Studies in Fuzziness and Soft Computing 293

Editor-in-Chief

Prof. Janusz Kacprzyk
Systems Research Institute
Polish Academy of Sciences
ul. Newelska 6
01-447 Warsaw
Poland
E-mail: kacprzyk@ibspan.waw.pl

For further volumes:
http://www.springer.com/series/2941

Rafik Aziz Aliev

Fundamentals of the Fuzzy Logic-Based Generalized Theory of Decisions

Springer

Author
Rafik Aziz Aliev
Joint Business Administration
Program (USA, Azerbaijan)
Baku
Azerbaijan

ISSN 1434-9922 e-ISSN 1860-0808
ISBN 978-3-642-44343-5 ISBN 978-3-642-34895-2 (eBook)
DOI 10.1007/978-3-642-34895-2
Springer Heidelberg New York Dordrecht London

Printed on acid-free paper

Springer is part of Springer Science+Business Media (www.springer.com)

Dedication

Dedicated to the memory of my wife Aida Alieva and my son Rahib and parents Aziz and Sayad Aliev.

Foreword

The path-breaking work of van Neumann-Morgenstern, published in 1946, opened the door to a massive intrusion of sophisticated mathematical theories into economics, games and decision analysis. In the years that followed, many great minds and Nobel Prize winners, among them John Nash, Gerard Debreu, John Harsanyi, Herbert Simon and Kahneman and Tversky, contributed greatly to a better understanding of decision analysis and economic behavior. Today, we are in possession over vast resource, call it R for short, of mathematical concepts, methods and theories for addressing problems and issues in economics, decision analysis and related fields. Viewed against this backdrop, a puzzling question arises. Why are sophisticated mathematical theories of limited use in dealing with problems in realistic settings?

There is a reason, in large measure R is based on classical, Aristotelian, bivalent logic. Bivalent logic is intolerant of imprecision and partiality of truth. What follows is that bivalent logic is not the right logic to serve as a foundation for mathematical theories of economic behavior and decision analysis—realms in which uncertainty, imprecision and partiality of truth is the norm rather than exception. Here is a simple example. Consider the standard, bivalent-logic-based definition of recession. An economic system is in a state of recession if there is a decline in GDP in two successive quarters. Based on this definition, the National Bureau of Economic Research announced in September 2010 that the recession came to an end in June 2009. The millions of people who have lost jobs and homes since then would find it hard to agree with this conclusion. The problem is that recession is not a bivalent concept—it is a matter of degree, as are many other concepts in economics and decision analysis. Realistically, recession should be associated with a Richter-like scale. What is widely unrecognized within the economics community is that bivalent logic is intrinsically unsuitable for construction of realistic models of economic systems. To deal with uncertainty, imprecision and partiality of truth, what is needed is fuzzy logic—a logic in which everything is or is allowed to be a matter of degree. Viewed in this perspective, Professor Aliev's work, "Fundamentals of the Fuzzy-Logic-Based Generalized Theory of Decisions" is a major contribution which shifts the foundations of decision analysis and economic behavior from bivalent logic to fuzzy logic. This shift opens the door to construction of much more realistic models of economic behavior and decision systems. The importance of Professor Aliev's generalized theory of decision-making is hard to exaggerate.

Professor Aliev's theory breaks away from traditional approaches. It contains many new concepts and ideas. To facilitate understanding of his theory, Professor Aliev includes in his book two introductory chapters. The first chapter is a succinct exposition of the basics of fuzzy logic, with emphasis on those parts of fuzzy logic which are of prime relevance to decision analysis. Professor Aliev is a prominent contributor to fuzzy logic and soft computing, and is a highly skilled expositor. His skill and expertise are reflected in this chapter.

As a preliminary to exposition of his theory, Professor Aliev presents in Chapter 2 a highly insightful review and critique of existing approaches—approaches which are based on bivalent logic. Among the theories which are discussed and critiqued in this chapter is the classical van Neumann-Morgenstern expected utility theory, and Kahneman-Tversky prospect theory. The principal problem with these theories is that they employ unrealistic models. In particular, they do not address a pivotal problem—decision-making with imprecise probabilities. Such probabilities are the norm rather than exception in realistic settings.

Exposition of Professor Aliev's theory begins in Chapter 3. A key concept in this theory is what Professor Aliev calls vague preference—preference which involves a mix of fuzzy and probabilistic uncertainties. In this context, a significant role is played by linguistic preference relations, exemplified by A is strongly preferred to B. The concept of a linguistic preference relation was introduced in my 1976 paper, "The linguistic approach and its application to decision analysis," followed by 1977 paper, "Linguistic characterization of preference relations as a basis for choice in social systems," but Professor Aliev's treatment goes far beyond what I had to say in those papers. An example of a question which arises: How does the concept of transitivity apply to linguistic preference relations? The concept of a vague preference relation is intended to serve as a better model of preference in realistic settings.

The core of Professor Aliev's generalized fuzzy-logic-based decision theory is described in Chapter 4. Here one finds detailed analyses of decision-making with various kinds of imperfect information. An important issue which is addressed is decision-making based on information which is described in natural language. Another important issue is decision-making with imprecise probabilities. Imprecise probabilities is a subject which is on the periphery of probability theory. There is a literature but there is a paucity of papers in which the problem of decision-making with imprecise probabilities is addressed. There is a reason. The problem of decision-making with imprecise probabilities does not lend itself to solution within the conceptual structure of theories based on bivalent logic.

In an entirely new direction, Professor Aliev describes an application of f-geometry to decision analysis. The concept of f-geometry was introduced in my 2009 paper "Toward extended fuzzy logic—A first step." In f-geometry, figures are drawn by hand with a spray-pen, with no drawing instruments such as a ruler and compass allowed. I am greatly impressed by Professor Aliev's skillful application of f-geometry to decision analysis.

Venturing beyond decision analysis, Professor Aliev develops a theory of fuzzy stability. In itself, the theory is an important contribution to the analysis of behavior of complex systems. Professor Aliev concludes his work with applying his theory to real-world problems in medicine, production and economics. This part of his book reflects his extensive experience in dealing with problems in planning and control of large-scale systems.

In conclusion, Professor Aliev's work may be viewed as a major paradigm shift—a paradigm shift which involves moving the foundation of decision analysis from bivalent logic to fuzzy logic. This move opens the door to construction of much better models of reality—reality in which uncertainty and imprecision lie at the center rather than on the periphery. To say that Professor Aliev's work is a major contribution is an understatement. Professor Aliev, the editor of the Series, Professor Kacprzyk, and Springer deserve a loud applause for producing a work which is certain to have a major and long-lasting impact.

Berkeley, California
August 28, 2012

Lotfi A. Zadeh

Preface

Every day decision making and decision making in complex human-centric systems are characterized by imperfect decision-relevant information. One main source of imperfect information is uncertainty of future, particularly, impossibility to exactly predict values of the variables of interest, actual trends, partially known present objective conditions etc. Even when it is sufficiently clear how events are developing, unforeseen contingencies may always essentially shift their trend. As a result, the prevailing amount of relevant information is carried not at a level of measurements but at a level of subjective perceptions which are intrinsically imprecise and are often described in a natural language (NL). The other source of imperfect information is behavior of a decision maker (DM) influenced by mental, psychological and other aspects like feelings, emotions, views etc. The latter are not to be described by numbers, information about them is imprecise, partially true and partially reliable, approximate, and, in general, is also described in NL. Due to imperfectness of information and complexity of decision problems, preferences of a DM in the real world are vague. Main drawback of the existing decision theories is namely incapability to deal with imperfect information and modeling vague preferences. Actually, a paradigm of non-numerical probabilities in decision making has a long history and arose also in Keynes's analysis of uncertainty. There is a need for further generalization – a move to decision theories with perception-based imperfect information described in NL on all the elements of a decision problem. Nowadays there are no economic models that provide sufficiently feasible descriptions of reality and new generation of decision theories is needed. Development of new theories is now possible due to an increased computational power of information processing systems which allows for computations with imperfect information, particularly, imprecise and partially true information, which are much more complex than computations over precise numbers and probabilities.

Thus, a new generation of decision theories should to some extent model this outstanding capability of humans. This means that the languages of new decision models for human-centric systems should be not languages based on binary logic and probability theory, but human-centric computational schemes able to operate on NL-described information.

In this book we suggest the developed decision theory with imperfect decision-relevant information on environment and a DM's behavior. This theory is based on the synthesis of the fuzzy sets theory as a mathematical tool for description and reasoning with perception-based information and the probability theory. The main difference of this theory from the existing theories is that it is based on a general statement of a decision problem taking into account imperfect information on all its elements used to describe environment and a behavior of a DM.

The book is composed of 8 chapters. The first chapter covers foundations of fuzzy sets theory, fuzzy logic and fuzzy mathematics which are the formal basis of the decision theory suggested in the present book.

Chapter 2 is devoted to review of the main existing decision theories. A description of a decision making problem including alternatives, states of nature, outcomes, preferences is explained. The key decision theories as von Neumann-Morgenstern Expected Utility, Subjective Expected Utility of Savage, Maximin Expected utility and other multiple priors-based utility models, Choquet Expected Utility, Cumulative Prospect theory and others are discussed in terms of the underlying motivations, features and disadvantages to deal with real-world imprecise and vague information.

Chapter 3 is devoted to uncertain (vague) preferences and imperfect information commonly faced with in real-world decision problems. The existing approaches like fuzzy preference relations, linguistic preference relations, probabilistic approaches to modeling uncertain preferences, incomplete preferences and basic types of imperfect information are explained. The motivation to use the fuzzy set theory for modeling uncertain preferences and imperfect information is provided.

Fuzzy logic-based theories of decision making with information described in natural language (NL) are suggested in Chapter 4. In Section 4.1 we suggest the theory in which vague preferences and a mix of fuzzy and probabilistic uncertainties are treated by a fuzzy-valued Choquet integral-based utility function. In section 4.2 two models of fuzzy logic-based multi-agent decision making are suggested. Section 4.3 presents alternative basics of operational approaches to decision making under interval and fuzzy uncertainty. Section 4.4 is devoted to a fuzzy optimality principle based approach of decision making under imperfect information without utility function. The approach utilizes a direct comparison of alternatives that may be more intuitive and reduces a sufficient loss of information related to the use of utility function for encoding preferences.

In economics, as in any complex humanistic system, motivations, intuition, human knowledge, and human behaviour, such as perception, emotions and norms, play dominant roles. In Chapter 5 we suggest a new approach to behavioral decision making which is based on a joint consideration of an environment conditions and a human behavior at the same fundamental level. In contrast to the basics of the existing behavioral decision theories, this allows for a transparent analysis of behavioral decision making. There are deeper and more subtle uncertain relationships between well-being and its determinants. In this chapter we consider two approaches which are based on fuzzy and probability theories to modeling of economic agents.

Sometimes perception-based information is not sufficiently clear to be modeled by means of fuzzy sets. In contrast, it remains at a level of some cloud images which are difficult to be caught by words. However, humans are able to make decisions based on visual perceptions. Modeling of this outstanding capability of humans, even to some limited extent, becomes a difficult yet a highly promising research area. This arises as a motivation of the methods of decision making suggested in Chapter 6. In this chapter we use Fuzzy Geometry and the extended fuzzy logic to cope with uncertain situations coming with unprecisiated decision-relevant information.

Many economic dynamical systems naturally become fuzzy due to the uncertain initial conditions and parameters. Stability is one of the fundamental concepts of such type of complex dynamical systems as physical, economical, socio-economical, and technical systems. In Chapter 7 we introduce and develop a Generalized Theory of Stability (GTS) for analysis of complex dynamical systems described by fuzzy differential equations (FDE). Different human-centric definitions of stability of dynamical systems are introduced. We also discuss and contrast several fundamental concepts of fuzzy stability, namely fuzzy stability of systems (FS), binary stability of fuzzy system (BFS), and binary stability of systems (BS) by showing that all of them arise as special cases of the proposed GTS. We also apply the obtained results to investigate stability of an economical system, including decision making in macroeconomic system.

Applications of the suggested fuzzy logic-based generalized theory of decisions to solving benchmark and real-world problems under imperfect information in production, medicine, economics and business are presented in Chapter 8.

We hope that the present book will be useful for researchers, practitioners and anyone who is interested in application of expressive power of fuzzy logic to solving real-world decision problems. The book is self containing and represents in a systematic way the suggested decision theory into the educational systems. The book will be helpful for teachers and students of universities and colleges, for managers and specialists from various fields of business and economics, production and social sphere.

We are grateful to Professor Lotfi Zadeh, for his suggestion to write this book, for his permanent support, invaluable ideas and advices.

I would like to express my thanks to Professors J. Kacprzyk, W. Pedrycz, V. Kreinovich, I. Turksen and B. Fazlollahi for helpful discussions on various topics of decision theories. Special thanks are due to Dr. O. Huseynov, A. Alizadeh and B. Guirimov for many enjoyable and productive conversations and collaborations.

Rafik Aziz Aliev

Contents

Chapter 1
Fuzzy Sets and Fuzzy Logic

1.1 Fuzzy Sets and Operations on Fuzzy Sets

Definition 1.1 Fuzzy Sets. Let X be a classical set of objects, called the universe, whose generic elements are denoted x. Membership in a classical subset A of X is **often** viewed as a characteristic function μ_A from A to $\{0,1\}$ such that

$$\mu_A(x) = \begin{cases} 1 \; iff \; x \in A \\ 0 \; iff \; x \notin A \end{cases}$$

where $\{0,1\}$ is called a valuation set; 1 indicates membership while 0 - non-membership.

If the valuation set is allowed to be in the real interval $[0,1]$, then A is called a fuzzy set denoted $\tilde{A}$ [2,3,6,8,57,58,114,133], $\mu_{\tilde{A}}(x)$ is the grade of membership of x in $\tilde{A}$

$$\mu_{\tilde{A}} : X \to [0,1]$$

As closer the value of $\mu_{\tilde{A}}(x)$ is to 1, so much x belongs to $\tilde{A}$.

$\tilde{A}$ is completely characterized by the set of pairs.

$$\tilde{A} = \{(x, \mu_{\tilde{A}}(x)), \; x \in X\}$$

Fuzzy sets with crisply defined membership functions are called ordinary fuzzy sets.

Properties of Fuzzy Sets

Definition 1.2. Equality of Fuzzy Sets. Two fuzzy sets $\tilde{A}$ and $\tilde{B}$ are said to be equal if and only if

$$\forall x \in X \quad \mu_{\tilde{A}}(x) = \mu_{\tilde{B}}(x) \quad \tilde{A} = \tilde{B}.$$

R.A. Aliev: *Fuzzy Logic-Based Generalized Theory of Decisions*, STUDFUZZ 293, pp. 1–64.
DOI: 10.1007/ 978-3-642-34895-2_1

Definition 1.3. The Support and the Crossover Point of a Fuzzy Set. The Singleton. The support of a fuzzy set $\tilde{A}$ is the ordinary subset of X that has nonzero membership in $\tilde{A}$:

$$\text{supp}\tilde{A}=\tilde{A}^{+0}=\{x\in X,\mu_A(x)>0\}$$

The elements of x such as $\mu_{\tilde{A}}(x)=1/2$ are the crossover points of $\tilde{A}$.

A fuzzy set that has only one point in X with $\mu_{\tilde{A}}=1$ as its support is called a singleton.

Definition 1.4. The Height of a Fuzzy Set. Normal and Subnormal Sets. The height of $\tilde{A}$ is

$$hgt(\tilde{A})=\sup_{x\in X}\mu_{\tilde{A}}(X)$$

i.e., the least upper bound of $\mu_{\tilde{A}}(x)$.

$\tilde{A}$ is said to be normalized iff $\exists x\in X\ ,\mu_{\tilde{A}}(x)=1$. This definition implies $hgt(\tilde{A})=1$. Otherwise $\tilde{A}$ is called subnormal fuzzy set.

The empty set $\varnothing$ is defined as

$$x\in X,\mu_{\varnothing}(x)=0,\text{of course}\,\forall x\in X\ \mu_X(x)=1$$

Definition 1.5. α-Level Fuzzy Sets. One of important way of representation of fuzzy sets is α-cut method. Such type of representation allows us to use properties of crisp sets and operations on crisp sets in fuzzy set theory.

The (crisp) set of elements that belongs to the fuzzy set $\tilde{A}$ at least to the degree α is called the α-level set:

$$A^{\alpha}=\{x\in X,\mu_{\tilde{A}}(x)\geq\alpha\}$$

$A^{\alpha}=\{x\in X,\mu_{\tilde{A}}(x)>\alpha\}$ is called "strong α-level set" or "strong α-cut".

Now we introduce fuzzy set A_{α}, defined as

$$A_{\alpha}(x)=\alpha A^{\alpha}(x) \tag{1.1}$$

Then the original fuzzy set $\tilde{A}$ may be defined as $\tilde{A}=\bigcup_{\alpha\in[0,1]}A_{\alpha}$. $\bigcup$ denotes the standard fuzzy union.

Definition 1.6. Convexity of Fuzzy Sets. A fuzzy set $\tilde{A}$ is convex iff

$$\mu_{\tilde{A}}(\lambda x_1+(1-\lambda)x_2)>\min(\mu_{\tilde{A}}(x_1),\mu_{\tilde{A}}(x_2)) \tag{1.2}$$

for all $x_1, x_2 \in R$, $\lambda \in [0,1]$, min denotes the minimum operator.

Alternatively, a fuzzy set $\tilde{A}$ on R is convex iff all its α-level sets are convex in the classical sense.

Definition 1.7. The Cardinality of a Fuzzy Set. When X is a finite set, the scalar cardinality $|\tilde{A}|$ of a fuzzy set $\tilde{A}$ on X is defined as

$$|\tilde{A}| = \sum_{x \in \tilde{A}} \mu_{\tilde{A}}(x)$$

Sometimes $|\tilde{A}|$ is called the power of $\tilde{A}$. $\|\tilde{A}\| = |\tilde{A}| / |X|$ is the relative cardinality.

When X is infinite, $|\tilde{A}|$ is defined as

$$|\tilde{A}| = \int_X \mu_{\tilde{A}}(x)\,dx$$

Definition 1.8. Fuzzy Set Inclusion. Given fuzzy sets $\tilde{A}, \tilde{B} \in \tilde{F}(X)$ $\tilde{A}$ is said to be included in $\tilde{B}\left(\tilde{A} \subseteq \tilde{B}\right)$ or $\tilde{A}$ is a subset of $\tilde{B}$ if $\forall x \in X, \mu_{\tilde{A}}(x) \le \mu_{\tilde{B}}(x)$.

When the inequality is strict, the inclusion is said to be strict and is denoted as $\tilde{A} < \tilde{B}$.

Let consider representations and constructing of fuzzy sets. It was mentioned above that each fuzzy set is uniquely defined by a membership function. In the literature one can find different ways in which membership functions are represented.

List Representation. If universal set $X = \{x_1, x_2, \ldots, x_n\}$ is a finite set, membership function of a fuzzy set $\tilde{A}$ on X $\mu_{\tilde{A}}(x)$ can be represented as table. Such table lists all elements in the universe X and the corresponding membership grades as shown below

$$\tilde{A} = \mu_{\tilde{A}}(x_1)/x_1 + \ldots + \mu_{\tilde{A}}(x_n)/x_n = \sum_{i=1}^{n} \mu_{\tilde{A}}(x_i)/x_i$$

Here symbol / (slash) does not denote division, it is used for correspondence between an element in the universe X (after slash) and its membership grade in the fuzzy set $\tilde{A}$ (before slash). The symbol + connects the elements (does not denote summation).

If X is a finite set then

$$\tilde{A} = \int_X \mu_{\tilde{A}}(x)/x\,.$$

Here symbol $\int_X$ is used for denoting a union of elements of set X.

Graphical Representation. Graphical description of a fuzzy set $\tilde{A}$ on the universe X is suitable in case when X is one or two-dimensional Euclidean space. Simple typical shapes of membership functions are usually used in fuzzy set theory and practice (Table 1.1).

Fuzzy n Cube Representation. All fuzzy sets on universe X with n elements can be represented by points in the n-dimensional unit cube – n-cube. Assume that universe X contains n elements $X = \{x_1, x_2, ..., x_n\}$. Each element x_i, $i = \overline{1,n}$ can be viewed as a coordinate in the n dimensional Euclidean space. A subset of this space for which values of each coordinate are restricted in $[0,1]$ is called n-cube. Vertices of the cube, i.e. bit list $(0,1,...,0)$ represent crisp sets. The points inside the cube define fuzzy subsets.

Analytical Representation. In case if universe X is infinite, it is not effective to use the above considered methods for representation of membership functions of a fuzzy sets. In this case it is suitable to represent fuzzy sets in an analytical form, which describes the shape of membership functions.

There are some typical formulas describing frequently used membership functions in fuzzy set theory and practice.

For example, bell-shaped membership functions often are used for representation of fuzzy sets. These functions are described by the formula:

$$\mu_{\tilde{A}}(x) = c \exp\left(-\frac{(x-a)^2}{b}\right)$$

which is defined by three parameters, a, b and c.

In general it is effective to represent the important typical membership functions by a parametrized family of functions.The following are formulas for describing the 6 families of membership functions

$$\mu_{\tilde{A}}(x, c_1) = [1 + c_1(x-a)^2]^{-1} \quad (1.3)$$

$$\mu_{\tilde{A}}(x, c_2) = \left[1 + c_2 |x-a|\right]^{-1} \quad (1.4)$$

$$\mu_{\tilde{A}}(x, c_3, d) = \left[1 + c_3 |x-a|^d\right]^{-1} \quad (1.5)$$

$$\mu_{\tilde{A}}(x, c_4, d) = \exp\left[-c_4 |x-a|^d\right] \quad (1.6)$$

Table 1.1 Typical membership functions

Type of Membership function	Graphical Representation	Analytical Representation
Triangular MF	$\mu_{\tilde{A}}(x)$ 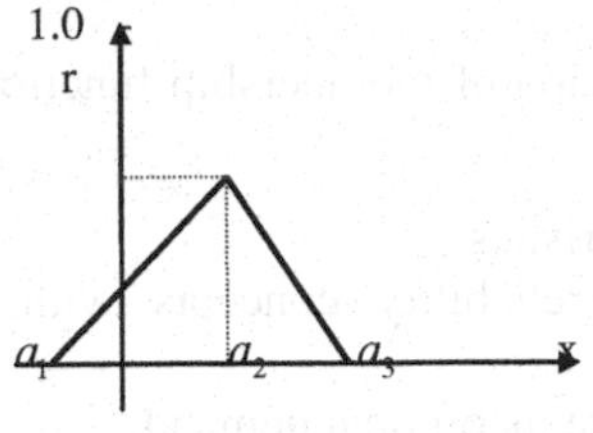	$\mu_{\tilde{A}}(x)=\begin{cases}\frac{x-a_1}{a_2-a_1}r, & \text{if } a_1\le x\le a_2\ ,\\ \frac{a_3-x}{a_3-a_2}r, & \text{if } a_2\le x\le a_3\ ,\\ 0\ , & otherwise\end{cases}$
Trapezoidal MF	$\mu_{\tilde{A}}(x)$ 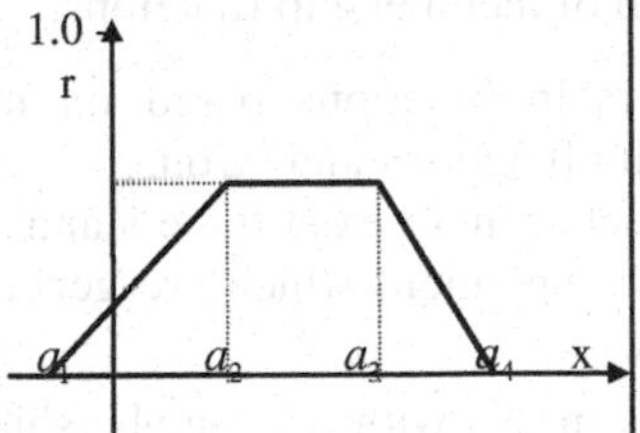	$\mu_{\tilde{A}}(x)=\begin{cases}\frac{x-a_1}{a_2-a_1}r, & \text{if } a_1\le x\le a_2\ ,\\ r\ , & \text{if } a_2\le x\le a_3\ ,\\ \frac{a_4-x}{a_4-a_3}r, & \text{if } a_3\le x\le a_4\ ,\\ 0\ , & otherwise\end{cases}$
S - shaped MF	$\mu_{\tilde{A}}(x)$ 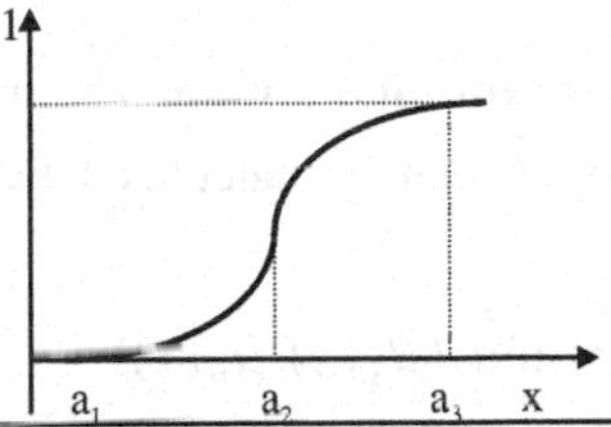	$\mu_{\tilde{A}}(x)=\begin{cases}0\ , & \text{if } x\le a_1\ ,\\ 2\left(\frac{x-a_1}{a_3-a_1}\right)^2, & \text{if } a_1<x<a_2\ ,\\ 1-2\left(\frac{x-a_1}{a_3-a_1}\right)^2, & \text{if } a_2\le x<a_3,\\ 1\ , & \text{if } a_3\le x\end{cases}$
Bell - shaped MF	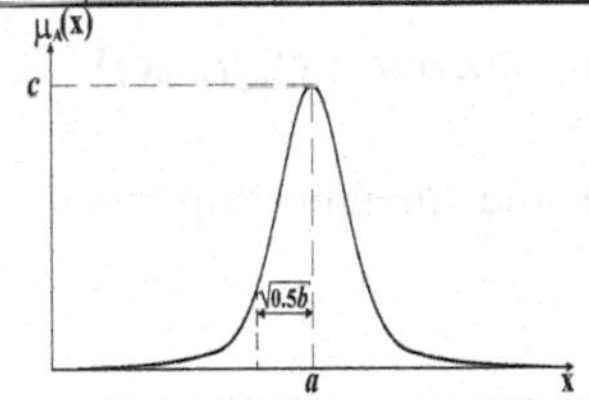	$\mu_{\tilde{A}}(x)=c\cdot\ \exp\left(-\frac{(x-a)^2}{b}\right)$

$$\mu_{\tilde{A}}(x,c_5)=\max\{0,[1-c_5\lvert x-a\rvert]\} \tag{1.7}$$

$$\mu_{\tilde{A}}(x,c_6)=c_6\exp\left[-\frac{(x-a)^2}{b}\right] \tag{1.8}$$

Here $c_i > 0,\ i = \overline{1,6},\ d > 1$ are parameters, a denotes the elements of corresponding fuzzy sets with the membership grade equal to unity.

Table 1.1 summarizes the graphical and analytical representations of frequently used membership functions (MF).

The problem of constructing membership functions is problem of knowledge engineering.

There are many methods for estimation of membership functions. They can be classified as follows:

1. Membership functions based on heuristics.
2. Membership functions based on reliability concepts with respect to the particular problem.
3. Membership functions based on more theoretical demand.
4. Membership functions as a model for human concepts.
5. Neural networks based construction of membership functions.

The estimation methods of membership functions based on more theoretical demand use axioms, probability density functions and so on.

Let consider operations on fuzzy sets. There exist three standard fuzzy operations: fuzzy intersection, union and complement which are generalization of the corresponding classical set operations.

Let's $\tilde{A}$ and $\tilde{B}$ be two fuzzy sets in X with the membership functions $\mu_{\tilde{A}}$ and $\mu_{\tilde{B}}$ respectively. Then the operations of intersection, union and complement are defined as given below.

Definition 1.9. Fuzzy Standard Intersection and Union. The intersection ($\bigcap$) and union ($\bigcup$) of fuzzy sets $\tilde{A}$ and $\tilde{B}$ can be calculated by the following formulas:

$$\forall x \in X \quad \mu_{\tilde{A}\cap\tilde{B}}(x) = \min\ (\mu_{\tilde{A}}(x), \mu_{\tilde{B}}(x))$$

$$\forall x \in X \quad \mu_{\tilde{A}\cup\tilde{B}}(x) = \max\ (\mu_{\tilde{A}}(x), \mu_{\tilde{B}}(x))$$

where $\mu_{\tilde{A}\cap\tilde{B}}(x)$ and $\mu_{\tilde{A}\cup\tilde{B}}(x)$ are the membership functions of $\tilde{A}\bigcap\tilde{B}$ and $\tilde{A}\bigcup\tilde{B}$, respectively.

Definition 1.10. Standard Fuzzy Complement. The complement $\tilde{A}^c$ of $\tilde{A}$ is defined by the membership function:

$$\forall x \in X \quad \mu_{\tilde{A}^c}(x) = 1 - \mu_{\tilde{A}}(x).$$

As already mentioned $\mu_{\tilde{A}}(x)$ is interpreted as the degree to which x belongs to $\tilde{A}$. Then by the definition $\mu_{\tilde{A}^c}(x)$ can be interpreted as the degree to which x does not belong to $\tilde{A}$.

The standard fuzzy operations do not satisfy the law of excluded middle $\tilde{A} \cup \tilde{A}^c = X$ and the law of contradiction $\tilde{A} \cap \tilde{A}^c = \varnothing$ of classical set theory. But commutativity, associativity, idempotency, distributivity, and De Morgan laws are held for standard fuzzy operations.

For fuzzy union, intersection and complement operations there exist a broad class of functions. Function that qualify as fuzzy intersections and fuzzy unions are defined as t-norms and t-conorms.

Definition 1.11. t-Norms. t-norm is a binary operation in [0,1], i.e. a binary function t from [0,1] into [0,1] that satisfies the following axioms

$$t\left(\mu_{\tilde{A}}(x),1\right) = \mu_{\tilde{A}}(x) \tag{1.9}$$

if $\mu_{\tilde{A}}(x) \le \mu_{\tilde{C}}(x)$ and $\mu_{\tilde{B}}(x) \le \mu_{\tilde{D}}(x)$ then

$$t(\mu_{\tilde{A}}(x),\mu_{\tilde{B}}(x)) \le t(\mu_{\tilde{C}}(x),\mu_{\tilde{D}}(x)) \tag{1.10}$$

$$t(\mu_{\tilde{A}}(x),\mu_{\tilde{B}}(x)) = t(\mu_{\tilde{B}}(x),\mu_{\tilde{A}}(x)) \tag{1.11}$$

$$t(\mu_{\tilde{A}}(x),t(\mu_{\tilde{B}}(x),\mu_{\tilde{C}}(x))) = t(t(\mu_{\tilde{A}}(x),\mu_{\tilde{B}}(x),\mu_{\tilde{C}}(x))) \tag{1.12}$$

Here (1.9) is boundary condition, (1.10)-(1.12) are conditions of monotonicity, commutativity and associativity, respectively.

The function t takes as its arguments the pair consisting of the element membership grades in set $\tilde{A}$ and in set $\tilde{B}$, and yields membership grades of the element in the $\tilde{A} \cap \tilde{B}$

$$(\tilde{A} \cap \tilde{B})(x) = t[\tilde{A}(x),\tilde{B}(x)] \quad \forall x \in X.$$

The following are frequently used t-norm-based fuzzy intersection operations:

Standard Intersection

$$t_0(\mu_{\tilde{A}}(x),\mu_{\tilde{B}}(x)) = \min\{\mu_{\tilde{A}}(x),\mu_{\tilde{B}}(x)\} \tag{1.13}$$

Algebraic Product

$$t_1(\mu_{\tilde{A}}(x),\mu_{\tilde{B}}(x)) = \mu_{\tilde{A}}(x)\cdot\mu_{\tilde{B}}(x) \tag{1.14}$$

Bounded Difference

$$t_2(\mu_{\tilde{A}}(x),\mu_{\tilde{B}}(x)) = \mu_{\tilde{A}\cap\tilde{B}}(x) = \max(0,\mu_{\tilde{A}}(x)+\mu_{\tilde{B}}(x)-1) \tag{1.15}$$

Drastic Intersection

$$t_3(\mu_{\tilde{A}}(x),\mu_{\tilde{B}}(x)) = \begin{cases} \min\{\mu_{\tilde{A}}(x),\mu_{\tilde{B}}(x)\} & \text{if } \mu_{\tilde{A}}(x)=1 \\ & \text{or } \mu_{\tilde{B}}(x)=1 \\ 0 & \text{otherwise} \end{cases} \tag{1.16}$$

For four fuzzy intersections the following is true

$$t_3(\mu_{\tilde{A}}(x),\mu_{\tilde{B}}(x)) \le t_2(\mu_{\tilde{A}}(x),\mu_{\tilde{B}}(x)) \le t_1(\mu_{\tilde{A}}(x),\mu_{\tilde{B}}(x)) \le t_0(\mu_{\tilde{A}}(x),\mu_{\tilde{B}}(x)) \quad (1.17)$$

Definition 1.12. t-Conorms. t-conorm is a binary operation in $[0,1]$, i.e. a binary function $S:[0,1]\times[0,1]\rightarrow[0,1]$ that satisfies the following axioms

$$S(\mu_{\tilde{A}}(x),0)=\mu_{\tilde{A}}(x) \text{ ;(boundary condition)} \quad (1.18)$$

if $\mu_{\tilde{A}}(x)\le\mu_{\tilde{C}}(x)$ and $\mu_{\tilde{B}}(x)\le\mu_{\tilde{D}}(x)$ then

$$S(\mu_{\tilde{A}}(x),\mu_{\tilde{B}}(x))\le S(\mu_{\tilde{C}}(x),\mu_{\tilde{D}}(x)) \text{ ; (monotonicity)} \quad (1.19)$$

$$S(\mu_{\tilde{A}}(x),\mu_{\tilde{B}}(x))=S(\mu_{\tilde{B}}(x),\mu_{\tilde{A}}(x)) \text{ ; (commutativity)} \quad (1.20)$$

$$S(\mu_{\tilde{A}}(x),S(\mu_{\tilde{B}}(x),\mu_{\tilde{C}}(x)))=S(S(\mu_{\tilde{A}}(x),\mu_{\tilde{B}}(x),\mu_{\tilde{C}}(x))) \text{ ; (associativity).} \quad (1.21)$$

The function S yields membership grade of the element in the set $\tilde{A}\cup\tilde{B}$ on the argument which is pair consisting of the same elements membership grades in set $\tilde{A}$ and $\tilde{B}$

$$(A\cup B)(X)=S[A(x),B(x)] \quad (1.22)$$

The following are frequently used t-conorm based fuzzy union operations.

Standard Union

$$S_0(\mu_{\tilde{A}}(x),\mu_{\tilde{B}}(x))=\max\{\mu_{\tilde{A}}(x),\mu_{\tilde{B}}(x)\} \quad (1.23)$$

Algebraic Sum

$$S_1(\mu_{\tilde{A}}(x),\mu_{\tilde{B}}(x))=\mu_{\tilde{A}}(x)+\mu_{\tilde{B}}(x)-\mu_{\tilde{A}}(x)\cdot\mu_{\tilde{B}}(x) \quad (1.24)$$

Drastic Union

$$S_3(\mu_{\tilde{A}}(x),\mu_{\tilde{B}}(x))=\begin{cases}\max\{\mu_{\tilde{A}}(x),\mu_{\tilde{B}}(x)\} & \text{if } \mu_{\tilde{A}}(x)=0 \\ & \text{or } \mu_{\tilde{B}}(x)=0 \\ 1 & \text{otherwise}\end{cases} \quad (1.25)$$

For four fuzzy union operations the following is true

$$S_0(\mu_{\tilde{A}}(x),\mu_{\tilde{B}}(x))\le S_1(\mu_{\tilde{A}}(x),\mu_{\tilde{B}}(x))\le S_2(\mu_{\tilde{A}}(x),\mu_{\tilde{B}}(x))\le S_3(\mu_{\tilde{A}}(x),\mu_{\tilde{B}}(x)) \quad (1.26)$$

Definition 1.13. Cartesian Product of Fuzzy Sets. The Cartesian product of fuzzy sets $\tilde{A}_1,\tilde{A}_2,\ldots,\tilde{A}_n$ on universes $X_1,X_2,\ldots,X_n$ respectively is a fuzzy set in the product space $X_1 \times X_2 \times \ldots \times X_n$ with the membership function $\mu_{\tilde{A}_1 \times \tilde{A}_2 \times \ldots \times \tilde{A}_n}(x) = \min\left\{\mu_{\tilde{A}_i}(x_i) \mid x = (x_1, x_2 \ldots, x_n), x_i \in X_i\right\}$.

Definition 1.14. Power of Fuzzy Sets. m-th power of a fuzzy set $\tilde{A}^m$ is defined as

$$\mu_{\tilde{A}^m}(x) = [\mu_{\tilde{A}}(x)]^m \quad , \ \forall \mathrm{x} \in \mathrm{X}, \ \forall \mathrm{m} \in \mathrm{R}^+ \tag{1.27}$$

where R^+ is positively defined set of real numbers.

Definition 1.15. Concentration and Dilation of Fuzzy Sets

Let $\tilde{A}$ be fuzzy set on the universe:

$$\tilde{A} = \left\{x, \mu_{\tilde{A}}(x) / x \in X\right\}$$

Then the operator $Con_m \tilde{A} = \{(x, [\mu_{\tilde{A}}(x)]^m) / \mathrm{x} \in \mathrm{X}\}$ is called concentration of $\tilde{A}$ and the operator $Dil_n \tilde{A} = \{(x, \sqrt{\mu_{\tilde{A}}(x)}\ ) / \mathrm{x} \in \mathrm{X}\}$ is called dilation of $\tilde{A}$.

Definition 1.16. Difference of Fuzzy Sets. Difference of fuzzy sets is defined by the formula:

$$\forall x \in X, \ \mu_{\tilde{A}|-|\tilde{B}}(x) = \max(0, \mu_{\tilde{A}}(x) - \mu_{\tilde{B}}(x)) \tag{1.28}$$

$\tilde{A}|-|\tilde{B}$ is the fuzzy set of elements that belong to $\tilde{A}$ more than to $\tilde{\mathrm{B}}$.

Symmetrical difference of fuzzy sets $\tilde{A}$ and $\tilde{\mathrm{B}}$ is the fuzzy set $\tilde{\mathrm{A}} \nabla \tilde{\mathrm{B}}$ of elements that belong more to $\tilde{A}$ than to $\tilde{\mathrm{B}}$:

$$\forall x \in X \quad \mu_{\tilde{A} \nabla \tilde{B}}(x) = \left|\mu_{\tilde{A}}(x) - \mu_{\tilde{B}}(x)\right| \tag{1.29}$$

Definition 1.17. Fuzzy Number. A fuzzy number is a fuzzy set $\tilde{A}$ on R which possesses the following properties: a) $\tilde{A}$ is a normal fuzzy set; b) $\tilde{A}$ is a convex fuzzy set; c) α-cut of $\tilde{A}$, A^α is a closed interval for every $\alpha \in (0,1]$; d) the support of $\tilde{A}$, A^{+0} is bounded.

In Fig. 1.1 some basic types of fuzzy numbers are shown. For comparison of a fuzzy number with a crisp number in Fig. 1.2 crisp number 2 is given.

Let consider arithmetic operation on fuzzy numbers. There are different methods for developing fuzzy arithmetic. In this section we present three methods.

Method based on the extension principle. By this method basic arithmetic operations on real numbers are extende to operations on fuzzy numbers. Let $\tilde{A}$ and $\tilde{B}$ be two fuzzy numbers and $*$ denote any of four arithmetic operations $\{+, -, \cdot, :\}$.

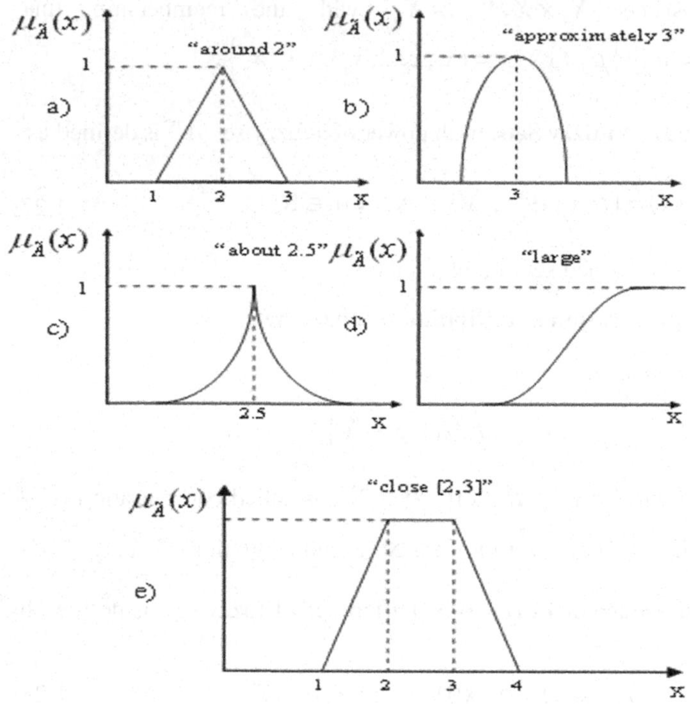

Fig. 1.1 Types of fuzzy numbers

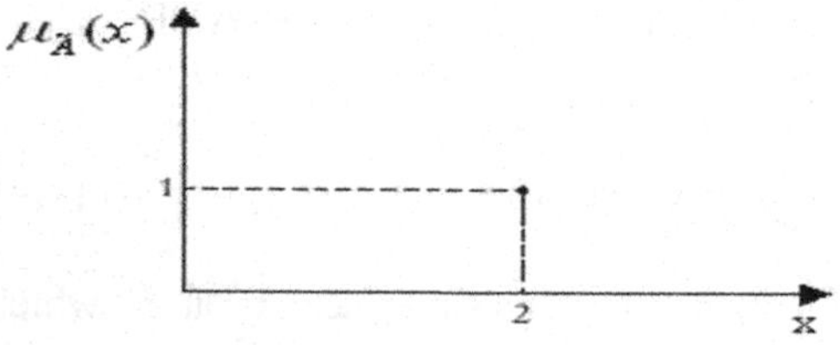

Fig. 1.2 Crisp number 2

A fuzzy set $\tilde{A} * \tilde{B}$ on R can be defined by the equation

$$\forall z \in \mathrm{R} \quad \mu_{(\tilde{A}*\tilde{B})}(z) = \sup_{z=x*y} \min[\mu_{\tilde{A}}(x), \mu_{\tilde{B}}(y)] \tag{1.30}$$

It is shown in [57] that $\tilde{A} * \tilde{B}$ is fuzzy number and the following theorem has been formulated and proved.

Theorem 1.1. *Let* $* \in \{+, -, \cdot, :\}$, *and let* $\tilde{A}$, $\tilde{B}$ *denote continuous fuzzy numbers. Then, the fuzzy set* $\tilde{A} * \tilde{B}$ *defined by (1.30) is a continuous fuzzy number.*

Then for four basic arithmetic operations on fuzzy numbers we can write

$$\mu_{(\tilde{A}+\tilde{B})}(z) = \sup_{z=x+y} \min[\mu_{\tilde{A}}(x), \mu_{\tilde{B}}(y)] \tag{1.31}$$

$$\mu_{(\tilde{A}-\tilde{B})}(z) = \sup_{z=x-y} \min[\mu_{\tilde{A}}(x), \mu_{\tilde{B}}(y)] \tag{1.32}$$

$$\mu_{(\tilde{A}\cdot\tilde{B})}(z) = \sup_{z=x\cdot y} \min[\mu_{\tilde{A}}(x), \mu_{\tilde{B}}(y)] \tag{1.33}$$

$$\mu_{(\tilde{A}:\tilde{B})}(z) = \sup_{z=x:y} \min[\mu_{\tilde{A}}(x), \mu_{\tilde{B}}(y)] \tag{1.34}$$

Method Based on Interval Arithmetic and α-Cuts. This method is based on representation of arbitrary fuzzy numbers by their α-cuts and use interval arithmetic to the α-cuts. Let $\tilde{A}, \tilde{B} \subset R$ be fuzzy numbers and $*$ denote any of four operations. For each $\alpha \in (0,1]$, the α-cut of $\tilde{A} * \tilde{B}$ is expressed as

$$(\tilde{A} * \tilde{B})^{\alpha} = A^{\alpha} * B^{\alpha} \tag{1.35}$$

For $*$ we assume $0 \notin \mathrm{supp}(\tilde{B})$.

The resulting fuzzy number $\tilde{A} * \tilde{B}$ can be defined as

$$\tilde{A} * \tilde{B} = \bigcup_{\alpha \in [0,1]} \alpha (A * B)^{\alpha} \tag{1.36}$$

Next we using (1.35), (1.36) illustrate four arithmetic operations on fuzzy numbers.

Addition. Let $\tilde{A}$ and $\tilde{B}$ be two fuzzy numbers and A^{α} and B^{α} their α-cuts

$$A^{\alpha} = [a_1^{\alpha}, a_2^{\alpha}]; B^{\alpha} = [b_1^{\alpha}, b_2^{\alpha}] \tag{1.37}$$

Then we can write

$$A^{\alpha} + B^{\alpha} = [a_1^{\alpha}, a_2^{\alpha}] + [b_1^{\alpha}, b_2^{\alpha}] = [a_1^{\alpha} + b_1^{\alpha}, a_2^{\alpha} + b_2^{\alpha}], \forall \alpha \in [0,1] \tag{1.38}$$

here

$$A^{\alpha} = \{x / \mu_{\tilde{A}}(x) \geq \alpha\}; B^{\alpha} = \{x / \mu_{\tilde{B}}(x) \geq \alpha\} \tag{1.39}$$

Subtraction. Subtraction of given fuzzy numbers $\tilde{A}$ and $\tilde{B}$ can be defined as

$$(A-B)^{\alpha} = A^{\alpha} - B^{\alpha} = [a_1^{\alpha} - b_2^{\alpha}, a_2^{\alpha} - b_1^{\alpha}], \forall \alpha \in [0,1] \tag{1.40}$$

We can determine (1.40) by addition of the image $\tilde{B}^{-}$ to $\tilde{A}$

$$\forall\alpha\in[0,1], B^{\alpha^-}=[-b_2^\alpha,-b_1^\alpha] \quad (1.41)$$

Multiplication. Let two fuzzy numbers $\tilde{A}$ and $\tilde{B}$ be given. Multiplication $\tilde{A}\cdot\tilde{B}$ is defined as

$$(A\cdot B)^\alpha = A^\alpha\cdot B^\alpha=[a_1^\alpha,a_2^\alpha]\cdot[b_1^\alpha,b_2^\alpha]\forall\alpha\in[0,1] \quad (1.42)$$

Multiplication of fuzzy number $\tilde{A}$ in R by ordinary numbers $k\in R^+$ is performed as follows

$$\forall\tilde{A}\subset R\ kA^\alpha=[ka_1^\alpha,ka_2^\alpha] \quad (1.43)$$

Division. Division of two fuzzy numbers $\tilde{A}$ and $\tilde{B}$ is defined by

$$A^\alpha : B^\alpha=[a_1^\alpha,a_2^\alpha]:[b_1^\alpha,b_2^\alpha]\ \forall\alpha\in[0,1] \quad (1.44)$$

Definition 1.18. Absolute Value of a Fuzzy Number. Absolute value of fuzzy number is defined as:

$$abs(\tilde{A})=\begin{cases}\max(\tilde{A},-\tilde{A}), & \text{for } R^+\\ 0, & \text{for } R^-\end{cases} \quad (1.45)$$

Let consider Z-number and operations on Z –numbers [128]. Decisions are based on decision-relevant information which must be reliable. Basically, the concept of a Z -number relates to the issue of reliability of information. A Z -number, Z, has two components, Z=($\tilde{A}$, $\tilde{B}$). The first component, $\tilde{A}$, is a restriction (constraint) on the values which a real-valued uncertain variable, X, is allowed to take. The second component, $\tilde{B}$, is a measure of reliability (confidence)of the first component. Typically, $\tilde{A}$ and $\tilde{B}$ are described in a natural language.

The concept of a Z -number has a potential for many applications, especially in the realms of economics and decision analysis.

Much of the information on which decisions are based is uncertain. Humans have a remarkable capability to make rational decisions based on information which is uncertain, imprecise and or incomplete. Formalization of this capability, at least to some degree motivates the concepts Z -number [128].

The ordered triple $(X,\tilde{A},\tilde{B})$ is referred to as a Z -valuation. A Z -valuation is equivalent to an assignment statement, X is $(\tilde{A},\tilde{B})$. X is an uncertain random variable. For convenience, $\tilde{A}$ is referred to as a value of X, with the understanding that, $\tilde{A}$ is not a value of X but a restriction on the values which X can take. The second component, $\tilde{B}$, is referred to as confidence(certainty). When X is a random variable, certainty may be equated to probability. Typically, $\tilde{A}$ and $\tilde{B}$

are perception-based and are described in NL. A collection of Z-valuations is referred to as Z-information. It should be noted that much of everyday reasoning and decision-making is based on Z-information. For purposes of computation, when $\tilde{A}$ and $\tilde{B}$ are described in NL, the meaning of $\tilde{A}$ and $\tilde{B}$ is precisiated through association with membership functions, $\mu_{\tilde{A}}$ and $\mu_{\tilde{B}}$, respectively. Simple examples of Z-valuations are:

(anticipated budget deficit, about 3 million dollars, likely);
(price of oil in the near future, significantly over 50 dollars/barrel, veri likely).

The Z-valuation $(X, \tilde{A}, \tilde{B})$ may be viewed as a restriction on X defined by:

Prob (X is $\tilde{A}$) is $\tilde{B}$.

In a Z-number, $(\tilde{A}, \tilde{B})$, the underlying probability distribution p_X, is not known. What is known is a restriction on p_X which may be expressed as [128]:

$$\int_R \mu_{\tilde{A}}(u) p_X(u) du \text{ is } \tilde{B}$$

An important qualitative attribute of a Z-number is informativeness. Generally, but not always, a Z-number is informative if its value has high specificity, that is, is tightly constrained [110], and its certainty is high. Informativeness is a desideratum when a Z-number is a basis for a decision. A basic question is: When is the informativeness of a Z-number sufficient to serve as a basis for an intelligent decision?

The concept of a Z-number is based on the concept of a fuzzy granule [120,121,124]. A concept which is closely related to the concept of a Z-number is the concept of a Z^+-number. Basically, a Z^+-number, Z^+, is a combination of a fuzzy number, $\tilde{A}$, and a random number, R, written as an ordered pair $Z^+ = (\tilde{A}, \tilde{B})$. In this pair, $\tilde{A}$ plays the same role as it does in a Z-number, and R is the probability distribution of a random number. Equivalently, R may be viewed as the underlying probability distribution of X in the Z-valuation $(X, \tilde{A}, \tilde{B})$. Alternatively, a Z^+-number may be expressed as $(\tilde{A}, p_X)$ or $(\mu_{\tilde{A}}, p_X)$, where $\mu_{\tilde{A}}$ is the membership function of $\tilde{A}$. A Z^+-valuation is expressed as $(X, \tilde{A}, p_X)$ or, equivalently, as $(X, \mu_{\tilde{A}}, p_X)$, where p_X is the probability distribution (density) of X.

The scalar product of $\mu_{\tilde{A}}$ and p_X, $\mu_{\tilde{A}} p_X$ is the probability measure, $P_{\tilde{A}}$, of $\tilde{A}$. More concretely,

$$\mu_{\tilde{A}} p_X = P_{\tilde{A}} = \int_R \mu_{\tilde{A}}(u) p_X(u) du \tag{1.46}$$

It is this relation that links the concept of a Z-number to that of a Z^+-number.

More concretely,

$$Z(\tilde{A},\tilde{B}) = Z^{+}(\tilde{A}, \mu_{\tilde{A}}\, p_X \; is \; \tilde{B})$$

What should be underscored is that in the case of a Z-number what is known is not p_X but a restriction on p_X expressed as $\mu_{\tilde{A}}\, p_X$ is $\tilde{B}$.

Let X be a real-valued variable taking values in U. For our purposes, it will be convenient to assume that U is a finite set $U=\{u_1,...,u_n\}$. We can associate with X a possibility distribution μ, and a probability distribution p, expressed as:

$$\mu = \mu_1 / u_1 + ..., + \mu_n / u_n$$

$$p = \mu_1 \setminus p_1 + ..., + \mu_n \setminus p_n$$

Here μ_i / u_i means that μ_i, $i=1,...n$, is the possibility that $X = u_i$. Similarly, $p_i \setminus u_i$ means that p_i is the probability that $X = u_i$.

The possibility distribution, μ, may be combined with the probability distribution, p, through what is referred to as confluence. More concretely,

$$\mu : p = (\mu_1, p_1)/u_1 + ... + (\mu_n, p_n)/u_n$$

As was noted earlier, the scalar product, expressed as $\mu \cdot p$, is the probability measure of $\tilde{A}$. In terms of the bimodal distribution, the Z^{+}-valuation and the Z-valuation associated with X may be expressed as:

$$(X, \tilde{A}, p_X)$$

$$(X, \tilde{A}, \tilde{B}),\; \mu_{\tilde{A}}\, p_X \; is \; \tilde{B},$$

respectively, with the understanding that $\tilde{B}$ is a possibilistic restriction on $\mu_{\tilde{A}}\, p_X$.

A key idea which underlies the concept of a Z-mouse [128] is that visual interpretation of uncertainty is much more natural than its description in natural language or as a membership function of a fuzzy set. This idea is closely related to the remarkable human capability to precisiate (graduate) perceptions, that is, to associate perceptions with degrees.

Using a Z-mouse, a Z-number is represented as two f-marks on two different scales. The trapezoidal fuzzy sets which are associated with the f-marks serve as objects of computation.

Let us consider computation with Z-numbers. Computation with Z^{+}-numbers is much simpler than computation with Z-numbers. Assume that $*$ is a binary operation whose operands are Z^+-numbers, $Z^+{}_X = (\tilde{A}_X, R_X)$ and $Z^+{}_Y = (\tilde{A}_Y, R_Y)$ By definition,

$$Z^+{}_X * Z^+{}_Y = (\tilde{A}_X * \tilde{A}_Y R_X * R_Y) \tag{1.47}$$

with the understanding that the meaning of $*$ in $R_X * R_Y$ is not the same as the meaning of $*$ in $\tilde{A}_X * \tilde{A}_Y$. In this expression, the operands of $*$ in $\tilde{A}_X * \tilde{A}_Y$ are fuzzy numbers; the operands of $*$ in $R_X * R_Y$ are probability distributions.

Assume that $*$ is sum. In this case, $\tilde{A}_X + \tilde{A}_Y$ is defined by:

$$\mu_{(\tilde{A}_X + \tilde{A}_Y)}(v) = \sup_u(\mu_{\tilde{A}_X}(u) \wedge \mu_{\tilde{A}_Y}(v-u)), \wedge = \min \tag{1.48}$$

Similarly, assuming that R_X and R_Y are independent, the probability density function of $R_X * R_Y$ is the convolution, $\circ$, of the probability density functions of R_X and R_Y. Denoting these probability density functions as p_{R_X} and p_{R_Y}, respectively, we have:

$$p_{R_X + R_Y}(v) = \int_R p_{R_X}(u) p_{R_Y}(v-u) du \tag{1.49}$$

Thus,

$$Z^+{}_X + Z^+{}_Y = (\tilde{A}_X + \tilde{A}_Y, p_{R_X} \circ p_{R_Y}) \tag{1.50}$$

More generally, to compute $Z_X * Z_Y$ what is needed is the extension principle of fuzzy logic [114,115].

Turning to computation with Z-numbers, assume for simplicity that $* = \text{sum}$. Assume that $Z_X = (\tilde{A}_X, \tilde{B}_X)$ and $Z_Y = (\tilde{A}_Y, \tilde{B}_Y)$. Our problem is to compute the sum Z=X+Y. Assume that the associated Z-valuations are $(X, \tilde{A}_X, \tilde{B}_X)$, $(Y, \tilde{A}_Y, \tilde{B}_Y)$ and $(Z, \tilde{A}_Z, \tilde{B}_Z)$.

The first step involves computation of p_Z. To begin with, let us assume that p_X and p_Y are known, and let us proceed as we did in computing the sum of Z^+-numbers. Then

$$p_Z = p_X \circ p_Y$$

or more concretely

$$p_Z(v) = \int_R p_X(u) p_Y(v-u) du$$

In the case of Z-numbers what we know are not p_X and p_Y but restrictions on p_X and p_Y

$$\int_R \mu_{\tilde{A}_X}(u) p_X(u) du \text{ is } \tilde{B}_X$$

$$\int_R \mu_{\tilde{A}_Y}(u) p_Y(u) du \text{ is } \tilde{B}_Y$$

In terms of the membership functions of $\tilde{B}_X$ and $\tilde{B}_Y$, these restrictions may be expressed as:

$$\mu_{\tilde{B}_X}\left(\int_R \mu_{\tilde{A}_X}(u) p_X(u) du\right)$$

$$\mu_{\tilde{B}_Y}\left(\int_R \mu_{\tilde{A}_Y}(u) p_Y(u) du\right)$$

Additional restrictions on p_X and p_Y are:

$$\int_R p_X(u) du = 1$$

$$\int_R p_Y(u) du = 1$$

$$\int_R u p_X(u) du = \frac{\int_R u \mu_{\tilde{A}_X}(u) du}{\int_R \mu_{\tilde{A}_X}(u) du} \quad \text{(compatibility)}$$

$$\int_R u p_Y(u) du = \frac{\int_R u \mu_{\tilde{A}_Y}(u) du}{\int_R \mu_{\tilde{A}_Y}(u) du} \quad \text{(compatibility)}$$

Applying the extension principle, the membership function of p_Z may be expressed as:

$$\mu_{p_Z}(p_Z) = \sup_{p_X, p_Y}\left(\mu_{\tilde{B}_X}\left(\int_R \mu_{\tilde{A}_X}(u) p_X(u) du\right) \wedge \mu_{\tilde{B}_Y}\left(\int_R \mu_{\tilde{A}_Y}(u) p_Y(u) du\right)\right)$$

subject to

$$p_Z = p_X \circ p_Y$$

$$\int_R p_X(u) du = 1$$

$$\int_R p_Y(u) du = 1$$

$$\int_R u p_X(u) du = \frac{\int_R u \mu_{\tilde{A}_X}(u) du}{\int_R \mu_{\tilde{A}_X}(u) du}$$

$$\int_R u p_Y(u)du = \frac{\int_R u\mu_{\tilde{A}_Y}(u)du}{\int_R \mu_{\tilde{A}_Y}(u)du}$$

The second step involves computation of the probability of the fuzzy event, Z is $\tilde{A}_Z$, given p_Z. As was noted earlier, in fuzzy logic the probability measure of the fuzzy event X is $\tilde{A}$, where $\tilde{A}$ is a fuzzy set and X is a random variable with probability density p_X, is defined as:

$$\int_R \mu_{\tilde{A}}(u) p_X(u)du$$

Using this expression, the probability measure of $\tilde{A}_Z$ may be expressed as:

$$B_Z = \int_R \mu_{\tilde{A}_Z}(u) p_Z(u)du,$$

where

$$\mu_{\tilde{A}_Z}(u) = \sup_v(\mu_{\tilde{A}_X}(v) \wedge \mu_{\tilde{A}_Y}(u-v))$$

It should be noted that B_Z is a number when p_Z is a known probability density function. Since what we know about p_Z is its possibility distribution, $\mu_{p_Z}(p_Z)$, $\tilde{B}_Z$ is a fuzzy set with membership function $\mu_{\tilde{B}_Z}$. Applying the extension principle, we arrive at an expression for $\mu_{\tilde{B}_Z}$. More specifically,

$$\mu_{\tilde{B}_Z}(w) = \sup_{p_Z} \mu_{p_Z}(p_Z)$$

subject to

$$w = \int_R \mu_{\tilde{A}_Z}(u) p_Z(u)du$$

Where $\mu_{p_Z}(p_Z)$ is the result of the first step. In principle, this completes computation of the sum of Z-numbers, Z_X and Z_Y.

In a similar way, we can compute various functions of Z-numbers. The basic idea which underlies these computations may be summarized as follows. Suppose that our problem is that of computing $f(Z_X, Z_Y)$, where Z_X and Z_Y are Z-numbers, $Z_X = (\tilde{A}_X, \tilde{B}_X)$ and $Z_Y = (\tilde{A}_Y, \tilde{B}_Y)$ respectively, and $f(Z_X, Z_Y) = (\tilde{A}_Z, \tilde{B}_Z)$. We begin by assuming that the underlying probability distributions p_X and p_Y are known. This assumption reduces the computation of $f(Z_X, Z_Y)$ to computation of $f(Z_X^+, Z_Y^+)$, which can be carried out through the use of the version of the extension principle which applies to restrictions which are Z^+-numbers. At this point, we recognize that what we know are not p_X and p_Y but restrictions on p_X and p_Y. Applying the version of the extension principle which relates to probabilistic restrictions, we are led to $f(Z_X, Z_Y)$. We can compute the restriction, $\tilde{B}_Z$, of the scalar product of $f(\tilde{A}_X, \tilde{A}_Y)$ and $f(p_X, p_Y)$. Since $\tilde{A}_Z = f(\tilde{A}_X, \tilde{A}_Y)$, computation of $\tilde{B}_Z$ completes the computation of $f(Z_X, Z_Y)$.

There are many important directions which remain to be explored, especially in the realm of calculi of Z-rules and their application to decision analysis and modeling of complex systems.

Computation with Z-numbers may be viewed as a generalization of computation with numbers, intervals, fuzzy numbers and random numbers. More concretely, the levels of generality are: computation with numbers (ground level1);computation with intervals (level1); computation with fuzzy numbers (level 2); and computation with Z-numbers (level3). The higher the level of generality, the greater is the capability to construct realistic models of real-world systems, especially in the realms of economics and decision analysis.

It should be noted that many numbers, especially in fields such as economics and decision analysis are in reality Z-numbers, but they are not treated as such because it is much simpler to compute with numbers than with Z-numbers. Basically, the concept of a Z-number is a step toward formalization of the remarkable human capability to make rational decisions in an environment of imprecision and uncertainty.

We now consider fuzzy relations, linguistic variables. In modeling systems the internal structure of a system must be described first. An internal structure is characterized by connections (associations) among the elements of system. As a rule these connections or associations are represented by means of relation. We will consider here fuzzy relations which gives us the representation about degree or strength of this connection.

There are several definitions of fuzzy relation [54,113,117]. Each of them depends on various factors and expresses different aspects of modeling systems.

Definition 1.19. Fuzzy Relation. Let $X_1, X_2, ..., X_n$ be nonempty crisp sets. Then, a $\tilde{R}(X_1, X_2, ..., X_n)$ is called a fuzzy relation of sets $X_1, X_2, ..., X_n$, if $\tilde{R}(X_1, X_2, ..., X_n)$ is the fuzzy subset given on Cartesian product $X_1 \times X_2 \times ... \times X_n$.

If $n = 2$, then fuzzy relation is called binary fuzzy relation, and is denoted as $\tilde{R}(X, Y)$. For three, four, or n sets the fuzzy relation is called ternary, quaternary, or n-ary, respectively.

In particular, if $X_1 = X_2 = ... = X_n = X$ we say that fuzzy relation R is given on set X among elements $x_1, x_2, ..., x_n \in X$.

Notice, that fuzzy relation can be defined in another way. Namely, by two ordered fuzzy sets.

Assume, two fuzzy sets $\mu_{\tilde{A}}(x)$ and $\mu_{\tilde{B}}(y)$ are given on crisp sets X and Y, respectively. Then, it is said, that fuzzy relation $R_{\tilde{A}\tilde{B}}(X,Y)$ is given on sets X and Y, if it is defined in the following way

$$\mu_{R_{\tilde{A}\tilde{B}}}(x,y) = \min_{x,y}[\mu_{\tilde{A}}(x), \mu_{\tilde{B}}(y)]$$

for all pairs (x,y), where $x \in X$ and $y \in Y$. As above, fuzzy relation $R_{\tilde{A}\tilde{B}}$ is defined on Cartesian product.

Let fuzzy binary relation on set X be given. Consider the following three properties of relation $\tilde{R}$:

1. Fuzzy relation $\tilde{R}$ is reflexive, if

$$\mu_{\tilde{R}}(x,x) = 1$$

for all $x \in X$. If there exist $x \in X$ such that this condition is violated, then relation $\tilde{R}$ is irreflexive, and if $\tilde{R}(x,x) = 0$ for all $x \in X$, the relation $\tilde{R}$ is antireflective;

2. A fuzzy relation $\tilde{R}$ is symmetric if it satisfies the following condition:

$$\mu_{\tilde{R}}(x,y) = \mu_{\tilde{R}}(y,x)$$

for all $x, y \in X$. If from $\tilde{R}(x,y) > 0$ and $\tilde{R}(y,x) > 0$ follows $x = y$ for all $x, y \in X$ relation $\tilde{R}$ is called antisymmetric;

3. A fuzzy relation $\tilde{R}$ is transitive (or, more specifically, max-min transitive) if

$$\mu_{\tilde{R}}(x,z) \geq \max_{y \in Y} \min(\mu_{\tilde{R}}(x,y),\ \mu_{\tilde{R}}(y,z))$$

is satisfied for all pairs $(x,z) \in X$.

Definition 1.20. Fuzzy Proximity. A fuzzy relation is called a fuzzy proximity or fuzzy tolerance relation if it is reflexive and symmetric. A fuzzy relation is called a fuzzy similarity relation if it is reflexive, symmetric, and transitive.

Definition 1.21. Fuzzy Composition. Let $\tilde{A}$ and $\tilde{B}$ be two fuzzy sets on $X \times Y$ and $Y \times Z$, respectively. A fuzzy relation $\tilde{R}$ on $X \times Z$ is defined as

$$\tilde{R} = \{((x,z), \mu_{\tilde{R}}(x,z) \mid (x,z) \in X \times Z\} \quad (1.51)$$

here

$$\mu_{\tilde{R}} : X \times Y \to [0,1]$$

$$(x,z) \mapsto \mu_{\tilde{R}}(x,z) = \mu_{\tilde{A}\circ\tilde{B}}(x,z) = \underset{y\in Y}{S}(\mathrm{T}(\mu_{\tilde{A}}(x,y), \mu_{\tilde{B}}(y,z))) \tag{1.52}$$

For $x \in X$ and $z \in Z$, T and S are triangular norms and triangular conorms, respectively.

Definition 1.22. Equivalence (Similarity) Relation. If fuzzy relation $\tilde{R}$ is reflexive, symmetric and transitive then relation $\tilde{R}$ is an equivalence relation or similarity relation.

A fuzzy relation $\tilde{R}$ is a fuzzy compatibility relation if it is reflexive and symmetric. This relation is cutworthy. Compatibility classes are defined by means of α-cut. In fact, using α-cut a class of compatibility relation is represented by means of crisp subset.

Therefore a compatibility relation can also be represented by reflexive undirected graph.

Now consider fuzzy partial ordering.

Let X be nonempty set. It is well known, that to order a set it is necessary to give an order relation on this set. But sometimes our knowledge and estimates of the elements of a set are not accurate and complete. Thus, to order such set the fuzzy order on set must be defined.

Definition 1.23. Fuzzy Partial Ordering Relation. Let $\tilde{R}$ be binary fuzzy relation on X . Then fuzzy relation $\tilde{R}$ is called fuzzy partial ordering, if it satisfies the following conditions:

1.Fuzzy relation $\tilde{R}$ is reflexive;
2.Fuzzy relation $\tilde{R}$ is antisymmetric;
3.Fuzzy relation $\tilde{R}$ is fuzzy transitive.

If fuzzy partial order is given on set X then we will say that set X is fuzzy partially ordered.

Next we consider projections and cylindric extension.

Let $\tilde{R}$ be n-dimensional fuzzy relation on Cartesian product $X = X_1 \times X_2 \times \ldots \times X_n$ of nonempty sets $X_1, X_2, \ldots, X_n$ and $(i_1, i_2, \ldots, i_k)$ be a subsequence of $(1, 2, \ldots, n)$.

The practice and experimental evidence have shown that decision theories developed for a perfect decision-relevant information and 'well-defined' preferences are not capable of adequate modeling of real-world decision making. The reason is that real decision problems are characterized by imperfect decision-relevant information and vaguely defined preferences. This leads to the fact that when solving real-world decision problems we need to move away from traditional decision approaches based on exact modeling which is good rather for decision analysis of thought experiments.

More concretely, the necessity to sacrifice the precision and determinacy is by the fact that real-world problems are characterized by perception-based information and choices, for which natural language is more covinient and close than precise formal approaches. Modeling decision making from this perspective is impossible without dealing with fuzzy categories near to human notions and imaginations. In this connection, it is valuable to use the notion of linguistic variable first introduced by L.Zadeh [119]. Linguistic variables allow an adequate reflection of approximate in-word descriptions of objects and phenomena in the case if there is no any precise deterministic description. It should be noted as well that many fuzzy categories described linguistically even appear to be more informative than precise descriptions.

Definition 1.24. Linguistic Variable. A linguistic variable is characterized by the set (u,T,X,G,M), where u is the name of variable; T denotes the term-set of u that refer to as base variable whose values range over a universe X ; G is a syntactic rule (usually in form of a grammar) generating linguistic terms; M is a semantic rule that assigns to each linguistic term its meaning, which is a fuzzy set on X .

A certain $t \in T$ generated by the syntactic rule G is called a term. A term consisting of one or more words, the words being always used together, is named an atomary term. A term consisting of several atomary terms is named a composite term. The concatenation of some components of a composite term (i.e. the result of linking the chains of components of the composite term) is called a subterm. Here $t_1, t_2, ...$ are terms in the following expression

$$T = t_1 + t_2 + ...$$

The meaning of $M(t)$ of the term t is defined as a restriction $R(t;x)$ on the basis variable x conditioned by the fuzzy variable $\tilde{X}$:

$$M(t) \equiv R(t;x)$$

it is assumed here that $R(t;x)$ and, consequently, $M(t)$ can be considered as a fuzzy subset of the set X named as t.

The assignment equation in case of linguistic variable takes the form in which t-terms in T are names generated by the grammar G, where the meaning assigned to the term t is expressed by the equality

$$M(t) = R(term\,in\,T)$$

In other words the meaning of the term t is found by the application of the semantic rule M to the value of term t assigned according to the right part of equation.

Moreover, it follows that $M(t)$ is identical to the restriction associated with the term t.

It should be noted that the number of elements in T can be unlimited and then for both generating elements of the set T and for calculating their meaning, the application of the algorithm, not simply the procedure for watching term-set, is necessary.

We will say that a linguistic variable u is structured if its term-set T and the function M , which maps each element from the term-set into its meaning, can be given by means of algorithm. Then both syntactic and semantic rules connected with the structured linguistic variable can be considered algorithmic procedures for generating elements of the set T and calculating the meaning of each term in T , respectively.

However in practice we often encounter term-sets consisting of a small number of terms. This makes it easier to list the elements of term-set T and establishes a direct mapping from each element to its meaning. For axample, an intuitive description of possible economic conditions may be represented by linguistic terms like "strong econonmic growth", "weak economic growth" etc. Then the term set of linguistic variable "state of economy" can be written as follows:

T(state of economy) = "strong growth" + "moderate growth" + "static situation" + "recession".

The variety of economic conditions may also be described by ranges of the important economic indicators. However, numerical values of indicators may not be sufficiently clear even for experts and may arise questions and doubts. In contrast, linguistic description is well perceived by human intuition as qualitative and fuzzy.

1.2 Classical and Extented Fuzzy Logic

First we consider classical fuzzy logic. We will consider the logics with multi-valued and continuous values (fuzzy logic). Let's define the semantic truth function of this logic. Let P be statement and $T(P)$ its truth value, where

$$T(P)\in[0,1]$$

Negation values of the statement P are defined as:

$$T(\neg P)=1-T(P).$$

Hence

$$T(\neg\neg P)=T(P).$$

The implication connective is always defined as follows:

$$T(P\to Q)=T(\neg P\vee Q),$$

and the equivalence as

$$T(P \leftrightarrow Q) = T\left[(P \rightarrow Q) \wedge (Q \rightarrow P)\right].$$

It should be noted that exclusive disjunction ex, disjunction of negations (Shiffer's connective) $|$, conjunction of negations $\downarrow$ and $\sim\!\rightarrow$ (has no common name) are defined as negation of equivalence $\leftrightarrow$, conjunction $\wedge$, disjunction $\vee$, and implication $\rightarrow$, respectively.

The tautology denoted $\bullet$ and contradiction denoted $\circ$ will be, respectively:

$$T\left(\overset{\bullet}{P}\right) = T(P \vee \neg P); T\left(\overset{\circ}{P}\right) = T(P \wedge \neg P).$$

More generally

$$T\left(\overset{\bullet}{PQ}\right) = T((P \vee \neg P) \vee (Q \vee Q))$$

$$T\left(\overset{\circ}{PQ}\right) = T((P \wedge \neg P) \wedge (Q \wedge Q))$$

Semantic Analysis of Different Fuzzy Logics. Let $\tilde{A}$ and $\tilde{B}$ be fuzzy sets of the subsets of non-fuzzy universe U ; in fuzzy set theory it is known that $\tilde{A}$ is a subset of $\tilde{B}$ iff

$$\mu_{\tilde{A}} \leq \mu_{\tilde{B}}, \text{i.e. } \forall x \in U, \quad \mu_{\tilde{A}}(x) \leq \mu_{\tilde{B}}(x).$$

Definition 1.25. Power Fuzzy Set. For given fuzzy implication $\rightarrow$ and fuzzy set $\tilde{B}$ from the universe U , the power fuzzy set $\tilde{P}\tilde{B}$ from $\tilde{B}$ is given by membership function $\mu_{\tilde{P}\tilde{B}}$ [3,19]:

$$\mu_{\tilde{P}\tilde{B}}\tilde{A} = \underset{X \in U}{\wedge} (\mu_{\tilde{A}}(x) \rightarrow \mu_{\tilde{B}}(x))$$

Then the degree to which $\tilde{A}$ is subset of $\tilde{B}$, is

$$\pi(\tilde{A} \subseteq \tilde{B}) = \mu_{\tilde{P}\tilde{B}}\tilde{A}$$

Definition 1.26. If fuzzy implication operator [3,19] is given on the closed unit interval [0,1] then

$$a \leftarrow b = b \rightarrow a$$

$$a \leftrightarrow b = (a \rightarrow b) \wedge (a \leftarrow b) = (a \rightarrow b) \wedge (a \leftarrow b)$$

Definition 1.27. Degree of "Equivalency". Under the conditions of the definition $P\tilde{B}$ the degree to which fuzzy sets $\tilde{A}$ and $\tilde{B}$ are equivalent is:

$$\pi\left(\tilde{A} \equiv \tilde{B}\right) = \pi\left(\tilde{A} \subseteq \tilde{B}\right) \wedge \pi\left(\tilde{B} \subseteq \tilde{A}\right);$$

or

$$\pi\left(\tilde{A} \equiv \tilde{B}\right) = \wedge_{x \in U} (\mu_{\tilde{A}} x \rightarrow \mu_{\tilde{B}} x)$$

For practical purposes [3,19] in most cases it is advisable to work with multi-valued logics in which logical variable takes values from the real interval $I = [0,1]$ divided into 10 subintervals, i.e. by using set $V_{11} = [0, 0.1, 0.2, ..., 1]$.

We denote the truth values of premises $\tilde{A}$ and $\tilde{B}$ through $T(\tilde{A}) = a$ and $T(\tilde{B}) = b$. The implication operation in analyzed logics [2,3,88] has the following form:

1) min-logic

$$a \underset{\min}{\rightarrow} b = \begin{cases} a, & if \ \ a \leq b \\ b, & otherwise. \end{cases}$$

2) $S^{\#}$ - logic

$$a \underset{S^{\#}}{\rightarrow} b = \begin{cases} 1, & if \ \ a \neq 1 \ \ or \ \ b = 1, \\ 0 & , otherwise. \end{cases}$$

3) S - logic ("Standard sequence")

$$a \underset{S}{\rightarrow} b = \begin{cases} 1, & if \ \ a \leq b, \\ 0, & otherwise. \end{cases}$$

4) G - logic ("Gödelian sequence")

$$a \underset{G}{\rightarrow} b = \begin{cases} 1, & if \ \ a \leq b, \\ b, & otherwise. \end{cases}$$

5) $G43$ - logic

$$a \underset{G43}{\rightarrow} b = \begin{cases} 1, & if \ \ a = 0, \\ \min(1, b/a), & otherwise. \end{cases}$$

6) L- logic (Lukasiewicz's logic)

$$a \underset{L}{\rightarrow} b = \min(1, 1-a+b).$$

7) KD - logic

$$a \underset{KD}{\rightarrow} b = ((1-a) \vee b = \max(1-a, b).$$

In turn ALI1-ALI4 - logics, suggested by us, which will be used in further chapters are characterized by the following implication operations [4,5]:

8) ALI1 – logic

$$a \underset{ALI1}{\rightarrow} b = \begin{cases} 1-a, & if \ a < b, \\ 1, & if \ a = b, \\ b, & if \ a > b \end{cases}$$

9) ALI2 - logic

$$a \underset{ALI2}{\rightarrow} b = \begin{cases} 1, & if \ a \le b, \\ (1-a) \wedge b, & if \ a > b \end{cases}$$

10) ALI3 - logic

$$a \underset{ALI3}{\rightarrow} b = \begin{cases} 1, & if \ a \le b, \\ b/[a+(1-b)], & otherwise. \end{cases}$$

11) ALI4 – logic

$$a \xrightarrow[ALI4]{} b = \begin{cases} \dfrac{1-\mathrm{a}+b}{2}, & \mathrm{a} > \mathrm{b}, \\ 1, & \mathrm{a} \le \mathrm{b}. \end{cases}$$

A necessary observation to be made in the context of this discussion is that with the only few exceptions for S-*logic* (3) and G-*logic* (4), and ALI1-ALI4 (8)-(11), all other known fuzzy logics (1)-(2), (5)-(7) do not satisfy either the classical "modus-ponens" principle, or other criteria which appeal to the human perception of mechanisms of a decision making process being formulated in [74]. The proposed fuzzy logics ALI1-ALI4 come with an implication operators, which satisfy the classical principle of "modus-ponens" and meets some additional criteria being in line with human intuition.

The comparative analysis of the first seven logics has been given in [19]. The analysis of these seven logics has shown that only S- and G- logics satisfy the classical principle of Modus Ponens and allow development of improved rule of fuzzy conditional inference. At the same time the value of truthness of the implication operation in G-logic is equal either to 0 or 1; and only the value of truthness of logical conclusion is used in the definition of the implication operation in S-logic. Thus the degree of "fuzziness" of implication is decreased, which is a considerable disadvantage and restricts the use of these logics in approximate reasoning.

Definition 1.28. Top of a Fuzzy Set. The top of fuzzy set $\tilde{B}$ is

$$H\tilde{B} = \vee_U \mu_{\tilde{B}}(x).$$

Definition 1.29. Bottom of a Fuzzy Set. The bottom of fuzzy set $\tilde{B}$ is

$$p\tilde{B} = \wedge_U \mu_{\tilde{B}}(x).$$

Definition 1.30. Nonfuzziness. Nonfuzziness $a \in U$ is $ka = a \vee (1-a)$. Then nonfuzziness of fuzzy set $\tilde{B}$ is defined as:

$$k\tilde{B} = \wedge_U k\mu_{\tilde{B}}(x)$$

Let us give a brief semantic analysis of the proposed fuzzy logics ALI1-ALI3 by using the terminology accepted in the theory of power fuzzy sets. For this purpose we formulate the following.

Proposal. Possibility degree of the inclusion of set $\pi(\tilde{A} \subseteq \tilde{B})$ in fuzzy logic ALI1-ALI3 is determined as:

$$\pi_1\left(\tilde{A} \subseteq \tilde{B}\right) = \begin{cases} 1-\mu_{\tilde{A}}(x), & if \quad \mu_{\tilde{A}}(x) < \mu_{\tilde{B}}(x), \\ 1, & if \quad \mu_{\tilde{A}}(x) = \mu_{\tilde{B}}(x), \\ \mu_{\tilde{B}}(x), & if \quad \mu_{\tilde{A}}(x) > \mu_{\tilde{B}}(x); \end{cases}$$

$$\pi_2\left(\tilde{A} \subseteq \tilde{B}\right) = \begin{cases} 1, & if \ \mu_{\tilde{A}}(x) \le \mu_{\tilde{B}}(x), \\ \left(1-\mu_{\tilde{A}}(x)\right) \wedge \mu_{\tilde{B}}(x), & if \ \mu_{\tilde{A}}(x) > \mu_{\tilde{B}}(x); \end{cases}$$

$$\pi_3\left(\tilde{A} \subseteq \tilde{B}\right) = \begin{cases} 1, & if \quad \mu_{\tilde{A}}(x) \le \mu_{\tilde{B}}(x), \\ \dfrac{\mu_{\tilde{B}}(x)}{\mu_{\tilde{A}}(x) + \left(1-\mu_{\tilde{B}}(x)\right)}, & if \quad \mu_{\tilde{A}}(x) > \mu_{\tilde{B}}(x). \end{cases}$$

We note, that if $\mu_{\tilde{A}}(x) = 0$ or $\tilde{A} \neq \varnothing$, then the crisp inclusion is possible for fuzzy logic ALI1. Below we consider the equivalence of fuzzy sets.

Proposal. Possibility degree of the equivalence of the sets $\pi(\tilde{A} \equiv \tilde{B})$ is determined as:

$$\pi_1\left(\tilde{A} \equiv \tilde{B}\right) = \begin{cases} 1-\left[\left(1-\mu_{\tilde{A}}(x)\right) \vee \mu_{\tilde{B}}(x)\right], & if \quad \mu_{\tilde{A}}(x) < \mu_{\tilde{B}}(x), \\ 1, & if \quad \tilde{A} = \tilde{B}, \\ 1-\left[\left(1-\mu_{\tilde{B}}(x)\right) \vee \mu_{\tilde{A}}(x)\right], & if \quad \mu_{\tilde{A}}(x) > \mu_{\tilde{B}}(x), \end{cases}$$

$$\pi_2\left(\tilde{A} \equiv \tilde{B}\right) = \begin{cases} 1, & if \quad \tilde{A} = \tilde{B}, \\ \wedge_T \left\{\left[\left(1-\mu_{\tilde{A}}(x)\right) \wedge \mu_{\tilde{B}}(x)\right], \left[\left(1-\mu_{\tilde{B}}(x)\right) \wedge \mu_{\tilde{A}}(x)\right]\right\} & if \quad \tilde{A} \neq \tilde{B}, \\ 0, \quad if\ \exists x \,|||\, \mu_{\tilde{A}}(x) = 0, \mu_{\tilde{B}}(x) \neq 0 & (or\ vice\ versa), \\ and\ \ also\ \exists x \,|||\, \mu_{\tilde{A}}(x) = 1,\ \mu_{\tilde{B}}(x) \neq 1 & (or\ vice\ versa), \end{cases}$$

$$\pi_3\left(\tilde{A} \subseteq \tilde{B}\right) = \begin{cases} 1, & if \quad \mu_{\tilde{A}}(x) \leq \mu_{\tilde{B}}(x), \\ \dfrac{\mu_{\tilde{B}}(x)}{\mu_{\tilde{A}}(x) + \left(1-\mu_{\tilde{B}}(x)\right)} & ,if \quad \mu_{\tilde{A}}(x) > \mu_{\tilde{B}}(x). \end{cases}$$

Here the set $T = \{x \in U \,|\, \mu_{\tilde{A}} x \neq \mu_{\tilde{B}} x\}$ and $\tilde{A} = \tilde{B}$ means that $\forall x$

$\mu_{\tilde{A}}(x) = \mu_{\tilde{B}}(x)$ or in other words, $T = \varnothing$.

The symbol $|||$ means "such as ". From the expression $\pi_i\left(\tilde{A} \equiv \tilde{B}\right)$, $i = \overline{1,3}$, it follows that for ALI1 fuzzy logic the equivalency $\pi_1\left(\tilde{A} \equiv \tilde{B}\right) = 1$ takes place only when $\tilde{A} = \tilde{B}$. It is obvious that the equivalence possibility is equal to 0 only in those cases when one of the statements is crisp, i.e. either true or false, while the other is fuzzy.

Proposal. Degree to which fuzzy set $\tilde{B}$ is empty $\pi(\tilde{B} \equiv \varnothing)$ is determined as

$$\pi_1\left(\tilde{B} \equiv \varnothing\right) = \begin{cases} 1, & if \quad \tilde{B} = \varnothing, \\ 0, & otherwise; \end{cases}$$

$$\pi_2\left(\tilde{B} \equiv \varnothing\right) = \begin{cases} 1, \ if \quad H\tilde{B} < 1 \ or \ \tilde{B} = \varnothing, \\ 0, \ otherwise; \end{cases}$$

$$\pi_3\left(\tilde{B} \equiv \varnothing\right) = \begin{cases} 1, & if \quad \tilde{B} = \varnothing, \\ 0, & otherwise. \end{cases}$$

Here $\tilde{B}=\varnothing$ means that for $\forall x\, \mu_{\tilde{B}}(x)=0$, or equivalently $H\tilde{B}=0$.

We introduce the concept of disjointness of fuzzy sets. There are two kinds of the disjointness. For a set $\tilde{A}$ the first kind is defined by degree to which set $\tilde{A}$ is a subset of the complement of $\tilde{B}^c$. The second kind is the degree to which the intersection of sets is empty. Therefore, we formulate the following.

Proposal. Degree of disjointness of sets $\tilde{A}$ and $\tilde{B}$ is degree to which $\tilde{A}$ and $\tilde{B}$ are disjoint

$$\pi\left(\tilde{A}\ disj_1\ \tilde{B}\right)=\pi\left(\tilde{A}\subseteq\tilde{B}^C\right)\wedge\pi\left(\tilde{B}\subseteq\tilde{A}^C\right),$$

$$\pi\left(\tilde{A}\ disj_2\ \tilde{B}\right)=\pi\left(\left(\tilde{A}\cap\tilde{B}\right)=\varnothing\right).$$

Proposal. Disjointness grade of sets $\tilde{A}$ and $\tilde{B}$ is determined as

$$\pi_1\left(\tilde{A}\ disj_1\ \tilde{B}\right)=\begin{cases}1, & if\ \ \exists x\,|||\,\mu_{\tilde{A}}(x)=1-\mu_{\tilde{B}}(x),\\ \left(1-\mu_{\tilde{A}}(x)\right)\wedge\left(1-\mu_{\tilde{B}}(x)\right), & otherwise,\\ 0, & never;\end{cases}$$

$$\pi_2\left(\tilde{A}\ disj_1\ \tilde{B}\right)=\begin{cases}1, & if\ \mu_{\tilde{A}}(x)\le 1-\mu_{\tilde{B}}(x),\\ 0, & if\ \exists x\,|||\,\mu_{\tilde{A}}(x)=1,\ \ but\ \ \mu_{\tilde{B}}(x)\ne 0,\\ & or\ \mu_{\tilde{B}}(x)=1,\ but\ \mu_{\tilde{A}}(x)\ne 0,\\ \underset{T}{\wedge}\left[\left(1-\mu_{\tilde{A}}(x)\right),\left(1-\mu_{\tilde{B}}(x)\right)\right], & otherwise;\end{cases}$$

$$\pi_3\left(\tilde{A}\ disj_1\ \tilde{B}\right)=\begin{cases}1, & if\ \ \mu_{\tilde{A}}(x)=\mu_{\tilde{B}}(x)\ \ or\ \mu_{\tilde{B}}(x)=0,\\ \underset{T}{\wedge}\left[\dfrac{1-\mu_{\tilde{B}}(x)}{\mu_{\tilde{A}}(x)+\left(1-\mu_{\tilde{B}}(x)\right)},\dfrac{1-\mu_{\tilde{A}}(x)}{\mu_{\tilde{B}}(x)+\left(1-\mu_{\tilde{A}}(x)\right)}\right], & otherwise,\\ 0, & never.\end{cases}$$

here $T=\{x\,|||\,\mu_{\tilde{A}}(x)>1-\mu_{\tilde{B}}(x)\}$.

We note that, the disjointness degree of the set is equal to 0 only for fuzzy logic ALI2, when under the condition that one of the considered fuzzy sets is normal, the other is subnormal.

Proposal. Degree to which set is a subset of its complement for the considered fuzzy logics $\pi_i(\tilde{A}\subseteq\tilde{B}^C)$ takes the following form

$$\pi_1(\tilde{A} \subseteq \tilde{A}^C) = \begin{cases} 1, & if\ H\tilde{A} = 0, \\ 0, & if\ H\tilde{A} = 1, \\ 1 - H\tilde{A}, & otherwise; \end{cases}$$

$$\pi_2(\tilde{A} \subseteq \tilde{A}^C) = \begin{cases} 1, & if\ H\tilde{A} \leq 0, \\ 0, & if\ H\tilde{A} = 1, \\ 1 - H\tilde{A}, & otherwise; \end{cases}$$

$$\pi_3(\tilde{A} \subseteq \tilde{A}^C) = \begin{cases} 1, & if\ H\tilde{A} \leq 0.5, \\ 0, & if\ H\tilde{A} = 1, \\ (1 - H\tilde{A})/(2H\tilde{A}), & otherwise; \end{cases}$$

It is obvious that for the fuzzy logic ALI1 the degree to which a set is the subset of its complement is equal to the degree to which this set is empty. It should also be mentioned that the semantic analysis given in [6,8,9] as well as the analysis given above show a significant analogy between features of fuzzy logics ALI1 and KD. However, the fuzzy logic ALI1, unlike the KD logic, has a number of advantages. For example, ALI1 logic satisfies the condition $\mu_{\tilde{A}}x \wedge (\mu_{\tilde{A}}x \rightarrow \mu_{\tilde{B}}x) \leq \mu_{\tilde{B}}x$ necessary for development of fuzzy conditional inference rules. ALI2 and ALI3 logics satisfy this inequality as well. This allows them to be used for the formalization of improved rules of fuzzy conditional inference and for the modeling of relations between main elements of a decision problem under uncertainty and interaction among behavioral factors.

Extended Fuzzy Logic [127]

Fuzzy logic adds to bivalent logic an important capability—a capability to reason precisely with imperfect information. In classical fuzzy logic, results of reasoning are expected to be provably valid, or p-valid for short. Extended fuzzy logic adds an equally important capability—a capability to reason imprecisely with imperfect information. This capability comes into play when precise reasoning is infeasible, excessively costly or unneeded. In extended fuzzy logic, p-validity of results is desirable but not required. What is admissible is a mode of reasoning which is fuzzily valid, or f-valid for short. Actually, much of everyday human reasoning is f-valid reasoning. What is important to note is that f-valid reasoning based on a realistic model may be more useful than p-valid reasoning based on an unrealistic model. As John Maynard Keynes states, "*It is better to be roughly right than precisely wrong*" In constructing better models of reality, a problem that has to be faced is that as the complexity of a system, increases, it becomes increasingly difficult to construct a model, which is both cointensive, that is, close-fitting, and precise. This applies, in particular, to systems in which human judgment, perceptions and emotions play a prominent role. Economic systems, legal systems and political systems are cases in point. As the complexity of a system increases further, a point is reached at which construction of a model which is both cointensive and precise is not merely difficult—it is impossible. It is at this point that extended fuzzy logic comes into play. Actually, extended fuzzy logic is not the only formalism that comes into play at this point. The issue of what to do when an exact solution cannot be found or is excessively costly is associated with a vast literature. Prominent in this literature are various approximation theories [16], theories centered on bounded rationality [100], qualitative reasoning [106], commonsense reasoning [65,78] and theories of argumentation [101]. Extended fuzzy logic differs from these and related theories both in spirit and in substance. The difference will become apparent in Section 1.3, in which the so-called f -geometry is used as an illustration. To develop an understanding of extended fuzzy logic, it is expedient to start with the following definition of classical fuzzy logic. Classical fuzzy logic is a precise conceptual system of reasoning, deduction and computation in which the objects of discourse and analysis are, or are allowed to be, associated with imperfect information. In fuzzy logic, the results of reasoning, deduction and computation are expected to be provably valid (p -valid) within the conceptual structure of fuzzy logic. In fuzzy logic precision is achieved through association of fuzzy sets with membership functions and, more generally, association of granules with generalized constraints [126]. What this implies is that classical fuzzy logic is what may be called precisiated logic.

At this point, a key idea comes into play. The idea is that of constructing a fuzzy logic, which, in contrast to classical, is unprecisiated. What this means is that in unprecisiated fuzzy logic UFL membership functions and generalized constraints are not specified, and are a matter of perception rather than measurement. A question which arises is: What is the point of constructing UFL - a logic in which provable validity is off the table? But what is not off the table is what may be called fuzzy validity, or f -validity for short. As will be shown in section 1.3 a model of UFL is f -geometry. Actually, everyday human reasoning is preponderantly f -valid reasoning. Humans have a remarkable capability to perform a wide variety of physical and mental tasks without any measurements and any computations. In this context, f -valid reasoning is perception-based. The concept of unprecisiated fuzzy logic provides a basis for the concept of extended fuzzy logic, EFL. More specifically, EFL is the result of adding UFL to classical fuzzy logic. Basically, extended fuzzy logic. Effect, extended fuzzy logic adds to fuzzy logic a capability to deal imprecisely with imperfect information when precision is infeasible, carries a high cost or is unneeded. This capability is a necessity when repeated attempts at constructing a theory which is both realistic and precise fail to achieve success. Cases in point are the theories of rationality, causality and decision-making under second order uncertainty, that is, uncertainty about uncertainty. There is an important point to be made. f -Validity is a fuzzy concept and hence is a matter of degree. When a chain of reasoning leads to a conclusion, a natural question is: What is the possibly fuzzy degree of validity, call it the validity index, of the conclusion? In most applications involving f -valid reasoning a high validity index is a desideratum. How can it be achieved? Achievement of a high validity index is one of the principal objectives of extended fuzzy logic. The importance of extended fuzzy logic derives from the fact that it adds to fuzzy logic an essential capability—the capability to deal with unprecisiated imperfect information.

1.3 Fuzzy Analyses and Fuzzy Geometry

In this section we concern with the necessary concepts related to the calculus of fuzzy set-valued mappings, for short fuzzy functions. Let X be an arbitrary set.A family τ of fuzzy sets in X is called a fuzzy topology for X and the pair (X,τ) a fuzzy topological space if: (i) $\mu_X \in \tau$ and $\mu_\phi \in \tau$; (ii) $\cup_{i\subset I} \tilde{A}_i \in \tau$ whenever each $\tilde{A}_i \in \tau (i \in I)$; and (iii) $\tilde{A} \cap \tilde{B} \in \tau$ whenever $\tilde{A}, \tilde{B} \in \tau$ [25].

Definition 1.31. Fuzzy Function [25]. A fuzzy function $\tilde{f}$ from a set X into a set Y assigns to each x in X a fuzzy subset $\tilde{f}(x)$ of Y. We denote it by $\tilde{f}: X \to Y$. We can identify $\tilde{f}$ with a fuzzy subset $G_{\tilde{f}}$ of $X \times Y$ and $\tilde{f}(x)(y) = G_{\tilde{f}}(x,y)$.

If $\tilde{A}$ is a fuzzy subset of X, then the fuzzy set $\tilde{f}(\tilde{A})$ in Y is defined by

$$\tilde{f}(\tilde{A})(y) = \sup_{x\in X}[G_{\tilde{f}}(x,y) \wedge \tilde{A}(x)]$$

The graph $\tilde{G}_{\tilde{f}}$ of $\tilde{f}$ is the fuzzy subset of X×Y associated with $\tilde{f}$,

$$\tilde{G}_{\tilde{f}} = \{(x,y) \in X \times Y : [\tilde{f}(x)](y) \neq 0 \ \}$$

Let X be a fuzzy topological space. Neighborhood of a fuzzy set $\tilde{A} \subset X$ is any fuzzy set $\tilde{B}$ for which there is an open fuzzy set $\tilde{V}$ satisfying $\tilde{A} \subset \tilde{V} \subset \tilde{B}$. Any open fuzzy set $\tilde{V}$ that satisfies $\tilde{A} \subset \tilde{V}$ is called an open neighborhood of $\tilde{A}$.

A fuzzy function $\tilde{f}: X \to Y$ between two fuzzy topological spaces X and Y is: upper semicontinuous at the point x, if for every open neighborhood U of $\tilde{f}(x), \tilde{f}^u(U)$ is a neighborhood of x in X; lower semicontinuous at x, if for every open fuzzy set $\tilde{V}$ which intersects $\tilde{f}(x)$, $f^l(U)$ is a neighborhood of x; and continuous if it is both upper and lower semicontinuous.

Let $\mathcal{E}^n$ [34,62] be a space of all fuzzy subsets of $\mathcal{R}^n$. These subsets satisfy the conditions of normality, convexity, and are upper semicontinuous with compact support.

Definition 1.32. Fuzzy Closeness [25]. A function $\tilde{f}: X \to Y$ between two fuzzy topological spaces is fuzzy closed or has fuzzy closed graph if its graph is a closed fuzzy subset of X×Y

Definition 1.33. Composition [25]. Let $\tilde{f}: X \to Y$ and $\tilde{g}: Y \to Z$ be two fuzzy functions. The composition $\tilde{g} \circ \tilde{f}: X \to Z$ $\tilde{f}: X \to Z$ is defined by $(g \circ f)(x) = \; = \cup\{g(y): [f(x)](y) \neq 0\}$.

Theorem 1.2. Convex Hull of a Fuzzy Set [25]. *Let X, Y and Z be three fuzzy topological spaces. Let $\tilde{f}: X \to Y$ and $\tilde{g}: Y \to Z$ be two fuzzy functions. Then*

$$(i)\ (\tilde{g}_0\tilde{f})^u(\tilde{A}) = \tilde{f}^u(\tilde{g}^u(\tilde{A}))$$

and

$$(ii)\ (\tilde{g}_0\tilde{f})^l(\tilde{A}) = \tilde{f}^l(\tilde{g}^l(\tilde{A}))$$

where $\tilde{A}$ is an open fuzzy subset of Z.

A fuzzy set $\tilde{A}$ in $\tilde{E}$ is called convex if for each $t \in [0,1], [tA + (1-t)A](x) \le A(x)$. The convex hull of a fuzzy set $\tilde{B}$ is smallest convex fuzzy set containing $\tilde{B}$ and is denoted by $\tilde{c}_0(\tilde{B})$.

Definition 1.34. Fuzzy Topological Vector Space [25]. A fuzzy linear topology on a vector space E over K is a fuzzy topology τ on E such that the two mappings:

$$f: E \times E \to E,\ f(x,y) = x + y,$$

$$h: K \times E \to E,\ h(t,x) = tx,$$

are continuous when K has the usual fuzzy topology and $K \times E, E \times E$ the corresponding product fuzzy topologies. A linear space with a fuzzy linear topology is called a fuzzy topological vector space. A fuzzy topological vector space E is called locally convex if it has a base at origin of convex fuzzy sets.

Definition 1.35. Fuzzy Multivalued Functions [25]. If $\tilde{f}, \tilde{g}: X \to Y$ are two fuzzy multivalued functions, where Y is a vector space, then we define:

(1) The sum fuzzy multivalued function $\tilde{f} + \tilde{g}$ by

$$(\tilde{f} + \tilde{g})(x) = \tilde{f}(x) + \tilde{g}(x) = \{y + z : y \in \tilde{f}(x) \text{ and } z \in \tilde{g}(x)\}$$

(2) The convex hull of a fuzzy multivalued function $\tilde{c}_0(\tilde{f})$ of $\tilde{f}$ by

$$(\tilde{c}_0(\tilde{f}))(x) = \tilde{c}_0(\tilde{f}(x)).$$

(3) If Y is a fuzzy topological vector space, the closed convex hull of a fuzzy multivalued function $cl(\tilde{c}_0(\tilde{f}))$ of $\tilde{f}$ by

$$(cl(\tilde{c}_0(\tilde{f})))(x) = cl(\tilde{c}_0(\tilde{f}(x)))$$

Below a definition of measurability of fuzzy mapping $\tilde{F}: T \to \mathcal{E}^n$ is given.

Definition 1.36. Measurability of Fuzzy Mapping [34,62]. We say that a mapping $\tilde{F}: T \to \mathcal{E}^n$ is strongly measurable if for all $\alpha \in [0,1]$ the set-valued mapping $F_\alpha : T \to P_K(\mathcal{R}^n)$ defined by

$$F_\alpha(t) = [F(t)]^\alpha$$

is (Lebesgue) measurable , when $P_K(\mathcal{R}^n)$ is endowed with the topology generated by the Hausdorff metric d_H .

If $\tilde{F}: T \to \mathcal{E}^n$ is continuous with respect to the metric d_H then it is strongly measurable [34,62].

A mapping $\tilde{F}: T \to \mathcal{E}^n$ is called integrably bounded if there exists an integrable function h such that $\| x \| \le h(t)$ for all $x \in \tilde{F}_0(t)$.

Definition 1.37. Integrability of Fuzzy Mapping [34,62]. Let $\tilde{F}: T \to \mathcal{E}^n$. The integral of $\tilde{F}$ over T , denoted $\int_T \tilde{F}(t)dt$ or $\int_a^b \tilde{F}(t)dt$, is defined levelwise by the equation

$$[\int_T \tilde{F}(t)dt] = \int_T F_\alpha(t)dt = \{ \int_T f(t)dt \mid f: T \to \mathcal{R}^n \text{ is a measurable selection for } F_\alpha \}$$

for all $0 < \alpha \le 1$. A strongly measurable and interably bounded mapping $\tilde{F}: T \to \mathcal{E}^n$ is said to be integrable over T if $\int_T \tilde{F}(t)dt \in \mathcal{E}^n$.

Hausdorff Distance [34,62]. Let $P_K(R^n)$ denote the family of all nonempty compact convex subsets of R^n and define the addition and scalar multiplication in $P_K(R^n)$ as usual. Let C and D be two nonempty bounded subsets of R^n . The distance between C and D is defined by using the Hausdorff metric

$$d_H(C,D) = \left\{ \max(\sup_{c \in C} \inf_{d \in D} \|c-d\|, \sup_{d \in D} \inf_{c \in C} \|c-d\|) \right\} \tag{1.53}$$

where $\|\cdot\|$ denotes the usual Euclidean norm in R^n. Then it becomes clear that $(P_K(R^n), d_H)$ becomes a metric space.

The next necessary concept that will be used in the sequel is the concept of difference of two elements of $\mathcal{E}^n$ referred to as Hukuhara difference:

Definition 1.38. Hukuhara Difference [34,62]. Let $\tilde{X}, \tilde{Y} \in \mathcal{E}^n$. If there exists $\tilde{Z} \in \mathcal{E}^n$ such that $\tilde{X} = \tilde{Y} + \tilde{Z}$, then $\tilde{Z}$ is called a Hukuhara difference of $\tilde{X}$ and $\tilde{Y}$ and is denoted as $\tilde{X} -_h \tilde{Y}$.

Note that with the standard fuzzy difference for $\tilde{Z}$ produced of $\tilde{X}$ and $\tilde{Y}$, $\tilde{X} \neq \tilde{Y} + \tilde{Z}$. We use Hukuhara difference when we need $\tilde{X} = \tilde{Y} + \tilde{Z}$.

Let us consider and example. Let $\tilde{X}$ and $\tilde{Y}$ be triangular fuzzy sets $\tilde{X} = (3,7,11)$ and $\tilde{Y} = (1,2,3)$. Then Hukuhara difference of $\tilde{X}$ and $\tilde{Y}$ is $\tilde{X} -_h \tilde{Y} = (3,7,11) -_h (1,2,3) = (3-1, 7-2, 11-3) = (2,5,8)$ Indeed, $\tilde{Y} + (\tilde{X} -_h \tilde{Y}) = (1,2,3) + (2,5,8) = (3,7,11) = \tilde{X}$.

Definition 1.39. Fuzzy Hausdorff Distance [10,11]. Let $\tilde{A}, \tilde{B} \in \mathcal{E}^n$. The fuzzy Hausdorff distance $\tilde{d}_{fH}$ between $\tilde{A}$ and $\tilde{B}$ is defined as

$$\tilde{d}_{fH}(\tilde{A}, \tilde{B}) = \bigcup_{\alpha \in [0,1]} \alpha \left[d_H(A^1, B^1), \sup_{\alpha \le \bar{\alpha} \le 1} d_H(A^{\bar{\alpha}}, B^{\bar{\alpha}}) \right], \tag{1.54}$$

where d_H is the Hausdorff distance [34,62] and A^1, B^1 denote the cores ($\alpha = 1$ level sets) of fuzzy sets $\tilde{A}, \tilde{B}$ respectively. Let us denote by $\|\tilde{A} -_h \tilde{B}\| = d_{fH}(\tilde{A} -_h \tilde{B}, \hat{0})$ a fuzzy norm of the Hukuhara difference. We note that $d_{fH}(\tilde{A} -_h \tilde{B}, \hat{0}) = d_{fH}(\tilde{A}, \tilde{B})$. We will be using this difference in further considerations.

Let us consider a small example. Let $\tilde{A}$ and $\tilde{B}$ be triangular fuzzy sets $\tilde{A} = (2,3,4)$ and $\tilde{B} = (6,8,12)$. Then the fuzzy Hausdorff distance d_{fH} between $\tilde{A}$ and $\tilde{B}$ is defined as a triangular fuzzy set $\tilde{d}_{fH}(\tilde{A}, \tilde{B}) = (5,5,8)$.

Fuzzy Norms. Let $\tilde{x}, \tilde{y} \in E^n$. We denote by $\|\tilde{x} -_h \tilde{y}\|_{fH}$ a fuzzy norm defined as

$$\|\tilde{x} -_h \tilde{y}\|_{fH} = d_{fH}(\tilde{x}, \tilde{y}). \tag{1.55}$$

It is the fuzzy Haussdorf distance mentioned above.

Let $\tilde{u} = (\tilde{u}_1, \tilde{u}_2, ..., \tilde{u}_n) \in E^n$. We denote by $\|\tilde{u}\|_f$ a fuzzy norm defined by the formula

$$\|\tilde{u}\|_f = |\tilde{u}_1| + |\tilde{u}_2| + ... + |\tilde{u}_n|. \tag{1.56}$$

where $|.|$ is the absolute value of a fuzzy number [3,7].

Derivatives of Fuzzy Functions and Fuzzy Derivatives [46,52]. It is necessary to distinguish between the following cases:

–we are given a fuzzy function and our interest is to determine its derivative at a particular point a (see Fig 1.3 (a));

–we have a function but the information about the point $\tilde{a}$ at which we are to consider the derivative is vague (uncertain) (see Fig 1.3 (b));

–we have a fuzzy function and we are interested in its derivative at a vague point $\tilde{a}$ (see Fig 1.3 (c)).

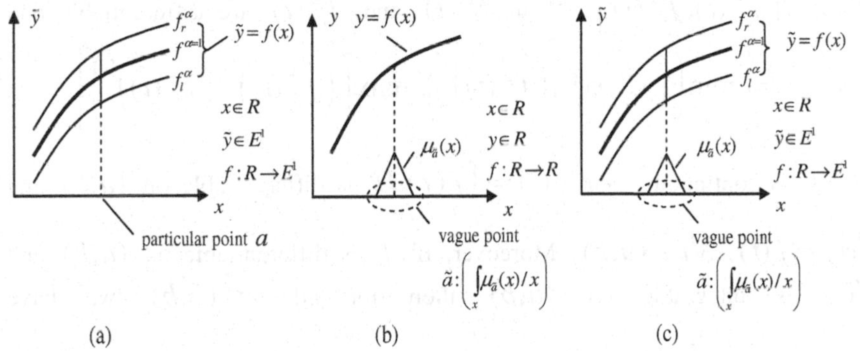

Fig. 1.3 Derivatives of fuzzy functions and fuzzy derivatives

In this paper, we analyze the situations in which the points are not exactly known, and therefore they need to be substituted by subjective and vague estimates, viz. could be treated as fuzzy sets (numbers) defined over some interval.

Strongly Generalized Differentiability [24]. Let $f:(a, b) \to E^n$ and $t_0 \in (a,b)$. We say that f is strongly generalized differentiable at t_0 if there exists an element $f'(t_0) \in E^n$, such that

a) for all $h>0$ sufficiently small, $\exists\ f(t_0+h)-_h f(t_0)$, $f(t_0)-_h f(t_0-h)$ (i.e. the length of $diam\left((f(t))^\alpha\right)$ increases) and the limits (in the supremum metric [34])

$$\lim_{h\to 0^+}\frac{f(t_0+h)-_h f(t_0)}{h}=\lim_{h\to 0^+}\frac{f(t_0)-_h f(t_0-h)}{h}=f'(t_0),$$

or

b) for all $h>0$ sufficiently small, $\exists\ f(t_0)-_h f(t_0+h)$, $f(t_0-h)-_h f(t_0)$ (i.e. the length of $diam\left((f(t))^\alpha\right)$ decreases) and the limits (in the supremum metric [34]) interval

$$\lim_{h\to 0^+}\frac{f(t_0)-_h f(t_0+h)}{(-h)}=\lim_{h\to 0^+}\frac{f(t_0-h)-_h f(t_0)}{(-h)}=f'(t_0),$$

(h and $(-h)$ shown in the denominators mean $1/h$ and $1/(-h)$ respectively).

Let $f:(a,b)\to E^1$ be a differentiable function. We introduce the notation $f^\alpha(t)=\left[f_l^\alpha(t),f_r^\alpha(t)\right]$. Then $f_l^\alpha(t)$ and $f_r^\alpha(t)$ are differentiable and

$$(f'(t))^\alpha=\left[\min\left(\left(f_l^\alpha(t)\right)',\left(f_r^\alpha(t)\right)'\right),\max\left(\left(f_l^\alpha(t)\right)',\left(f_r^\alpha(t)\right)'\right)\right].$$

If f is continuous then $g(t)=\int_a^t f(\tau)d\tau$ is differentiable on (a,b) and $g'(t)=f(t)$, $\forall t\in(a,b)$. Moreover, if f is differentiable on (a,b) and $f'(\cdot)$ is integrable on (a,b) then for all $t\in(a,b)$ we have $f(t)=f(t_0)+\int_{t_0}^t f'(\tau)d\tau$, $a<t_0\le t<b$.

Possibility Measure [3,110,121]. Given two fuzzy sets defined in the same universe of discourse $\mathbf{X}$, a fundamental question arises as to their similarity or proximity. There are several well-documented approaches covered in the literature. One of them concerns a *possibility measure*. The possibility measure, denoted by $Poss(\tilde{A},\tilde{X})$ describes a level of overlap between two fuzzy sets and is expressed in the form

$$Poss(\tilde{A},\tilde{X}) = \sup_{x\in X}\left[\tilde{A}(x)t\tilde{X}(x)\right],$$

where t is a t-norm. Computationally, we note that the possibility measure is concerned with the determination of the intersection between $\tilde{A}$ and $\tilde{X}$, $\tilde{A}(x)t\tilde{X}(x)$, that is followed by the optimistic assessment of this intersection. It is done by picking up the highest values among the intersection grades of $\tilde{A}$ and $\tilde{X}$ that are taken over all elements of the universe of discourse $\mathbf{X}$. For example, let $\tilde{a}$ and $\tilde{b}$ be fuzzy sets with trapezoidal membership functions:

$$\mu_{\tilde{a}}(x)=\begin{cases}1-\dfrac{a_1-x}{\alpha_l}, & \text{if } a_1-\alpha_l\le x\le a_1\\ 1, & \text{if } a_1\le x\le a_2\\ 1-\dfrac{x-a_2}{\alpha_r}, & \text{if } a_2-\alpha_r\le x\le a_2\\ 0, & \text{otherwise}\end{cases}\qquad \mu_{\tilde{b}}(x)=\begin{cases}1-\dfrac{b_1-x}{\beta_l}, & \text{if } b_1-\beta_l\le x\le b_1\\ 1, & \text{if } b_1\le x\le b_2\\ 1-\dfrac{x-b_2}{\beta_r}, & \text{if } b_2-\beta_r\le x\le b_r\\ 0, & \text{otherwise}\end{cases}$$

The graphs of the corresponding membership functions $\mu_{\tilde{a}}(x)$ and $\mu_{\tilde{b}}(x)$ are shown in Fig.1.4.

Then the possibility measure of the proposition "$\tilde{a}$ is equal to $\tilde{b}$" is defined as follows:

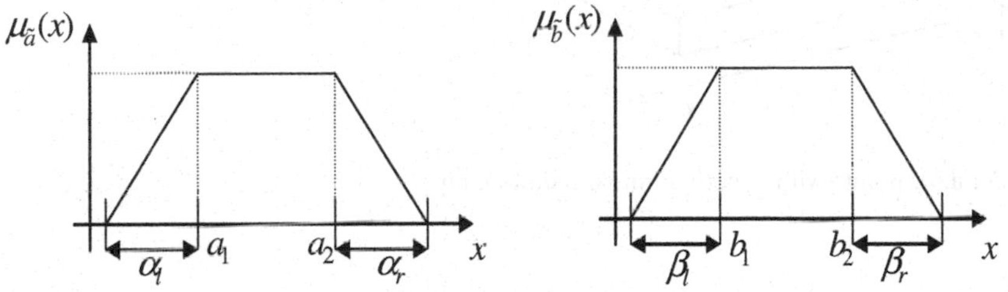

Fig. 1.4 Trapezoidal fuzzy numbers $\tilde{a}$ and $\tilde{b}$

$$(\tilde{a}=\tilde{b})=Poss(\tilde{a}/\tilde{b})=\max\min(\mu_{\tilde{a}}(x),\mu_{\tilde{b}}(x))=\begin{cases}1-\dfrac{a_1-b_2}{\alpha_l+\beta_r}, & \text{if } 0<a_1-b_2<\alpha_l+\beta_r\\ 1, & \text{if } \max(a_1,b_1)\le\min(a_2,b_2)\\ 1-\dfrac{b_1-a_2}{\alpha_r+\beta_l}, & \text{if } 0<b_1-a_2<\alpha_r+\beta_l\\ 0, & \text{otherwise}\end{cases} \tag{1.57}$$

Fuzzy Geometry

In general, fuzzy geometry may be considered as extension of conventional geometry to the fuzzy case [29,77,90-94,96,107]. Fuzzy geometry includes the topological concepts of area, perimeter, compactness, length, adjacency etc. These measures can be used to reflect the ambiguity in decision relevant information.

Definition 1.40. Fuzzy Point. Fuzzy point $\tilde{x}_0$ is a convex fuzzy subset of R^i. Fuzzy point in R is characterized by kernel x_0 whose precise location is only approximately known.

A crisp point $x_0 \in R^i$ is the kernel, from which membership function decreases in all directions monotonically [17]. In Fig. 1.5 and Fig. 1.6 fuzzy points with hyperpyramidal (Fig.1.5) and hyperparaboloidal (Fig. 1.6) membership functions are shown. In first case imprecision of location of fuzzy point is expressed by intervals for the components, in second case by definite matrix in all directions of the space.

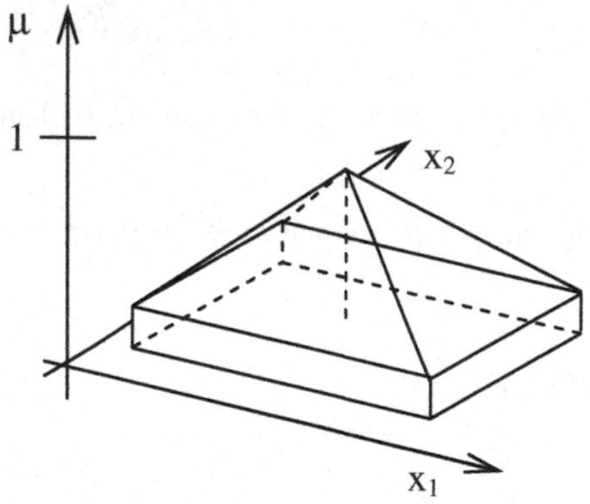

Fig. 1.5 Fuzzy points with hyperpyramidal membership

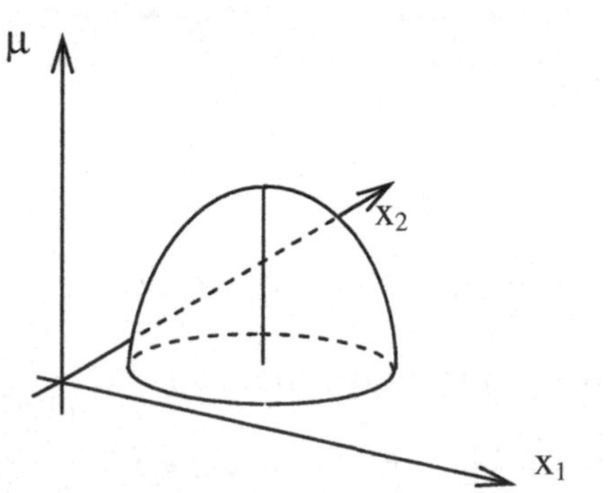

Fig. 1.6 Fuzzy points with hyperparaboloidal membership

Definition 1.41. Fuzzy Interval. If fuzzy domain I of the real line R is bounded by two normalized convex fuzzy sets then it is called fuzzy interval. In Fig. 1.7 fuzzy interval with fuzzy ends $\mu_{\tilde{a}}(x)$ and $\mu_{\tilde{b}}(x)$ is given. A crisp interval [a, b] is the kernel, from which the membership function decreases to zero [17]. Analogously, fuzzy region in R^i is represented as a crisp region, which is surrounded by a fuzzy transition zone, in which the membership function decreases monotonically to zero [18].

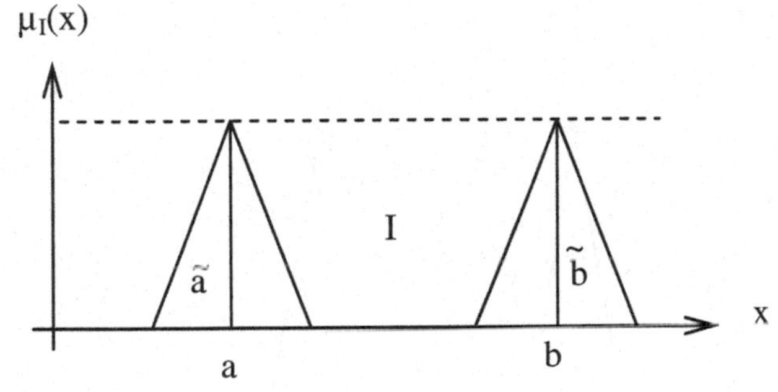

Fig. 1.7 Fuzzy interval

Definition 1.42. Length of a Fuzzy Interval. Length of fuzzy interval I is defined as

$$L(I) = \int_{I_0} \mu_I(x)dx$$

Here $I_0 = cl\{x \mid \mu_I(x) > 0\}$ is support of fuzzy interval.

Definition 1.43. Distance between Fuzzy Points. A distance between two points $d(x_1, x_2)$ by using the extension principle translates to a fuzzy distance between fuzzy sets.

The fuzzy distance between two fuzzy sets $\tilde{A}$ and $\tilde{B}$ on X (X is metric space) is defined as [18,81]

$$\mu_{d(\tilde{A},\tilde{B})}(y) = \sup_{\substack{(x_1,x_2)\in X\times X \\ d(x_1,x_2)=y}} \min(\mu_{\tilde{A}}(x_1), \mu_{\tilde{B}}(x_2))$$

Example. Fuzzy distance in case when $X = R^1$, $d(x_1, x_2) = |x_1 - x_2|$ is shown in Fig 1.8.

Some distances frequently used in practical problems are given below (for R^1):

$$d(x_1,x_2)=(x_1-x_2)^2 \quad \text{Euclidean distance}$$
$$d(x_1,x_2)=(|x_1-x_2|^p)^{1/p} \quad \text{Minkowski metric}$$
$$d(x_1,x_2)=c\,|x_1-x_2| \quad \text{Tschebyscheff metric}$$
$$d(x_1,x_2)=|x_1-x_2| \quad \text{Hamming distance}$$

Definition 1.44. Fuzzy Area. The area of fuzzy subset is defined as the area of fuzzy subset $\tilde{A}$ given on R^2 is defined as

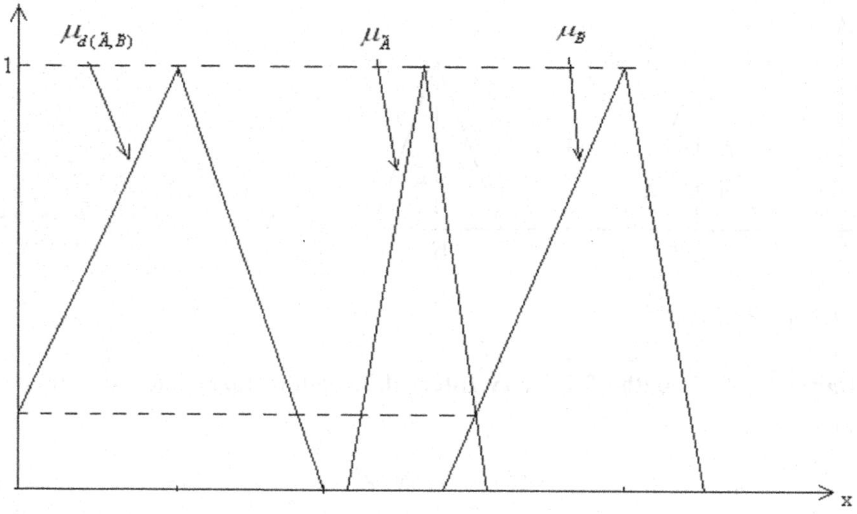

Fig 1.8 Fuzzy distance between fuzzy sets

$$S(\tilde{A})=\iint_{\tilde{A}_0}\mu_{\tilde{A}}(x,y)dxdy \tag{1.58}$$

Here $\tilde{A}_0=\{(x,y)\,|\,m_{\tilde{A}}(x,y)>0\}$ is support of fuzzy region $\tilde{A}$.

For fuzzy region, represented by piecewise membership function, the area is defined as [96]

$$S(\tilde{A})=\sum_i \mu(i) \tag{1.59}$$

Definition 1.45. Perimeter. In case of fuzzy set $\tilde{A}$ when $\mu_{\tilde{A}}$ is piecewise constant, the perimeter of fuzzy set $\tilde{A}$ is defined as

$$P_{\tilde{A}}=\sum_{i,j,k}\left|\mu(i)-\mu(j)\right| * L(i,j,k) \tag{1.60}$$

Here $\mu(i)$ and $\mu(j)$ are the membership values of two adjacent regions, $L(i, j, k)$ is length of a k-th arc of these regions.

Definition 1.46. Compactness. The compactness of a fuzzy set $\tilde{A}$ with area $S_{\tilde{A}}$ and perimeter $P_{\tilde{A}}$ is defined as

$$C(\tilde{A}) = \frac{S_{\tilde{A}}}{\mathrm{P}_{\tilde{A}}^{2}} \tag{1.61}$$

Definition 1.47. Length and Breadth of a Fuzzy Set. The length of a fuzzy set $\tilde{A}$ is defined as

$$l(\tilde{A}) = \max_{x}\left\{ \int \mu_{\tilde{A}}(x, y)dy \right\}, \tag{1.62}$$

where the integral is taken over the region with $\mu_{\tilde{A}}(x, y) > 0$. For discrete case formula (1.62) takes form

$$l(\tilde{A}) = \max_{x}\left\{ \sum_{y} \mu_{\tilde{A}}(x, y) \right\} \tag{1.63}$$

The breadth of a fuzzy set $\tilde{A}$ is defined as

$$b(\tilde{A}) = \max_{y}\left\{ \int \mu_{\tilde{A}}(x, y)dx \right\} \tag{1.64}$$

or

$$b(\tilde{A}) = \max_{y}\left\{ \sum_{x} \mu_{\tilde{A}}(x, y) \right\} \tag{1.65}$$

Definition 1.48. Index of Area Coverage (IOAC). IOAC of a fuzzy set $\tilde{A}$ is defined as

$$IOAC(\tilde{A}) = \frac{S_{\tilde{A}}}{l(\tilde{A}) \cdot b(\tilde{A})} \tag{1.66}$$

This index for fuzzy region represents the fraction of the maximum area (covered by the length and breadth of the region) actually covered by the region.

Now let us consider f -Geometry and f-transformation suggested by Zadeh [127].

In the described above geometry the underlying logic is precisiated fuzzy logic. In the world of f -geometry, suggested by Zadeh [127] the underlying logic is unprecisiated fuzzy logic, UFL. This f -Geometry differs both in spirit and in substance from Poston's fuzzy geometry [87], coarse geometry [89], fuzzy geometry of Rosenfeld [94], fuzzy geometry of Buckley and Eslami [29], fuzzy geometry of Mayburov [71], and fuzzy geometry of Tzafestas [102].

The counterpart of a crisp concept in Euclidean geometry is a fuzzy concept in this fuzzy geometry. Fuzzy concept may be obtained by fuzzy transformation (f -transform) of a crisp concept.

For example, the f -transform of a point is an f -point, the f -transform of a line is an f -line, the f-transform of a triangle is an f -triangle, the f -transform of a circle is an f -circle and the f -transform of parallel is f -parallel (Fig. 1.9). In summary, f -geometry may be viewed as the result of application of f -transformation to Euclidean geometry.

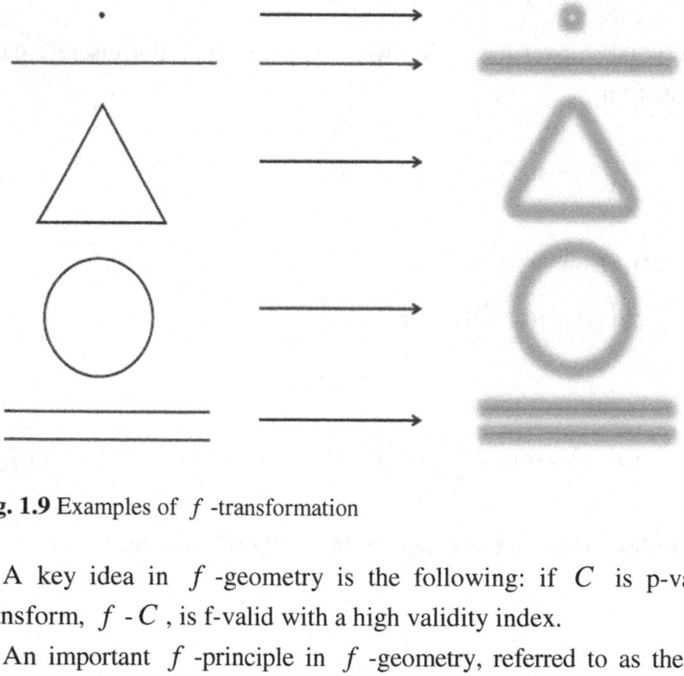

Fig. 1.9 Examples of f -transformation

A key idea in f -geometry is the following: if C is p-valid then its f -transform, f - C , is f-valid with a high validity index.

An important f -principle in f -geometry, referred to as the validation principle, is the following. Let p be a p-valid conclusion drawn from a chain of premises $*p_1,...,*p_n$. Then, using the star notation, $*p$ is an f -valid conclusion drawn from $*p_1,...,*p_n$, and $*p$ has a high validity index. It is this principle that is employed to derive f -valid conclusions from a collection of f-premises.

A basic problem which arises in computation of f -transforms is the following. Let g be a function, a functional or an operator. Using the star notation, let an f-transform , $*C$, be an argument of g . The problem is that of computing $g(*C)$. Generally, computing $g(*C)$ is not a trivial problem.

An f -valid approximation to $g(*C)$ may be derived through application of an f -principle which is referred to as precisiation / imprecisiation principle or P/I principle, for short [123]. More specifically, the principle may be expressed as

$$g(*C)* = *g(C)$$

where*=should be read as approximately equal. In words, $g(*C)$ is approximately equal to the f -transform of $g(C)$.

1.4 Approximate Reasoning

In our daily life we often make inferences where *antecedents* and *consequents* a represented by fuzzy sets. Such inferences cannot be realized adequately by the methods, which are based either on two-valued logic or many-valued logic. In order to facilitate such an inference, Zadeh [114,118,119,122,123,125] suggested an inference rule called a "compositional rule of inference". Using this inference rule, Zadeh, Mamdani [68], Mizumoto et al [38,74,75], R.Aliev and A.Tserkovny [7,9,12,13] suggested several methods for fuzzy reasoning in which the antecedent contain a conditional proposition involving fuzzy concepts:

$$\begin{array}{l} \text{Ant 1: If } x \text{ is } \tilde{P} \text{ then } y \text{ is } \tilde{Q} \\ \text{Ant2: } x \text{ is } \tilde{P}' \\ \hline \text{Cons: } y \text{ is } \tilde{Q}'. \end{array} \qquad (1.67)$$

Those methods are based on implication operators present in various fuzzy logics. This matter has been under a thorough discussion for the last couple decades. Some comparative analysis of such methods was presented in [20-23,38,40,47,50,51,53,69,70,74,75,98,111,112]. A number of authors proposed to use a certain suite of fuzzy implications to form fuzzy conditional inference rules [7,9,38,39,59,68,74,75]. The implication operators present in the theory of fuzzy sets were investigated in [7,9,14,26-28,30-33,35,36,41,42,45,48,55,60,61,63,66, 67,69,72,73,76,79,80,82,84,86,99,103,104,108,109,112,129,131,132].On the other hand, statistical features of fuzzy implication operators were studied in [83,105] In turn, the properties of stability and continuity of fuzzy conditional inference rules were investigated in [37,39,49,56]. We will begin with a *formation* of a fuzzy logic regarded as an *algebraic system closed under all its operations.* In the sequel an investigation of statistical characteristics of the proposed fuzzy logic will be presented. Special attention will be paid to building a set of fuzzy conditional inference rules on the basis of the fuzzy logic proposed in this study. Next, continuity and stability features of the formalized rules will be investigated. Lately in fuzzy sets research the great attention is paid to the development of Fuzzy Conditional Inference Rules (CIR) [1,5,36,56,64,72,80,95]. This is connected with the feature of the natural language to contain a certain number of fuzzy concepts (F-concepts), therefore we have to make logical inference in which the preconditions and conclusions contain such F-concepts. The practice shows that there is a huge variety of ways in which the formalization of rules for such kind of inferences can be made. However, such inferences cannot be satisfactorily formalized using the classical Boolean Logic, i.e. here we need to use multi-valued logical systems. The development of the conditional logic rules embraces mainly three types of fuzzy propositions:

$$P_1 = IF\ x\ is\ \tilde{A}\ THEN\ y\ is\ \tilde{B}$$

$$P_2 = IF\ x\ is\ \tilde{A}\ THEN\ y\ is\ \tilde{B}$$

$$OTHERWISE\ \tilde{C}$$

$$P_3 = IF\ x_1\ is\ \tilde{A}_1\ AND\ x_2\ is\ \tilde{A}_2 ... AND ... AND\ x_n\ is\ \tilde{A}_n$$

$$THEN\ y\ is\ \tilde{B}$$

The conceptual principle in the formalization of fuzzy rules is the Modus Ponens inference (separation) rule that states:

$IF(\alpha \rightarrow \beta)$ is true and α is true $THEN\ \beta$ is true.

The methodological base for this formalization is the compositional rule suggested by L.Zadeh [114,116]. Using this rule, he formulated some inference rules in which both the logical preconditions and consequences are conditional propositions including F-concepts. Later E.Mamdani [68] suggested inference rule, which like Zadeh's rule was developed for the logical proposition of type P_1. In other words the following type F-conditional inference is considered:

$$\begin{array}{l} \text{Proposition 1: } IF\ x\ is\ \tilde{A}\ THEN\ y\ is\ \tilde{B} \\ \text{Proposition 2: } x\ is\ \tilde{A}' \\ \hline \text{Conclusion: y is } \tilde{B}, \end{array} \tag{1.68}$$

where $\tilde{A}$ and $\tilde{A}'$ are F -concepts represented as F -sets in the universe U ; $\tilde{B}$ is F -conceptions or F -set in the universe V . It follows that B' is the consequence represented as a F-set in V . To obtain a logical conclusion based on the CIR, the Propositions 1 and 2 must be transformed accordingly to the form of binary F-relation $\tilde{R}(A_1(x)), A_2(y))$ and unary F -relation $\tilde{R}(A_1(x))$.

Here $A_1(x)$ and $A_2(y)$ are defined by the attributes x and y which take values from the universes U and V , respectively. Then

$$\tilde{R}\ (A_1(x)) = \tilde{A}' \tag{1.69}$$

According to Zadeh-Mamdani's inference rule $\tilde{R}(A_1(x)), A_2(y))$ is defined as follows.

The maximin conditional inference rule

$$\tilde{R}_m(A_1(x),\ A_2(y)) = (\tilde{A}\times\tilde{B})\cup(\neg\tilde{A}\times V) \tag{1.70}$$

The arithmetic conditional inference rule

$$\tilde{R}_a(A_1(x), A_2(y)) = (\neg\tilde{A}\times V)\oplus(U\times\tilde{B}) \tag{1.71}$$

The mini-functional conditional inference rule

$$\tilde{R}_c(A_1(x),\ A_2(y)) = \tilde{A}\times\tilde{B} \tag{1.72}$$

where $\times$, $\cup$ and $\neg$ are the Cartesian product, union, and complement operations, respectively; $\oplus$ is the limited summation.

Thus, in accordance with [68,114,116] the logical consequence $\tilde{R}(A_2(y))$, ($\tilde{B}$' in (1.72)) can be derived as follows:

$$\tilde{R}(A_2(y)) = \tilde{A}'\circ[(\tilde{A}\times\tilde{B})]\cup[\neg\tilde{A}\times U)]$$

$$\tilde{R}(A_2(y)) = \tilde{A}'\circ[(\neg\tilde{A}\times V)]\oplus[\neg U\times\tilde{B})]$$

or

$$\tilde{R}(A_2(y)) = \tilde{A}'\circ(\tilde{A}\times\tilde{B})$$

where $\circ$ – is the F -set maximin composition operator.

On the base of these rules the conditional inference rules for type P_2 were suggested in [15]:

$$\begin{aligned}&\tilde{R}_4(\tilde{A}_1(x), \tilde{A}_2(y)) =\\&= [(\tilde{A}\times V)\oplus(U\times\tilde{B})]\cap[(\tilde{A}\times V)\oplus(U\times\tilde{C})]\end{aligned} \tag{1.73}$$

$$\begin{aligned}&\tilde{R}_5(A_1(x), A_2(y)) =\\&= [(\neg\tilde{A}\times V)\cup(U\times\tilde{B})]\cap[(\tilde{A}\times V)\cup(U\times\tilde{C})]\end{aligned} \tag{1.74}$$

$$\tilde{R}_6(A_1(x),\ A_2(y)) = [(\tilde{A}\times\tilde{B})\cup(\neg\tilde{A}\times\tilde{C})] \tag{1.75}$$

Note that in [15] also the fuzzy conditional inference rules for type P_3 were suggested:

$$\tilde{R}_7(A_1(x),\ A_2(y)) = \left[\bigcap_{i=1,n}(\neg\tilde{A}_i \times V)\right] \oplus [(U \times \tilde{B})] \tag{1.76}$$

$$\tilde{R}_8(A_1(x),\ A_2(y)) = \left[\bigcap_{i=1,n}(\neg\tilde{A}_i \times V)\right] \cup [(U \times \tilde{B})] \tag{1.77}$$

$$\begin{aligned}\tilde{R}_9(A_1(x),\ A_2(y)) &= (\neg\tilde{A} \times V) \oplus (U \times \tilde{B}) = \\ &= \int_{U \times V} 1 \wedge (1 - \mu_{\tilde{A}}(u) + \mu_{\tilde{B}}(v))/(u,v)\end{aligned} \tag{1.78}$$

In order to analyze the effectiveness of rules (1.68)-(1.78) we use some criteria for F-conditional logical inference suggested in [38]. The idea of these criteria is to compare the degree of compatibility of some fuzzy conditional inference rules with the human intuition when making approximate reasoning. These criteria are the following:

Criterion I
Precondition 1: IF x is $\tilde{A}$ THEN y is $\tilde{B}$
Precondition 2: x is $\tilde{A}$

Conclusion: y is $\tilde{B}$

Criterion II-1 Precondition 1: IF x is $\tilde{A}$ THEN y is $\tilde{B}$
Precondition 2: x is very $\tilde{A}$

Conclusion: y is very $\tilde{B}$

Criterion II-2 Precondition 1: IF x is $\tilde{A}$ THEN y is $\tilde{B}$
Precondition 2: x is very $\tilde{A}$

Conclusion: y is B

Criterion III
Precondition 1: IF x is $\tilde{A}$ THEN y is $\tilde{B}$
Precondition 2: x is more or less $\tilde{A}$

Conclusion: y is more or less $\tilde{B}$

Criterion IV-1
Precondition 1: IF x is $\tilde{A}$ THEN y is $\tilde{B}$
Precondition 2: x is not $\tilde{A}$

Conclusion: y is unknown

Criterion IV-2
Precondition 1: IF x is $\tilde{A}$ THEN y is $\tilde{B}$
Precondition 2: x is not $\tilde{A}$

Conclusion: y is not $\tilde{B}$

In [38] it was shown that in Zadeh-Mamdani's rules the relations $\tilde{R}_m$, $\tilde{R}_c$ and $\tilde{R}_c$ do not always satisfy the above criteria. For the case of mini-operational rule $\tilde{R}_c$ it has been found that criteria I and II-2 are satisfied while criteria II-1 and III are not.

In [38] an important generalization was made that allows some improvement to the mentioned F-conditional logical inference rules. It was shown there that for the conditional proposition arithmetical rule defined by Zadeh

$$P_1 = IF\ x\ is\ \tilde{A}\ THEN\ y\ is\ \tilde{B}$$

the following takes place

$$\begin{aligned}&\tilde{R}_9(A_1(x),\ A_2(y)) = (\neg\tilde{A}\times V)\oplus(U\times\tilde{B}) = \\ &= \int_{U\times V} 1\wedge(1-\mu_{\tilde{A}}(u)+\mu_{\tilde{B}}(v))/(u,v)\end{aligned}$$

The membership function for this F-relation is

$$1\wedge(1-\mu_{\tilde{A}}(u)+\mu_{\tilde{B}}(v))$$

that obviously meets the implication operation or the Ply-operator for the multi-valued logic L (by Lukasiewicz), i.e.

$$T(P \underset{L}{\rightarrow} Q), T(P) \tag{1.79}$$

where $T(P \underset{L}{\rightarrow} Q), T(P)$ and $T(Q)$ - are the truth values for the logical propositions $P \underset{L}{\rightarrow} Q, P$ and Q respectively.

In other words, these expressions can be considered as adaptations of implication in the L-logical system to a conditional proposition.

Having considered this fact, the following expression was derived:

$$\begin{aligned}&\tilde{R}_a(A_1(x),\ A_2(y)) = (\neg\tilde{A}\times V)\oplus(U\times\tilde{B}) = \\ &= \int_{U\times V} 1\wedge(1-\mu_{\tilde{A}}(u)+\mu_{\tilde{B}}(v))/(u,v) = \\ &= \int_{U\in V} (\mu_{\tilde{A}}(u) \underset{L}{\rightarrow} \mu_{\tilde{B}}(v))/(u,v) = (\tilde{A}\times V)\rightarrow(U\times\tilde{B})\end{aligned} \tag{1.80}$$

In [38] an opinion was expressed that the implication operation or the Ply-operator in the expression (1.80) may belong to any multi-valued logical system.

The following are guidelines for deciding which logical system to select for developing F-conditional logical inference rules [38]. Let F-sets $\tilde{A}$ from U and B from V are given in the form:

$$\tilde{A} = \int_V \mu_{\tilde{A}}(u)/u, \tilde{B} = \int_V \mu_{\tilde{B}}(v)/v$$

Then, as mentioned above, the conditional logical proposition P_1 can be transformed to the F-relation $\tilde{R}(A_1(x), A_2(y))$ by adaptation of the Ply-operator in multi-valued logical system, i.e.

$$\tilde{R}(A_1(x), A_2(y)) = \tilde{A} \times V \to U \times \tilde{B} = \int_{U \times V} (\mu_{\tilde{A}}(u) \to \mu_{\tilde{B}}(v))/(u, v) \tag{1.81}$$

where the values $\mu_{\tilde{A}}(u) \to \mu_{\tilde{B}}(v)$ are depending on the selected logical system.

Assuming $\tilde{R}(A_1(x)) = \tilde{A}$ we can conclude a logical consequence $\tilde{R}(A_2(y))$, then using the CIR for $\tilde{R}(A_1(x))$ and $\tilde{R}(A_1(x), A_2(y))$, then

$$\begin{aligned} \tilde{R}(A_2(y)) &= \tilde{A} \circ \tilde{R}(A_1(x), A_2(y)) = \\ &= \int_U \mu_{\tilde{A}}(u)/u \circ \int_{U \times V} \mu_{\tilde{A}}(u) \to \mu_{\tilde{B}}(v))/(u, v) = \\ &= \int_V \vee_{u \in V} [\mu_{\tilde{A}}(u) \wedge (\mu_{\tilde{A}}(u) \to \mu_{\tilde{B}}(v))] \end{aligned} \tag{1.82}$$

For the criterion I to be satisfied, one of the following equalities must hold true

$$\tilde{R}(A_2(y)) = \tilde{B},$$

$$\vee_{u \in V} [\mu_{\tilde{A}}(u) \wedge (\mu_{\tilde{A}}(u) \to \mu_{\tilde{B}}(v))] = \mu_{\tilde{B}}(v),$$

or

$$[\mu_{\tilde{A}}(u) \wedge (\mu_{\tilde{A}}(u) \to \mu_{\tilde{B}}(v))] \le \mu_{\tilde{B}}(v) \tag{1.83}$$

the latter takes place for any u∈U and v∈V or in terms of truth values:

$$T(P \wedge (P \to Q)) \le T(Q) \tag{1.84}$$

The following two conditions are necessary for formalization of F-conditional logical inference rules: the conditional logical inference rules (CIR) must meet the criteria I-IV; the conditional logical inference rules (CIR) satisfy the inequality (1.84). Now we consider formalization of the fuzzy conditional inference for a different type of conditional propositions. As was shown above, the logical inference for conditional propositions of type P_1 is of the following form:

$$\begin{array}{l} \text{Proposition 1: } IF\ x\ is\ \tilde{A}\ THEN\ y\ is\ \tilde{B} \\ \text{Proposition 2: } x\ is\ \tilde{A}' \\ \hline \text{Conclusion: } y\ is\ \tilde{B}' \end{array} \tag{1.85}$$

where $\tilde{A}$, $\tilde{B}$, and $\tilde{A}'$ are F-concepts represented as F-sets in U, V, and V, respectively, which should satisfy the criteria I, II-1, III, and IV-1.

For this inference if the Proposition 2 is transformed to an unary F-relation in the form $\tilde{R}(A_1(x)) = \tilde{A}'$ and the Proposition 1 is transformed to an F-relation $\tilde{R}(A_1(x), \tilde{R}(A_2(y))$ defined below, then the conclusion $\tilde{R}(A_2(y))$ is derived by using the corresponding F-conditional logical inference rule, i.e.

$$\tilde{R}(A_2(y) = \tilde{R}(A_1(x)) \circ \tilde{R}(A_1(x)) \tag{1.86}$$

where $\tilde{R}(A_2(y))$ is equivalent to $\tilde{B}'$ in (1.85).

Fuzzy Conditional Inference Rule 1

Theorem 1.3. *If the F-sets $\tilde{A}$ from U and $\tilde{B}$ from V are given in the traditional form:*

$$\tilde{A} = \int_U \mu_{\tilde{A}}(u)/u, \tilde{B} = \int_V \mu_{\tilde{B}}(v)/v \tag{1.87}$$

and the relation for the multi-valued logical system ALI1

$$\begin{aligned} \tilde{R}_1(A_1(x), A_2(y)) &= \tilde{A} \times V \xrightarrow[ALI1]{} U \times \tilde{B} = \\ &= \int_{U \times V} \mu_{\tilde{A}}(u)/(u,v) \xrightarrow[ALI1]{} \int_{U \times V} \mu_{\tilde{B}}(v)/(u,v) = \\ &= \int_{U \times V} (\mu_{\tilde{A}}(u) \xrightarrow[ALI1]{} \mu_{\tilde{B}}(v))/(u,v) \end{aligned} \tag{1.88}$$

where

$$\mu_{\tilde{A}}(u) \xrightarrow[ALI1]{} \mu_{\tilde{B}}(v) = \begin{cases} 1 - \mu_{\tilde{A}}(u), & \mu_{\tilde{A}}(u) < \mu_{\tilde{B}}(v) \\ 1, & \mu_{\tilde{A}}(u) = \mu_{\tilde{B}}(v) \\ \mu_{\tilde{B}}(v), & \mu_{\tilde{A}}(u) > \mu_{\tilde{B}}(v) \end{cases}$$

then the criteria I-IV are satisfied.

We will consider ALI4 in detailes.

Consider a continuous function $F(p,q) = p - q$ which defines a distance between p and q where p, q assume values in the unit interval. Notice that $F(p,q) \in [-1,1]$, where $F(p,q)^{\min} = -1$ and $F(p,q)^{\max} = 1$. The normalized version of $F(p,q)$ is defined as follow

$$F(p,q)^{norm} = \frac{F(p,q) - F(p,q)^{\min}}{F(p,q)^{\max} - F(p,q)^{\min}} = \frac{F(p,q)+1}{2} = \frac{p-q+1}{2} \tag{1.89}$$

It is clear that $F(p,q)^{norm} \in [0,1]$. This function quantifies a concept of "closeness" between two values (potentially the ones for the truth values of *antecedent* and *consequent*), defined within unit interval, which therefore could play significant role in the formulation of the implication operator in a fuzzy logic.

Definition 1.49. An implication is a continuous function I from $[0,1]\times[0,1]$ into $[0,1]$ such that for $\forall p,\ p',\ q,\ q'\ r \in [0,1]$ the following properties are satisfied

(I1) If $p \le p'$, then $I(p,q) \ge I(p',q)$ (Antitone in first argument),
(I2) If $q \le q'$, then $I(p,q) \le I(p,q')$ (Monotone in second argument),
(I3) $I(0,q)=1$, (Falsity),
(I4) $I(1,q) \le q$ (Neutrality),
(I5) $I(p,I(q,r)) = I(q,I(p,r))$ (Exchange),
(I6) $I(p,q) = I(n(q),n(p))$ (Contra positive symmetry), where $n()$ - is a negation, which could be defined as $n(q) = T(\neg Q) = 1 - T(Q)$

Let us define the implication operation

$$I(p,q) = \begin{cases} 1 - F(p,q)^{norm}, & p > q \\ 1, & p \le q \end{cases} \tag{1.90}$$

where $F(p,q)^{norm}$ is expressed by (1.89). Before showing that operation $I(p,q)$ satisfies axioms *(I1)-(I6),* let us show some basic operations encountered in proposed fuzzy logic.

Let us designate the truth values of the *antecedent* P and *consequent* Q as $T(P) = p$ and $T(P) = q$, respectively. The relevant set of proposed fuzzy logic operators is shown in Table 1.2.

To obtain the truth values of these expressions, we use well known logical properties such as

$$p \to q = \neg p \vee q,\ p \wedge q = \neg(\neg p \vee \neg q) \text{ and alike.}$$

In other words, we propose a new many-valued system, characterized by the set of *union* ($\cup$) and *intersection* ($\cap$) operations with relevant *complement*, defined as $T(\neg Q) = 1 - T(Q)$. In addition, the operators $\downarrow$ and $\uparrow$ are expressed as negations of the $\cup$ and $\cup$, respectively. It is well known that the *implication* operation in fuzzy logic supports the foundations of decision-making exploited in numerous schemes of approximate reasoning. Therefore let us prove that the proposed *implication* operation in (1.90) satisfies axioms *(I1)-(I6)*. For this matter, let us emphasize that we are working with a many-valued system, whose values for our purposes are the elements of the real interval $R = [0,1]$. For our discussion the set of truth values $V_{11} = \{0, 0.1, 0.2, ..., 0.9, 1\}$ is sufficient. In further investigations, we use this particular set V_{11}.

Table 1.2 Fuzzy logic operators

Name	Designation	Value
Tautology	$\overset{\bullet}{P}$	1
Controversy	$\overset{\circ}{P}$	0
Negation	$\neg P$	$1-P$
Disjunction	$P \vee Q$	$\begin{cases} \frac{p+q}{2}, p+q \neq 1, \\ 1, p+q=1 \end{cases}$
Conjunction	$P \wedge Q$	$\begin{cases} \frac{p+q}{2}, p+q \neq 1, \\ 0, p+q=1 \end{cases}$
Implication	$P \rightarrow Q$	$\begin{cases} \frac{1-p+q}{2}, p \neq q, \\ 1, p=q \end{cases}$
Equivalence	$P \leftrightarrow Q$	$\begin{cases} \min((p-q),(q-p)), p \neq q, \\ 1, p=q \end{cases}$
Pierce Arrow	$P \downarrow Q$	$\begin{cases} 1-\frac{p+q}{2}, p+q \neq 1, \\ 0, p+q=1 \end{cases}$
Shaffer Stroke	$P \uparrow Q$	$\begin{cases} 1-\frac{p+q}{2}, p+q \neq 1, \\ 1, p+q=1 \end{cases}$

Theorem 1.4. *Let a continuous function $I(p,q)$ be defined by (1.90) i.e.*

$$I(p,q) = \begin{Bmatrix} 1-F(p,q)^{norm}, & p>q \\ 1, & p \leq q \end{Bmatrix}, p>q = \begin{cases} \frac{1-p+q}{2}, p>q \\ 1, p \leq q \end{cases} \tag{1.91}$$

where $F(p,q)^{norm}$ is defined by (1.89). Then axioms (I1)-(I6) are satisfied and, therefore (1.91) is an implication operation.

It should be mentioned that the proposed fuzzy logic could be characterized by yet some other three features:

$p \wedge 0 \equiv 0, p \leq 1,$ whereas $p \wedge 1 \equiv p, p \geq 0$ and $\neg\neg p = p$.

As a conclusion, we should admit that all above features confirm that *resulting system* can be applied to V_{11} for every finite and infinite n up to that $(V_{11}, \neg, \wedge, \vee, \rightarrow)$ is then *closed* under all its operations.

Let us investigate Statistical Properties of the Fuzzy Logic. In this section, we discuss some properties of the proposed fuzzy implication operator (1.91), assuming that the two propositions (*antecedent/consequent*) in a given compound proposition are independent of each other and the truth values of the propositions are uniformly distributed [64] in the unit interval. In other words, we assume that the propositions P and Q are independent from each other and the truth values $v(P)$ and $v(Q)$ are uniformly distributed across the interval $[0,1]$. Let $p = v(P)$ and $q = v(Q)$. Then the value of the implication $I = v(p \rightarrow q)$ could be represented as the function $I = I(p,q)$.

Because p and q are assumed to be uniformly and independently distributed across $[0,1]$, the expected value of the implication is

$$E(I) = \iint_R I(p,q)dpdq, \tag{1.92}$$

Its variance is equal to

$$Var(I) = E[(I - E(I))^2] = \iint_R (I(p,q) - E(I))^2 dpdq = E[I^2] - E[I^2] \tag{1.93}$$

where $R = \{(p,q): 0 \leq p \leq 1, 0 \leq q \leq 1\}$ From (1.92) and given (1.93) as well as the fact that

$I(p,q) = \begin{cases} I_1(p,q), p > q, \\ I_2(p,q), p \leq q, \end{cases}$ we have the following

$$E(I_1) = \iint_{\Re} I_1(p,q)dpdq, = \int_0^1\int_0^1 \frac{1-p+q}{2} dpdq = \frac{1}{2}\int_0^1 (\int_0^1 (1-p+q)dp)dp =$$
$$= \frac{1}{2}\left[\int_0^1 \left((p - \frac{p^2}{2} + p)\bigg|_{p=0}^{p=1}\right)dq\right] = \frac{1}{2}\left[\frac{1}{2} + \frac{q^2}{2}\bigg|_{q=0}^{q=1}\right] = \frac{1}{2} \tag{1.94}$$

Whereas $E(I_2) = 1$ Therefore $E(I) = (E(I_1) + E(I_2))/2 = 0.75$

From (1.93) we have

$$I_1^2(p,q) = \frac{1}{4}(1 - p + q)^2 = \frac{1}{4}(1 - 2p + 2q + p^2 - 2pq + q^2)$$

$$E(I_1^2) = \iint_{\Re} I_1^2(p,q)dpdq, = \frac{1}{4}\int_0^1(\int_0^1(1-2p+2q+p^2-2pq+q^2)dp)dq =$$

$$\frac{1}{4}\int_0^1[p-2\frac{p^2}{2}+\frac{p^3}{3}-2\frac{p^2}{2}q+2q+q^2]\Bigg|_{p=0}^{p=1}dq = \frac{1}{4}\int_0^1(\frac{1}{3}+q+q^2)dq =$$

$$= \frac{1}{4}\left[\frac{q}{3}+\frac{q^2}{2}+\frac{q^3}{3}\right]\Bigg|_{q=0}^{q=1} = \frac{7}{24}$$

Here $E(I_2^2) = 1$ Therefore $E(I^2) = (E(I_1^2) + E(I_2^2))/2 = \frac{31}{48}$ From (1.93) and (1.94) we have $Var(I) = \frac{1}{12} = 0.0833$

Both values of $E(I)$ and $Var(I)$ demonstrate that the proposed fuzzy implication operator could be considered as one of the fuzziest from the list of the exiting implications [45]. In addition, it satisfies the set of important Criteria I-IV, which is not the case for the most implication operators mentioned above.

As it was mentioned in [38] "in the semantics of natural language there exist a vast array of concepts and humans very often make inferences antecedents and consequences of which contain fuzzy concepts". A formalization of methods for such inferences is one of the most important issues in fuzzy sets theory. For this purpose, let U and V (from now on) be two *universes of discourses* and P and Q are corresponding fuzzy sets:

$$\tilde{P} = \int_U \mu_{\tilde{P}}(u)/u\,,\ \tilde{Q} = \int_V \mu_{\tilde{Q}}(v)/v \tag{1.95}$$

Given (1.95), a *binary relationship* for the fuzzy conditional proposition of the type: "*If* x *is* $\tilde{P}$ *then* y *is* $\tilde{Q}$ " for proposed fuzzy logic is defined as

$$\tilde{R}(A_1(x), A_2(y)) = \tilde{P}\times V \to U\times\tilde{B} = \int_{U\times V}\mu_{\tilde{P}}(u)/(u,v) \to \int_{U\times V}\mu_{\tilde{Q}}(v)/(u,v) =$$
$$= \int_{U\times V}(\mu_{\tilde{P}}(u) \to \mu_{\tilde{Q}}(v))/(u,v) \tag{1.96}$$

Given (1.90), expression (1.96) reads as

$$\mu_{\tilde{P}}(u) \to \mu_{\tilde{Q}}(v) = \begin{cases} \dfrac{1-\mu_{\tilde{P}}(u)+\mu_{\tilde{Q}}(v)}{2}, \mu_{\tilde{P}}(u) > \mu_{\tilde{Q}}(v) \\ 1, \mu_{\tilde{P}}(u) \le \mu_{\tilde{Q}}(v) \end{cases} \tag{1.97}$$

It is well known that given a *unary relationship* $\tilde{R}(A_1(x))$ one can obtain the consequence $\tilde{R}(A_2(y))$ by applying a compositional rule of inference (CRI) to $\tilde{R}(A_1(x))$ and $\tilde{R}(A_1(x), A_2(y))$ of type (1.91):

$$\tilde{R}(A_2(y)) = \tilde{P} \circ \tilde{R}(A_1(x), A_2(y)) = \int_U \mu_{\tilde{P}}(u)/u \circ \int_{U\times V} \mu_{\tilde{P}}(u) \to \mu_{\tilde{Q}}(v)/(u,v) =$$
$$\int_V \bigcup_{u\in V} [\mu_{\tilde{P}}(u) \wedge (\mu_{\tilde{P}}(u) \to \mu_{\tilde{Q}}(v))]/v \tag{1.98}$$

In order to have Criterion I satisfied, that is $\tilde{R}(A_2(y)) = \tilde{Q}$ from (1.98), the equality

$$\int_V \bigcup_{u\in V} [\mu_{\tilde{P}}(u) \wedge (\mu_{\tilde{P}}(u) \to \mu_{\tilde{Q}}(v))] = \mu_{\tilde{Q}}(v) \tag{1.99}$$

has to be satisfied for any arbitrary v in V. To satisfy (1.99), it becomes necessary that the inequality

$$\mu_{\tilde{P}}(u) \wedge (\mu_{\tilde{P}}(u) \to \mu_{\tilde{Q}}(v)) \le \mu_{\tilde{Q}}(v) \tag{1.100}$$

holds for arbitrary $u \in U$ and $v \in V$. Let us define a new method of fuzzy conditional inference of the following type:

$$\begin{array}{l} \text{Ant 1: If } x \text{ is } \tilde{P} \text{ then } y \text{ is } \tilde{Q} \\ \text{Ant 2: } x \text{ is } \tilde{P}' \\ \hline \text{Cons: } y \text{ is } \tilde{Q}'. \end{array} \tag{1.101}$$

where $\tilde{P}, \tilde{P}' \subseteq U$ and $\tilde{Q}, \tilde{Q}' \subseteq V$. Fuzzy conditional inference in the form given by (1.101) should satisfy Criteria I-IV. It is clear that the inference (1.100) is defined by the expression (1.98), when $\tilde{R}(A_2(y)) = \tilde{Q}'$.

Theorem 1.5. *If fuzzy sets* $\tilde{P} \subseteq U$ *and* $\tilde{Q} \subseteq V$ *are defined by (1.96) and (1.97), respectively and*

$\tilde{R}(A_1(x), A_2(y))$ *is expressed as*

$$\tilde{R}(A_1(x), A_2(y)) = \tilde{P} \times V \xrightarrow[ALI4]{} U \times \tilde{Q} =$$
$$= \int_{U\times V} \mu_{\tilde{P}}(u)/(u,v) \xrightarrow[ALI4]{} \int_{U\times V} \mu_{\tilde{Q}}(v)/(u,v) =$$
$$= \int_{U\times V} (\mu_{\tilde{P}}(u) \xrightarrow[ALI4]{} \mu_{\tilde{Q}}(v))/(u,v)$$

where

$$\mu_{\tilde{P}}(u) \xrightarrow[ALI4]{} \mu_{\tilde{Q}}(v) = \begin{cases} \dfrac{1 - \mu_{\tilde{P}}(u) + \mu_{\tilde{Q}}(v)}{2}, \mu_{\tilde{P}}(u) > \mu_{\tilde{Q}}(v) \\ 1, \mu_{\tilde{P}}(u) \le \mu_{\tilde{Q}}(v) \end{cases} \tag{1.102}$$

then Criteria I, II, III and IV-1 [38] are satisfied [13].

Theorem 1.6. *If fuzzy sets* $\tilde{P} \subseteq U$ *and* $\tilde{Q} \subseteq V$ *are defined by (1.96) and (1.97), respectively, and*
$\tilde{R}(A_1(x), A_2(y))$ *is defined as*

$$\tilde{R}_1(A_1(x), A_2(y)) = (\tilde{P} \times V \xrightarrow[ALI4]{} U \times \tilde{Q}) \cap (\neg\tilde{P} \times V \xrightarrow[ALI4]{} U \times \neg\tilde{Q}) = \\ = \int_{U \times V} (\mu_{\tilde{P}}(u) \xrightarrow[ALI4]{} \mu_{\tilde{Q}}(v)) \wedge ((1-\mu_{\tilde{P}}(u)) \xrightarrow[ALI4]{} (1-\mu_{\tilde{Q}}(v)))/(u,v) \tag{1.103}$$

where

$$(\mu_{\tilde{P}}(u) \xrightarrow[ALI4]{} \mu_{\tilde{Q}}(v)) \wedge ((1-\mu_{\tilde{P}}(u)) \xrightarrow[ALI4]{} (1-\mu_{\tilde{Q}}(v))) = \\ = \begin{cases} \dfrac{1-\mu_{\tilde{P}}(u)+\mu_{\tilde{Q}}(v)}{2}, \mu_{\tilde{P}}(u) > \mu_{\tilde{Q}}(v), \\ 1, \mu_{\tilde{P}}(u) = \mu_{\tilde{Q}}(v), \\ \dfrac{1-\mu_{\tilde{P}}(u)+\mu_{\tilde{Q}}(v)}{2}, \mu_{\tilde{P}}(u) < \mu_{\tilde{Q}}(v), \end{cases}$$

Then Criteria I, II, III and IV-2 [38] are satisfied.

Theorems 1.4 and 1.5 show that fuzzy conditional inference rules, defined in (1.103) could adhere with human intuition to the higher extent as the one defined by (1.102). The major difference between mentioned methods of inference might be explained by the difference between *Criteria IV-1* and *IV-2*. In particular, a satisfaction of the *Criterion IV-1* means that in case of logical negation of an original antecedent we achieve an ambiguous result of an inference, whereas for the case of the *Criterion IV-2* there is a certainty in a logical inference. Let us to investigate stability and continuity of fuzzy conditional inference in this section. We revisit the fuzzy conditional inference rule (1.101). It will be shown that when the membership function of the observation $\tilde{P}$ is continuous, then the conclusion $\tilde{Q}$ depends continuously on the observation; and when the membership function of the relation $\tilde{R}$ is continuous then the observation $\tilde{Q}$ has a continuous membership function. We start with some definitions. A fuzzy set $\tilde{A}$ with membership function $\mu_{\tilde{A}} : \mathcal{R} \to [0,1] = I$ is called a fuzzy number if $\tilde{A}$ is normal, continuous, and convex. The fuzzy numbers represent the continuous possibility distributions of fuzzy terms of the following type

$$\tilde{A} = \int_{\mathcal{R}} \mu_A(x)/x$$

Let $\tilde{A}$ be a fuzzy number, then for any $\theta \geq 0$ we define $\omega_A(\theta)$ the modulus of continuity of $\tilde{A}$ by

$$\omega_{\tilde{A}}(\theta) = \max_{|x_1 - x_2| \leq \theta} \left| \mu_{\tilde{A}}(x_1) - \mu_{\tilde{A}}(x_2) \right| \tag{1.104}$$

An α-level set of a fuzzy interval $\tilde{A}$ is a non-fuzzy set denoted by $[A]^\alpha$ and is defined by $[A]^\alpha = \{t \in R \,|\, \mu_{\tilde{A}}(t) \geq \alpha\}$ for $\alpha = (0,1]$ and $[A]^{\alpha=0} = cl\left(\bigcup_{\alpha \in (0,1]} [A]^\alpha \right)$ for $\alpha = 0$. Here we use a metric of the following type

$$D(\tilde{A}, \tilde{B}) = \sup_{\alpha \in [0,1]} d([A]^\alpha, [B]^\alpha) \tag{1.105}$$

where d denotes the classical Hausdorff metric expressed in the family of compact subsets of R^2, i.e.

$$d([A]^\alpha, [B]^\alpha) = \max\{|a_1(\alpha) - b_1(\alpha)|, |a_2(\alpha) - b_2(\alpha)|\},$$

whereas

$[A]^\alpha = [a_1(\alpha), a_2(\alpha)], [B]^\alpha = b_1(\alpha), b_2(\alpha)$. When the fuzzy sets $\tilde{A}$ and $\tilde{B}$ have finite support $\{x_1, \ldots, x_n\}$ then their Hamming distance is defined as

$$H(\tilde{A}, \tilde{B}) = \sum_{i=1}^{n} \left| \mu_{\tilde{A}}(x_i) - \mu_{\tilde{B}}(x_i) \right|$$

In the sequel we will use the following lemma.

Lemma 1.1 [28]. Let $\delta \geq 0$ be a real number and let $\tilde{A}$, $\tilde{B}$ be fuzzy intervals. If $D(\tilde{A}, \tilde{B}) \leq \delta$, *Then*

$$\sup_{t \in \mathcal{R}} \left| \mu_{\tilde{A}}(t) - \mu_{B'}(t) \right| \leq \max\{\omega_{\tilde{A}}(\delta), \omega_{\tilde{B}}(\delta)\}$$

Consider the fuzzy conditional inference rule with different observations $\tilde{P}$ and $\tilde{P}'$:

Ant 1: If x is $\tilde{P}$ then y is $\tilde{Q}$

Ant2: x is $\tilde{P}$

Cons: y is $\tilde{Q}$.

Ant 1: If x is $\tilde{P}$ then y is $\tilde{Q}$

Ant2: x is $\tilde{P}'$

Cons: y is $\tilde{Q}'$.

According to the fuzzy conditional inference rule, the membership functions of the conclusions are computed as

$$\mu_{\tilde{Q}}(v) = \bigcup_{u \in R} [\mu_{\tilde{P}}(u) \wedge (\mu_{\tilde{P}}(u) \rightarrow \mu_{\tilde{Q}}(v))],$$

$$\mu_{\tilde{Q}'}(v) = \bigcup_{u \in R} [\mu_{\tilde{P}'}(u) \wedge (\mu_{\tilde{P}}(u) \rightarrow \mu_{\tilde{Q}}(v))],$$

or

$$\mu_{\tilde{Q}}(v) = \sup[\mu_{\tilde{P}}(u) \wedge (\mu_{\tilde{P}}(u) \rightarrow \mu_{\tilde{Q}}(v))],$$

$$\mu_{\tilde{Q}'}(v) = \sup[\mu_{\tilde{P}'}(u) \wedge (\mu_{\tilde{P}}(u) \rightarrow \mu_{\tilde{Q}}(v))], \tag{1.106}$$

The following theorem shows the fact that when the observations are closed to each other in the metric $D(.)$ of (1.105) type, then there can be only a small deviation in the membership functions of the conclusions.

Theorem 1.7. *(Stability theorem) Let* $\delta \geq 0$ *and let* $\tilde{P}$, $\tilde{P}'$ *be fuzzy intervals and an implication operation in the fuzzy conditional inference rule (1.106) is of type (1.97). If* $D(\tilde{P}, \tilde{P}') \leq \delta$, *then*

$$\sup_{v \in R} \left| \mu_{\tilde{P}}(v) - \mu_{\tilde{P}'}(v) \right| \leq \max\{\omega_{\tilde{P}}(\delta), \omega_{\tilde{P}'}(\delta)\}$$

Theorem 1.8. *(Continuity theorem) Let binary relationship* $\tilde{R}(u,v) = = \mu_{\tilde{P}}(u) \xrightarrow[ALI4]{} \mu_{\tilde{Q}}(v)$ *be continuous. Then* $\tilde{Q}$ *is continuous and* $\omega_{\tilde{Q}}(\delta) \leq \omega_{\tilde{R}}(\delta)$ *for each* $\delta \geq 0$.

While we use extended fuzzy logic to reason with partially true statements we need to extend logics (6) for partial truth. We consider here only extension at the Lukasewicz logic for partial truth. In order to deal with partial truth Pavelka [85] extended this logic by adding truth constants for all reals in [0,1] Hajek [43] simplified it by adding these truth constants $\overline{r}$ only for each rational $r \in [0,1]$ (so $\overline{r}$ is an atomic formula with truth value *r*). They also added 'book - keeping axioms'

$$\overline{r \Rightarrow s} \equiv \overline{r} \rightarrow \overline{s} \text{ for r, s rational} \in [0,1] .$$

This logic is called Rational Pavelka logic (RPL). RPL was introduced in order to reason with partially true statements. In this section we note that this can already be done in Lukasiewicz logic, and that the conservative extension theorems allow us to lift the completeness theorem, that provability degree equals truth degree from RPL to Lukasiewicz logic. This may be regarded as an additional conservative extension theorem, confirming that, even for partial truth, Rational Pavelka logic deals with exactly the same logic as Lukasiewicz logic - but in a very much more convenient way. RPL extends the language of infinite valued Łukasiewicz logic by adding to the truth constants 0 and 1 all rational numbers r of the unit interval $[0,1]$ A *graded formula* is a pair (φ, r) consisting of a formula φ of Łukasiewicz logic and a rational element $r \in [0,1]$, indicating that the truth value of φ is at least r, $\varphi \geq r$ [107]. For example, $\left(p(x), \frac{1}{2}\right)$ expresses the fact that the truth value of $p(x)$, $x \in Dom$, is at least $\frac{1}{2}$. The inference rules of RPL are the *generalization rule*

$$\frac{\varphi}{(\forall x)(\varphi)}, \tag{1.107}$$

and a modified version of the *modus ponens rule*,

$$\frac{(\varphi, r), (\varphi \to \psi, s)}{(\psi, r \otimes s)} \tag{1.108}$$

Here $\otimes$ denotes the Łukasiewicz t-norm. Rule (1.108) says that if formula φ holds at least with truth value *r*, and the implication $\varphi \to \psi$ holds at least with truth value *s*, then formula ψ holds at least with truth value $r \otimes s$. The modified modus ponens rule is derived from the so-called *book-keeping axioms* for the rational truth constants *r*. The book-keeping axioms add to the axioms of Łukasiewicz logic and provide rules for evaluating compound formulas involving rational truth constants [44]. The use of fuzzy reasoning trades accuracy against speed, simplicity and interpretability for lay users. In the context of ubiquitous computing, these characteristics are clearly advantageous.

References

1. Aguilo, I., Suner, J., Torrens, J.: A characterization of residual implications derived from left-continuous uninorms. Information Sciences 180, 3992–4005 (2010)
2. Aliev, R.A.: Fuzzy knowledge based Intelligent Robots. Radio i svyaz, Moscow (1995) (in Russian)
3. Aliev, R.A., Aliev, R.R.: Soft Computing and its Application. World Scientific, New Jersey (2001)
4. Aliev, R.A., Aliev, F.T., Babaev, M.D.: Fuzzy process control and knowledge engineering. Verlag TUV Rheinland, Koln (1991)
5. Aliev, R.A., Aliev, R.R.: Soft Computing, vol. I, II, III. ASOA Press, Baku (1997-1998) (in Russian)
6. Aliev, R.A., Bonfig, K.W., Aliev, F.T.: Messen, Steuern und Regeln mit Fuzzy-Logik. Franzis-Verlag, München (1993)
7. Aliev, R.A., Fazlollahi, B., Aliev, R.R.: Soft Computing and its Application in Business and Economics. Springer, Heidelberg (2004)

8. Aliev, R.A., Mamedova, G.A., Aliev, R.R.: Fuzzy Sets Theory and its application. Tabriz University Press, Tabriz (1993)
9. Aliev, R.A., Mamedova, G.A., Tserkovny, A.E.: Fuzzy control systems. Energoatomizdat, Moscow (1991)
10. Aliev, R.A., Pedrycz, W.: Fundamentals of a fuzzy-logic-based generalized theory of stability. IEEE Transactions on Systems, Man, and Cybernetics, Part B: Cybernetic 39(4), 971–988 (2009)
11. Aliev, R.A., Pedrycz, W., Fazlollahi, B., Huseynov, O.H., Alizadeh, A.V., Guirimov, B.G.: Fuzzy logic-based generalized decision theory with imperfect information. Information Sciences 189, 18–42 (2012)
12. Aliev, R.A., Tserkovny, A.: The knowledge representation in intelligent robots based on fuzzy sets. Soviet Math. Doklady 37, 541–544 (1988)
13. Aliev, R.A., Tserkovny, A.E.: A systemic approach to fuzzy logic formalization for approximate reasoning. Information Sciences 181, 1045–1059 (2011)
14. Azadeh, I., Fam, I.M., Khoshnoud, M., Nikafrouz, M.: Design and implementation of a fuzzy expert system for performance assessment of an integrated health, safety, environment (HSE) and ergonomics system: The case of a gas refinery. Information Sciences 178(22), 4280–4300 (2008)
15. Baldwin, J.F., Pilsworth, B.W.: A model of fuzzy reasoning through multivalued logic and set theory. Int. J. Man-Machines Studies 11, 351–380 (1979)
16. Ban, A.I., Gal, S.G.: Defects of Properties in Mathematics. Quantitative Characterizations. Series on Concrete and Applicable Mathematics, vol. 5. World Scientific, Singapore (2002)
17. Bandemer, H., Gottwald, S.: Fuzzy sets, Fuzzy logic, Fuzzy methods with applications. John Wiley and Sons, England (1995)
18. Bandemer, H., Nather, W.: Fuzzy data analysis. Kluwer Academic Publishers, Boston (1992)
19. Bandler, W., Kohout, L.: Fuzzy power sets and fuzzy implications operators. Fuzzy Sets and Systems 1, 13–30 (1980)
20. Bandler, W., Kohout, L.J.: Fuzzy relational products as a tool for analysis of complex artificial and natural systems. In: Wang, P.P., Chang, S.K. (eds.) Fuzzy Sets; Theory and Applications to Policy Analysis and Information Systems, p. 311. Plenum Press, New York (1980)
21. Bandler, W., Kohout, L.J.: Semantics of fuzzy implication operators and relational products. International Journal of Man-Machine Studies 12(1), 89–116 (1980)
22. Bandler, W., Kohout, L.J.: The identification of hierarchies in symptoms and patients through computation of fuzzy relational products. In: Parslow, R.D. (ed.) BCS 1981: Information Technology for the Eighties, P. 191. Heyden & Sons (1980)
23. Bandler, W., Kohout, L.J.: The four modes of inference in fuzzy expert systems. In: Trappl, R. (ed.) Cybernetics and Systems Research 2, pp. 581–586. North Holland, Amsterdam (1984)
24. Bede, B., Gal, S.G.: Generalizations of the differentiability of fuzzy-number-valued functions with applications to fuzzy differential equations. Fuzzy Sets and Systems 151, 581–599 (2005)
25. Beg, I.: Fuzzy multivalued functions. Centre for Advanced Studies in Mathematics, and Department of Mathematics, Lahore University of Management Sciences (LUMS), 54792-Lahore, Pakistan (2012), `http://wenku.baidu.com/view/71a84c136c175f0e7cd1372d.html`

26. Belohlavek, R., Sigmund, E., Zacpal, J.: Evaluation of IPAQ questionnaires supported by formal concept analysis. Information Science 181(10), 1774–1786 (2011)
27. Bloch, I.: Lattices of fuzzy sets and bipolar fuzzy sets, and mathematical morphology. Information Sciences 181(10), 2002–2015 (2011)
28. Bobillo, F., Straccia, U.: Reasoning with the finitely many-valued Łukasiewicz fuzzy Description Logic SROIQ. Information Sciences 181(4), 758–778 (2011)
29. Buckley, J.J., Eslami, E.: Fuzzy plane geometry I: Points and lines. Fuzzy Sets and Systems 86(2), 179–187 (1997)
30. Bustince, H., Barrenechea, E., Fernandez, J., Pagola, M., Montero, J., Guerra, C.: Contrast of a fuzzy relation. Information Sciences 180, 1326–1344 (2010)
31. Chajda, I., Halas, R., Rosenberg, I.G.: On the role of logical connectives for primality and functional completeness of algebras of logics. Information Sciences 180(8), 1345–1353 (2010)
32. Chen, T.: Optimistic and pessimistic decision making with dissonance reduction using interval valued fuzzy sets. Information Sciences 181(3), 479–502 (2010)
33. Davvaz, B., Zhan, J., Shum, K.P.: Generalized fuzzy Hv-submodules endowed with interval valued membership functions. Information Sciences 178, Nature Inspired Problem-Solving 1, 3147–3159 (2008)
34. Diamond, P., Kloeden, P.: Metric spaces of fuzzy sets. Theory and applications. World Scientific, Singapoure (1994)
35. Dian, J.: A meaning based information theory - inform logical space: Basic concepts and convergence of information sequences. Information Sciences 180: Special Issue on Modelling Uncertainty 15, 984–994 (2010)
36. Fan, Z.-P., Feng, B.: A multiple attributes decision making method using individual and collaborative attribute data in a fuzzy environment. Information Sciences 179, 3603–3618 (2009)
37. Fedrizzi, M., Fuller, R.: Stability in Possibilistic Linear Programming Problems with Continuous Fuzzy Number Parameters. Fuzzy Sets and Systems 47, 187–191 (1992)
38. Fukami, S., Mizumoto, M., Tanaka, K.: Some considerations of fuzzy conditional inference. Fuzzy Sets and Systems 4, 243–273 (1980)
39. Fuller, R., Zimmermann, H.J.: On Zadeh's compositional rule of inference. In: Lowen, R., Roubens, M. (eds.) Fuzzy Logic: State of the Art, Theory and Decision Library, Series D, pp. 193–200. Kluwer Academic Publishers, Dordrecht (1993)
40. Gerhke, M., Walker, C.L., Walker, E.A.: Normal forms and truth tables for fuzzy logics. Fuzzy Sets and Systems 138, 25–51 (2003)
41. Grabisch, M., Marichal, J., Mesiar, R., Pap, E.: Aggregation functions: Construction methods, conjunctive, disjunctive and mixed classes. Information Sciences 181, 23 (2011)
42. Grzegorzewski, P.: On possible and necessary inclusion of intuitionistic fuzzy sets. Information Sciences 181, 342 (2011)
43. Hajek, P.: Fuzzy Logic from the Logical Point of View. In: Bartosek, M., Staudek, J., Wiedermann, J. (eds.) SOFSEM 1995. LNCS, vol. 1012, pp. 31–49. Springer, Heidelberg (1995)
44. Hajek, P.: Metamathematics of Fuzzy Logic. Trends in Logic. Kluwer Academic Publishers (1998)
45. Hu, Q., Yu, D., Guo, M.: Fuzzy preference based rough sets. Information Sciences 180: Special Issue on Intelligent Distributed Information Systems 15, 2003–2022 (2010)

46. Buckley, J.J., Feuring, T.: Fuzzy differential equations. Fuzzy Sets and Systems 151, 581–599 (2005)
47. Jantzen, J.: Array approach to fuzzy logic. Fuzzy Sets and Systems 70, 359–370 (1995)
48. Jayaram, B., Mesiar, R.: I-Fuzzy equivalence relations and I-fuzzy partitions. Information Sciences 179, 1278–1297 (2009)
49. Jenei, S.: Continuity in Zadeh's compositional rule of inference. Fuzzy Sets and Systems 104, 333–339 (1999)
50. Kallala, M., Kohout, L.J.: The use of fuzzy implication operators in clinical evaluation of neurological movement disorders. In: International Symposium on Fuzzy Information Processing in Artificial Intelligence and Operational Research, Christchurch College, Cambridge University (1984)
51. Kallala, M., Kohout, L.J.: A 2-stage method for automatic handwriting classification by means of norms and fuzzy relational inference. In: Proc. of NAFIPS 1986 (NAFIPS Congress), New Orleans (1986)
52. Kalina, M.: Derivatives of fuzzy functions and fuzzy derivatives. Tatra Mountains Mathematical Publications 12, 27–34 (1997)
53. Kandel, A., Last, M.: Special issue on advances in Fuzzy logic. Information Sciences 177, 329–331 (2007)
54. Kaufman, A.: Introduction to theory of fuzzy sets, vol. 1. Academic Press, Orlando (1973)
55. Kehagias, A.: Some remarks on the lattice of fuzzy intervals. Information Sciences 181(10), 1863–1873 (2010)
56. Kiszka, J.B., Kochanska, M.E., Sliwinska, D.S.: The influence of some fuzzy implication operators on the accuracy of a fuzzy model. Fuzzy Sets and Systems 15, (Part1) 111–128, (Part2) 223–240 (1985)
57. Klir, G.J., Yuan, B.: Fuzzy sets and fuzzy logic. Theory and Applications. PRT Prentice Hall, NJ (1995)
58. Klir, G.J., Clair, U.S., Yuan, B.: Fuzzy Set Theory. Foundations and Applications. PTR Prentice Hall, NJ (1997)
59. Kohout, L.J. (ed.): Perspectives on Intelligent Systems: A Framework for Analysis and Design. Abacus Press, Cambridge (1986)
60. Kolesarova, A., Mesiar, R.: Lipschitzian De Morgan triplets of fuzzy connectives. Information Sciences 180, 3488–3496 (2010)
61. Lai, J., Xu, Y.: Linguistic truth-valued lattice-valued propositional logic system lP(X) based on linguistic truth-valued lattice implication algebra. Information Sciences 180: Special Issue on Intelligent Distributed Information Systems, 1990–2002 (2010)
62. Lakshmikantham, V., Mohapatra, R.: Theory of fuzzy differential equations and inclusions. Taylor & Francis, London (2003)
63. Levy, P.: From social computing to reflexive collective intelligence: The IEML research program. Information Sciences 180: Special Issue on Collective Intelligence 2, 71 (2010)
64. Li, C., Yi, J.: Sirms based interval type–2 fuzzy inference systems properties and application. International Journal of Innovative Computing. Information and Control 6(9), 4019–4028 (2010)
65. Lifschitz, V. (ed.): Formalizing Common Sense, Papers by John McCarthy. Greenwood Publishing Group Inc., NJ (1990)

66. Long, Z., Liang, X., Yang, L.: Some approximation properties of adaptive fuzzy systems with variable universe of discourse. Information Sciences 180, 2991–3005 (2010)
67. Ma, H.: An analysis of the equilibrium of migration models for biogeography-based optimization. Information Sciences 180, 3444–3464 (2010)
68. Mamdani, E.H.: Application of fuzzy logic to approximate reasoning using linguistic syntheses. IEEE Transactions on Computers C-26(12), 1182–1191 (1977)
69. Mas, M., Monserrat, M., Torrens, J.: The law of importation for discrete implications. Information Sciences 179, 4208–4218 (2009)
70. Mas, M., Monserrat, M., Torrens, J., Trillas, E.: A Survey on Fuzzy Implication Functions. IEEE Transactions on Fuzzy Systems 15(6), 1107–1121 (2007)
71. Mayburov, S.: Fuzzy geometry of phase space and quantization of massive Fields. Journal of Physics A: Mathematical and Theoretical 41, 1–10 (2008)
72. Medina, J., Ojeda Aciego, M.: Multi-adjoint t-concept lattices. Information Sciences 180, 712–725 (2010)
73. Mendel, J.M.: On answering the question "Where do I start in order to solve a new problem involving interval type-2 fuzzy sets?". Information Sciences 179, 3418–3431 (2009)
74. Mizumoto, M., Fukami, S., Tanaka, K.: Some methods of fuzzy reasoning. In: Gupta, R., Yager, R. (eds.) Advances in Fuzzy Set Theory Applications. North-Holland, New York (1979)
75. Mizumoto, M., Zimmermann, H.-J.: Comparison of fuzzy reasoning methods. Fuzzy Sets and Systems 8, 253–283 (1982)
76. Molai, A.A., Khorram, E.: An algorithm for solving fuzzy relation equations with max-T composition operator. Information Sciences 178, 1293–1308 (2008)
77. Mordeson, J.N., Nair, P.S.: Fuzzy mathematics: an introduction for engineers and Scientists. Physica-Verlag, Heidelberg (2001)
78. Mueller, E.: Commonsense Reasoning. Morgan Kaufmann, San Francisco (2006)
79. Munoz-Hernandez, S., Pablos-Ceruelo, V., Strass, H.: R Fuzzy: Syntax, semantics and implementation details of a simple and expressive fuzzy tool over Prolog. Information Sciences 181(10), 1951–1970 (2011)
80. Nachtegael, M., Sussner, P., Melange, T., Kerre, E.E.: On the role of complete lattices in mathematical morphology: From tool to uncertainty model. Information Sciences (2010); corrected proof, available online 15 (in Press)
81. Nguyen, H.T., Walker, E.A.: A first Course in Fuzzy logic. CRC Press, Boca Raton (1996)
82. Noguera, C., Esteva, F., Godo, L.: Generalized continuous and left-continuous t-norms arising from algebraic semantics for fuzzy logics. Information Sciences 180, 1354–1372 (2010)
83. Oh, K.W., Bandler, W.: Properties of fuzzy implication operators, Florida State University, Tallahassee, FL, U. S. A. Department of Computer Science, pp. 24–33 (1988)
84. Ouyang, Y., Wang, Z., Zhang, H.: On fuzzy rough sets based on tolerance relations. Information Sciences 180, 532 (2010)
85. Pavelka, J.: On fuzzy logic I, II, III. Zeitschrift fur Mathematische Logik und Grundlagen der Mathematik 25(45-52), 119–134, 447–464 (1979)
86. Pei, D.: Unified full implication algorithms of fuzzy reasoning. Information Sciences 178, 520 (2008)
87. Poston, T.: Fuzzy geometry. Ph.D. Thesis, University of Warwick (1971)

88. Rescher, N.: Many-Valued Logic. McGraw–Hill, NY (1969)
89. Roe, J.: Index theory, coarse geometry, and topology of manifolds. In: CBMS: Regional Conf. Ser. in Mathematics. The American Mathematical Society, Rhode Island (1996)
90. Rosenfeld, A.: The diameter of a fuzzy set. Fuzzy Sets and Systems 13, 241–246 (1984)
91. Rosenfeld, A.: Distances between fuzzy sets. Pattern Recognition Letters 3(4), 229–233 (1985)
92. Rosenfeld, A.: Fuzzy rectangles. Pattern Recognition Letters 11(10), 677–679 (1990)
93. Rosenfeld, A.: Fuzzy plane geometry: triangles. Pattern Recognition Letters 15, 1261–1264 (1994)
94. Rosenfeld, A.: Fuzzy geometry: an updated overview. Information Science 110(3-4), 127–133 (1998)
95. Rutkowski, L., Cpalka, K.: Flexible Neuro-Fuzzy Systems. IEEE Transactions on Neural Networks 14(3), 554–573 (2003)
96. Sankar, K.P., Ghosh, A.: Fuzzy geometry in image analysis. Fuzzy Sets and Systems 48, 23–40 (1992)
97. Sankar, K.P.: Fuzzy geometry, entropy, and image information. In: Proceedings of the Second Join Technology Workshop on Neural Networks and Fuzzy Logic, vol. 2, pp. 211–232 (1991)
98. Serruier, M., Dubois, D., Prade, H., Sudkamp, T.: Learning fuzzy rules with their implication operator. Data & Knowledge Engineering 60, 71–89 (2007)
99. Shieh, B.S.: Infinite fuzzy relation equations with continuous t-norms. Information Sciences 178, 1961–1967 (2008)
100. Simon, H.: Models of Bounded Rationality: Empirically Grounded Economic Reason, vol. 3. MIT Press, Cambridge (1997)
101. Toulmin, S.: The Uses of Argument. Cambridge University Press, UK (2003)
102. Tzafestas, S.G., Chen, C.S., Fokuda, T., Harashima, F., Schmidt, G., Sinha, N.K., Tabak, D., Valavanis, K. (eds.): Fuzzy logic applications in engineering science. Microprocessor based and Intelligent Systems Engineering, vol. 29, pp. 11–30. Springer, Netherlands (2006)
103. Valle, M.E.: Permutation-based finite implicative fuzzy associative memories. Information Sciences 180, 4136–4152 (2010)
104. Valverde Albacete, F.J., Pelaez Moreno, C.: Extending conceptualization modes for generalized Formal Concept Analysis. Information Sciences 181(10), 1888–1909 (2011)
105. Wenstop, F.: Quantitative analysis with linguistic values. Fuzzy Sets and Systems 4(2), 99–115 (1980)
106. Werthner, H.: Qualitative Reasoning, Modeling and the Generation of Behavior. Springer (1994)
107. Wilke, G.: Approximate Geometric Reasoning with Extended Geographic Objects. In: Proceedings of the Workshop on Quality, Scale and Analysis Aspects of City Models, Lund, Sweden, December 3-4 (2009), `http://www.isprs.org/proceedings/XXXVIII/2W11/Wilke.pdf`
108. Xie, A., Qin, F.: Solutions to the functional equation I(x, y) = I(x, I(x, y)) for three types of fuzzy implications derived from uninorms. Information Sciences 186(1), 209–221 (2012)
109. Xu, Y., Liu, J., Ruan, D., Li, X.: Determination of [alpha]-resolution in lattice-valued first-order logic LF(X). Information Sciences 181(10), 1836–1862 (2010)

110. Yager, R.R.: On measures of specificity. In: Kaynak, O., Zadeh, L.A., Turksen, B., Rudas, I.J. (eds.) Computational Intelligence: Soft Computing and Fuzzy-Neuro Integration with Applications, pp. 94–113. Springer, Berlin (1998)
111. Yager, R.R.: On global requirements for implication operators in fuzzy modus ponens. Fuzzy Sets and Systems 106, 3–10 (1999)
112. Yager, R.R.: A framework for reasoning with soft information. Information Sciences 180(8), 1390–1406 (2010)
113. Yeh, R.T., Bang, S.Y.: Fuzzy relations, fuzzy graphs, and their applications to clustering analysis. In: Zadeh, L.A., Fu, K.S., Shimura, M.A. (eds.) Fuzzy Sets and Their Applications, pp. 125–149. Academic Press, NY (1975)
114. Zadeh, L.A.: Fuzzy Sets. Information and Control 8, 338–353 (1965)
115. Zadeh, L.A.: Probability measures of fuzzy events. Journal of Mathematical Analysis and Applications 23(2), 421–427 (1968)
116. Zadeh, L.A.: Fuzzy orderings. Information Sciences 3, 117–200 (1971)
117. Zadeh, L.A.: Similarity relations and Fuzzy orderings. Information Sciences 3, 177–200 (1971)
118. Zadeh, L.A.: Outline of a new approach to the analysis of complex system and decision processes. IEEE Trans. Systems, Man, and Cybernetics 3, 28–44 (1973)
119. Zadeh, L.A.: The concept of a linguistic variable and its applications in approximate reasoning. Information Sciences 8, 43–80, 301–357; 9, 199–251 (1975)
120. Zadeh, L.A.: Fuzzy sets and information granularity. In: Gupta, M., Ragade, R., Yager, R. (eds.) Advances in Fuzzy Set Theory and Applications, pp. 3–18. North-Holland Publishing Co., Amsterdam (1979)
121. Zadeh, L.A.: Possibility theory, soft data analysis. In: Cobb, L., Thrall, R.M. (eds.) Mathematical Frontiers of the Social and Policy Sciences, pp. 69–129. Westview Press, Boulder (1981)
122. Zadeh, L.A.: Fuzzy logic. IEEE Computer 21(4), 83–93 (1988)
123. Zadeh, L.A.: Toward a generalized theory of uncertainty — an outline. Information Sciences 172, 1–40 (2005)
124. Zadeh, L.A.: Generalized theory of uncertainty (GTU) – principal concepts and ideas. Computational statistics & Data Analysis 51, 15–46 (2006)
125. Zadeh, L.A.: Is there a need for fuzzy logic? Information Sciences 178, 2751–2779 (2008)
126. Zadeh, L.A.: Fuzzy logic. In: Encyclopedia of Complexity and Systems Science, pp. 3985–4009. Springer, Berlin (2009)
127. Zadeh, L.A.: Toward extended fuzzy logic. A first step. Fuzzy Sets and Systems 160, 3175–3181 (2009)
128. Zadeh, L.A.: A Note on Z-numbers. Information Sciences 181, 2923–2932 (2010)
129. Zhang, J., Yang, X.: Some properties of fuzzy reasoning in propositional fuzzy logic systems. Information Sciences 180, 4661–4671 (2010)
130. Zhang, L., Cai, K.Y.: Optimal fuzzy reasoning and its robustness analysis. Int. J. Intell. Syst. 19, 1033–1049 (2004)
131. Zhang, X., Yao, Y., Yu, H.: Rough implication operator based on strong topological rough algebras. Information Sciences 180, 3764–3780 (2010)
132. Zhao, S., Tsang, E.C.C.: On fuzzy approximation operators in attribute reduction with fuzzy rough sets. Information Sciences 178, Including Special Issue: Recent Advances in Granular Computing (2008); Fifth International Conference on Machine Learning and Cybernetics, vol. 15, pp. 3163–3176
133. Zimmermann, H.J.: Fuzzy Set Theory and its applications. Kluwer Academic Publishers, Norwell (1996)

Chapter 2
Brief Review of Theories of Decision Making

2.1 Existing Classical Theories of Decision

Making sound decisions requires to consistently take into account influence of various factors under uncertainty. Such analysis is not compatible with highly constrained computational ability of a human's brain. Thus, decision making needs to be based on the use of mathematical methods as a strong language of reasoning. This requires at first to formally define the problem of decision making. Depending on the decision relevant information on future, various routine decision making problems are used. However, all kinds of decision making problems have a common stem which is represented by the main elements of any decision making problem.

The first element is a set A of alternatives (sometimes called alternative actions, actions, acts, strategies etc) to choose from:

$$A=\{f_1,...,f_n\},n\geq 2$$

where $n\geq 2$ means that decision making may take place when at least two alternatives exist. In the problem of business development, f_1 may denote "to extend business", f_2– "to improve quality of services", f_3– "to open a new direction".

The next element of decision making problem is used to model objective conditions on which the results of any alternative action depend. This element is called *a set of states of nature*: $S=\{s_1,...,s_m\}$. A state of nature s_i is one possible objective condition. S is considered as a "*a space of mutually exclusive and exhaustive states*", according to a formulation suggested by L. Savage [43]. This means that all possible objective conditions (possible conditions of future) are known and only one of them $s_i, i=1,...,m$ will take place. The main problem is that it is not known for sure, which s_i will take place. For example, in the problem of business development, the set of states of nature may be considered as $S=\{s_1,s_2,s_3,s_4\}$, where s_1 denotes "high demand and low competition", s_2– "high demand and medium competition" s_3–"medium demand and low competition", s_4– "medium demand and high competition".

R.A. Aliev: *Fuzzy Logic-Based Generalized Theory of Decisions*, STUDFUZZ 293, pp. 65–88.
DOI: 10.1007/ 978-3-642-34895-2_2

The set S may also be infinite. For example, when considering inflation rate, S can be used as a continuous range.

The third element is results of actions in various states of nature. These results are called outcomes or consequences. Any action result in an outcome (lead to some consequence) in any state of nature. For example, if someone extends his business and high demand occurs then the profit will be high. If a low demand occurs, then extension results in a low profit, or even, in a loss. The outcomes may be of any type – quantitative or qualitative, monetary or non-monetary. A set of outcomes is commonly denoted X . As an outcome $x \in X$ is a result of an action f taken at a state of nature s, it is formalized as $x = f(s)$. So, an action f is formally a function whose domain is a set of states of nature S and the range is the set of outcomes $X : f : S \to X$. In order to formally compare actions $f \in A$, it is needed to formally measure all their outcomes $x \in X$, especially when the latter are qualitative. For this purpose a numeric function $u : X \to R$ is used to measure an outcome $x \in X$ in terms of its *utility* for a decision maker (DM). Utility $u(x)$ of an outcome $x \in X$ represents to what extent $x \in X$ is good, useful, or desirable for a DM. $u : X \to R$ is used to take into account various factors like reputation, health, mentality, psychology and others.

The fourth component is preferences of a DM. Given a set of alternatives A, the fact that a DM prefers an alternative $f \in A$ to an alternative $g \in A$ is denoted $f \succ g$. Indifference among f and g is denoted $f \sim g$. The fact that f is at least as good as g is denoted $f \succsim g$. Preferences are described as a binary relation $\succsim \in A \times A$.

In general, a decision making problem is formulated as follows:

Given
the set of alternatives A,
the set of states of nature S,
the set of outcomes X.

Determine an action $f^* \in A$ such that $f^* \succsim f$ for all $f \in A$.

In some models they formulate decision problem in a framework of outcomes and their probabilities only and don't use such concept as a set of states of nature. In this framework an alternative is described as a collection of its outcomes with the associated probabilities and is termed as lottery: $f = (x_1, p_1; ...; x_n, p_n)$.

The main issue here is to impose some reasonable assumptions on properties of DM's preferences. The latter critically depend on a type and an amount of information on S. Several typical cases exist with respect to this. In the idealized case, when it is known which state of nature will take place, we deal with *decision making under certainty*. In the case when objective (actual) probability of occurrence of each state of nature is known we deal with *decision making under risk.* In the situations when we find difficulties in assessment of unique precise probabilities to states of nature we deal with *decision making under ambiguity.* In the situations when there is no information on probabilities of states of nature we deal with *decision making under complete ignorance* (in the existing theories this is also referred to as *decision making under complete ignorance*). It is needed to mention that these four typical cases are rigorous descriptions of real-life decision situations. In general, real decision situations are more complex, diverse and ambiguous. In real-life, decisions are made under imperfect information on all elements of a decision problem. As Prof. L. Zadeh states, imperfect information is information which is in one or more respects is imprecise, uncertain, incomplete, unreliable, vague or partially true. For simplicity, imperfect information may be degenerated to one of the four typical cases described above.

From the other side, DM's preferences are determined by psychological, cognitive and other factors.

The solution of a decision making problem depends both on information and preference frameworks and consist in determination of the best action in terms of a DM's preferences. However, it is difficult to determine the best action by direct treatment of the preferences as a binary relation. For this purpose, a quantification of preferences is used. One approach to quantify preferences is to use a *utility function.* Utility function is a function $U:A \to R$ that for all $f,g \in A$ satisfies $U(f) \geq U(g)$ iff $f \succsim g$. Generally, any utility function is some aggregation $U(f)=\int_S u(f(s))ds$. The utility models differ on the type of aggregation $\int_S$. However, any utility function is a function existence of which is proven given the assumptions on properties of preferences. The use of utility function is more practical approach than direct treatment of preferences. However, the use of this approach leads to loss of information as any utility function transforms functions to numbers. From the other side, there exists some type of preferences for which a utility function does not exist.

In this chapter we will consider main categories of decision models suggested in the existing literature. Some of them are quite simple being based on idealistic assumptions on relevant information and preferences. Those which are based on more realistic assumptions are distinguished by an increased complexity.

Utility theory is one of the main parts of decision analysis and economics. The idea of a utility function consists in construction of a function that represents an individual's preferences defined over the set of possible alternatives [4,35,38,44, 48,49]. Formally speaking, a utility function $U(\cdot)$ is such a real-valued function that for any two possible alternatives f and g an inequality $U(f) \geq U(g)$ holds if and only if f is preferred or indifferent to g. For decision making under risk the first axiomatic foundation of the utility paradigm was the expected utility

(EU) theory of von Neumann and Morgenstern [48]. This model compares finite-outcome lotteries (alternatives) on the base of their utility values under conditions of exactly known utilities and probabilities of outcomes. Formally, let X be a set of outcomes without any additional structure imposed on it. The set of lotteries in the expected utility theory is the set of probability distributions over X with finite supports [24]:

$$L = \left\{ P : X \to [0,1] \middle| \sum_{x \in X} P(x) = 1 \right\}$$

Each probability distribution represents objective probabilities of possible outcomes. The EU model, as initially suggested is not based on a general decision problem framework which includes the concept of a set of states of nature. However, it can easily be applied in this framework also, once we consider an action f as a lottery $f = \left\{ P : f(S) \to [0,1] \middle| \sum_{s \in S} P(f(s)) = 1 \right\}$ as $f(S) \subset X$. Important issue to consider may be more complicated cases when various lotteries L can be faced with various probabilities. Such case is referred to as a compound, or a two-stage lottery, that is, a lottery which compose several lotteries as its possible results. To model this within L a convex combination is defined which reduces a compound lottery to a lottery in L as follows: for any $P, Q \in L$ and any $\alpha \in [0,1]$ $\alpha P + (1-\alpha)Q = R \in L$, where $R(x) = \alpha P(x) + (1-\alpha)Q(x)$. The axioms stating the assumptions on preference which underlie EU model are the following:

Weak-Order: (a) Completeness. Any two alternatives are comparable with respect to $\succsim$: for all f and g in A: $f \succsim g$ or $g \succsim f$. This means that for all f and g one has $f \succsim g$ or $g \succsim f$; (b) Transitivity. For all f, g and h in A: If $f \succsim g$ and $g \succsim h$ then $f \succsim h$.

Continuity: For all f, g and h in A: if $f \succ g$ and $g \succ h$ then there are α and β in $(0,1)$ such that $\alpha f + (1-\alpha)h \succ g \succ \beta f + (1-\beta)h$.

***Independence*:** For all acts f, g and h in A if $f \succsim g$, then $\alpha f + (1-\alpha)h \succsim \alpha g + (1-\alpha)h$ for all $\alpha \in (0,1)$.

The completeness property implies that despite the fact that each alternative f or g has its advantages and disadvantages with respect to the other, a DM supposed to be always able to compare two actions f and g on the base of his/her preferences either f is preferred to g or g is preferred to f or f and g are considered equivalent. The problem when f and g are absolutely not comparable should be resolved before the set of alternatives is completely determined. This problem may be solved by obtaining additional information to eliminate "ignorance" of preferences with respect to f and g. Alternatively, one of the alternatives should be disregarded.

Let us now present the utility representation of the vNM's axioms (i)-(iii). The vNM's EU representation theorem is given below:

Theorem 2.1 [48]. $\succsim \subset L \times L$ satisfies (i)-(iii) if and only if there exists $u : X \to \mathrm{R}$ such that for every P and Q in $\mathcal{L}$:

$$P \succsim Q \text{ iff } \sum_{x \in X} P(x)u(x) \geq \sum_{x \in X} Q(x)u(x),$$

Moreover, in this case, u is unique up to a positive linear transformation. So, a utility value $U(P)$ of a finite-outcome lottery $P = (x_1, p(x_1);...;x_n, p(x_n))$ is defined as $U(P) = \sum_{i=1}^{n} u(x_i)p(x_i)$. The problem of decision making is to find such

$$P^* \text{ that } U(P^*) = \max_{P \in A} U(P).$$

The assumptions of von Neumann and Morgenstern expected utility model stating that objective probabilities of events are known, makes this model unsuitable for majority of real-world applications. For example, what is an actual probability that a country will meet an economic crisis during a year? What is an actual probability that sales of a new product will bring profit next year? What is an actual probability that I will not get the flu upcoming winter? In any of these examples we deal either with the completely new phenomena, or phenomena which notably differs from the previous events, or phenomena that depends on uncertain or unforeseen factors. This means we have no representative experimental data or complete knowledge to determine objective probabilities. For such cases, L. Savage suggested a theory able to compare alternative actions on the base of a DM's experience or vision [45]. Savage's theory is based on a concept of subjective probability suggested by Ramsey [42] and de Finetti [13]. Subjective probability is DM's probabilistic belief concerning occurrence of an event and is assumed to be used by humans when no information on objective (actual) probabilities of outcomes is known. Savage's theory is called subjective expected utility (SEU) as it is based on the use of subjective probabilities in the expected utility paradigm of von Neumann and Morgenstern instead of objective probabilities. SEU became a base of almost all the utility models for decision making under uncertainty. The preferences in SEU model is formulated over acts as functions from S to X in terms of seven axioms.

The Savage's utility representation is as follows: provided that $\succsim$ satisfies all the axioms, there exists a unique probability measure μ on S and a function $u : X \to \mathrm{R}$ such that

$$f \succsim g \text{ iff } \int_S u(f(s))d\mu \geq \int_S u(g(s))d\mu$$

The problem of decision making consist in determination an action $f^* \in A$ such that

$$U(f^*) = \max_{f \in A} \int_S u(f(s))d\mu .$$

In quantitative sense, SEU model coincide with vNM's model – we again have expectation with respect to a probability measure. However, qualitatively, these theories differ. In vNM's model it is supposed that actual probabilities of outcomes are known. The main implications of SEU model are the following: 1) beliefs of a DM are probabilistic; 2) the beliefs are to be used linearly in utility representation of an alternative.

In von Neumann-Morgenstern and Savage theories it is assumed that individuals tend to maximize expected utility being motivated by material incentives (self-interest) [1] and make decisions in a rational way. In turn rationality means that individuals update their beliefs about probability of outcomes correctly (following Bayes' law) and that they can assign consistent subjective probabilities to each outcome. These theories are well-composed and have strong analytical power. However, they define human behavior as "ideal", i.e. inanimate. Experimental evidence has repeatedly shown that people violate the axioms of von Neumann-Morgenstern-Savage preferences in a systematic way. Indeed, these models are based on assumptions that people behave as "computational machines" functioning according to predefined mathematical algorithms. Of course, these don't correspond to the computational abilities of humans. From the other side, these models are developed for a perfect information framework, e.g. humans either know actual probabilities or they can assign subjective probabilities to each outcome. Really, actual probabilities are very seldom known in real life, and the use of subjective probabilities is very often questionable or not compatible with human choices.

Humans' decision activity is conditioned by psychological issues, mental, social and other aspects. These insights inspired a novel direction of studying how people actually behave when making decisions. This direction is called *behavioral economics* and takes it start in the Prospect Theory (PT) [32] of D.Kahneman and A. Tversky. PT [32,47] is the one of the most famous theories in the new view on a utility concept. This theory has a good success because it includes psychological aspects that form human behavior. Kahneman and Tversky uncovered a series of features of human behavior in decision making and used them to construct their utility model. The first feature is that people make decisions considering deviations from their initial wealth, i.e. gains and losses, rather than final wealth. To represent this, an alternative in their model is a lottery in which an outcome is considered as a change from a DM's current wealth called *reference point*, but not as a final wealth. They call such a lottery a prospect.

Furthermore, a so-called gain-loss asymmetry was observed: influence of losses on human choices dominates influence of gains. Humans' attitudes to risk depend on whether they deal with gains or losses.

For representing the features of perception of gains and losses, Kahneman and Tversky use value function $v()$ to model DM's tastes over monetary outcomes x (as changes from the current wealth), instead of a utility function (which is used in, for example, SEU to represent tastes over net wealth). Value function, as based on the experimental evidence on attitudes to gains and losses, has the following properties: 1) it is steeper in domain of gains than in domain of losses; 2) it is concave for gains and is convex for losses; 3) it is steepest in the reference point. Schematic view of the value function is given in fig. 2.1:

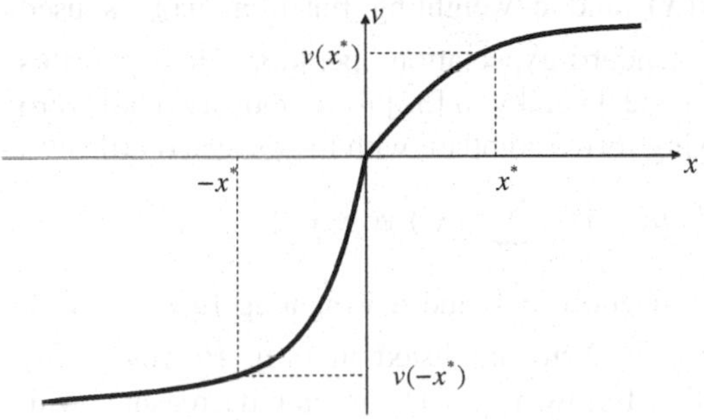

Fig. 2.1 Value function

The second main insight observed from experimental evidence is distorted perception of probabilities. In making decisions, people perceive the values of probabilities not exactly as they are but overestimate or underestimate them. This comes from the fact that the change of probability from, for example, 0 to 0.1 or from 0.9 to 1 is considered by people as notably more sufficient than the change from 0.3 to 0.4. The reason was explained as follows: in the first case the situation changes qualitatively – from an impossible outcome to some chance, in the second case the situation also change qualitatively – from very probable outcome to thecertain one. In other words, appearance of a chance or appearance of a guaranteed outcome is perceived more important than just the change of probabilty value. As a result, people overestimate low and underestimate high probabilities.

In order to model this evidence Kahneman and Tversky replace probabilities p with weights $w(p)$ as the values of so-called weighting function $w:[0,1]\rightarrow[0,1]$. This function non-linearly transforms an actual probability to represent distorted perception of the latter. Schematic view of the weighting function $w()$ is given in fig. 2.2:

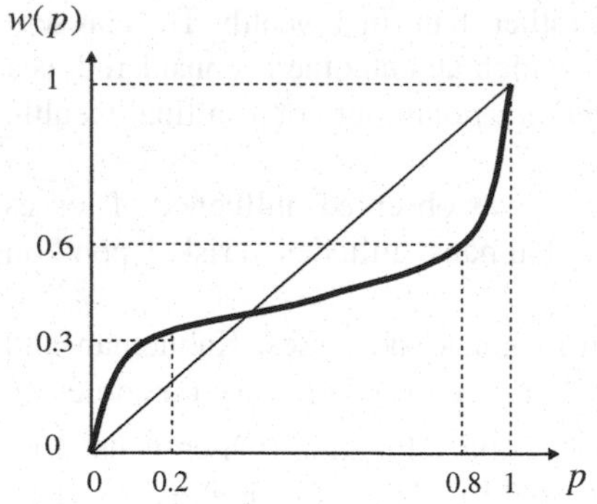

Fig. 2.2 Weighting function

In Prospect theory, Kahneman and Tversky suggested a new model of choice among prospects $(x_1, p(x_1);...;x_n, p(x_n))$. In this model, a value function $v()$ is used instead of utility function $u(x)$ and a weighting function $w()$ is used instead of probability measure in a standard expectation operation. However, this model was formulated by Kahneman and Tversky in [32] for at most two non-zero outcomes. The model has the following form (compare with the expected utility):

$$U((x_1, p(x_1);...;x_n, p(x_n)) = \sum_{i=1}^{n} v(x_i)w(p(x_i))$$

There are various forms of a value function $v()$ and a weighting function $w()$ suggested by various authors. Many of them are listed in [40]. However, any weighting function must satisfy the following: $w()$ is non-decreasing with $w(0) = 0, w(1) = 1$. Commonly, $w()$ is non-additive, that is, $w(p+q) \neq w(p) + w(q), p+q \leq 1$.

Prospect theory is a successful theory for decision making under risk and it was one of the first descriptive theories. This explains such phenomena as Allais paradox, certainty effect and framing effects [32].

Choquet Expected Utility (CEU) was suggested by Schmeidler [44] as a model with a new view on belief and representation of preferences in contrast to SEU model. In SEU model an overall utility is described as $U(f) = \int_S u(f(s))d\mu$ where μ is a probability measure. However, uncertainty as a vagueness of knowledge on occurrence of events may result in a non-additive belief for $s \in S$.

In CEU a belief is described by a capacity [12] – not necessarily additive measure v satisfying the following conditions [44]:

1. $v(\varnothing) = 0$
2. $\forall A, B \subset S, A \subset B$ implies $v(A) \leq v(B)$
3. $v(S) = 1$

Capacity is a model of a belief used for the cases when it is impossible to definitely separate states of nature by using probabilities of their occurrence. In CEU v is referred to as nonadditive probability.

The use of a capacity is not an only advantage of the CEU. The use of a capacity instead of an additive measure in Riemann integration adopted in SEU is not proper. Riemann integration of acts with respect to a capacity arises several unavoidable problems. Particularly, Riemann integration depends on the way the act is written. For example, consider an act f in expressed in two alternative forms: $(\$1,s_1;\$1,s_2;\$0,s_3)$ and $(\$1,\{s_1,s_2\};\$0,s_3)$. Riemann integration will give different results for these expressions as, in general, $v(\{s_1,s_2\})\neq v(\{s_1\})+v(\{s_2\})$. The other problems such as violation of continuity and monotonicity are well explained in [24]. In CEU, Choquet integral is used which is a generalization of Lebesgue integral obtained when a capacity is used instead of an additive measure. The use of Lebesgue integration removes the problems related to Riemann integration.

CEU is axiomatically developed for the Anscombe-Aumann framework, in which acts are functions from states to lotteries, i.e. to probabilistic outcomes. The main difference in the underlying assumptions on preferences in CEU from those in the expected utility models is a relaxation of the independence axiom. The independence property in this model is assumed only for comonotonic actions. Two acts f and g in A are said to be comonotonic if for no s and t in S, $f(s)\succ f(t)$ and $g(t)\succ g(s)$ hold. That is, as functions, f and g behave analogously.

Comonotonic Independence. For all pair wise comonotonic acts f,g and h in A if $f\succsim g$, then $\alpha f+(1-\alpha)h\succsim \alpha g+(1-\alpha)h$ for all $\alpha\in(0,1)$.

The other axioms of the CEU model are quite trivial. The utility representation in this model is $U(f)=\int_S u(f(s))dv$, where v is a nonadditive probability (capacity) and u is nonconstant and is unique up to a positive linear transformation. For finite S, i.e. for $S=\{s_1,\ldots,s_n\}$ CEU representation is as follows:

$$U(f)=\sum_{i=1}^{n}(u(f(s_{(i)}))-u(f(s_{(i+1)})))v(\{s_{(1)},\ldots,s_{(i)}\})$$

where (i) in the index of the states s implies that they are permuted such that $u(f(s_{(i)}))\geq u(f(s_{(i+1)}))$ and $u(f(s_{(n+1)}))=0$ by convention. Sometimes they use an equivalent expression for CEU representation:

$$U(f)=\sum_{i=1}^{n}u(f(s_{(i)}))(v(\{s_{(1)},...,s_{(i)}\})-v(\{s_{(1)},...,s_{(i-1)}\}))$$

with $\{s_{(1)},...,s_{(i-1)}\}=\varnothing$ for $i=1$.

One can see that SEU model is a special case of CEU model when v is additive. The disadvantages of CEU relates to difficulties of satisfactory interpretation of v. The typical cases are when v is taken as a lower envelope of a set of probability distributions possible for a considered problem (when a single distribution is unknown). Lower envelope is the measure that assigns minimal probability to an event among all the probabilities for this event each determined on the base of one possible probability distribution. This requires solving optimization problems which increases complexity of computations. In general, non-additive measure construction is difficult both from intuitive and computational points of view.

Cumulative Prospect Theory [47] (CPT) was suggested by Kahnemann and Tversky as a development of PT. PT was originally presented as a model for comparison of prospects with at most two non-zero outcomes. However, even for such primitive cases, the use of a non-additive weighting function in PT gives rise to at least one difficulty.

In order to overcome this difficulty, Quiggin [41] and Yaari [54] suggested using non-linear transformation of not a probability $p(x)$ of an outcome $x=a$, but of a cumulative probability $p(x\geq a)$ that an outcome x is not less than a predefined value a. Consider a prospect $f=(x_1,p(x_1);...;x_n,p(x_n))$ with non-negative outcomes, such that $x_1\geq...\geq x_n$ (this ordering does not lead to loss of generality, as this can always be achieved by permutation of indexes). EU of this prospect will be

$$U(f)=\sum_{i=1}^{n}p(x_i)u(x_i)$$

that can be rewritten as

$$U(f)=\sum_{i=1}^{n}\left(\sum_{j=1}^{i}p(x_j)\right)(u(x_i)-u(x_{i+1}))$$

where $x_{i+1}=0$ by convention. $\sum_{j=1}^{i}p(x_j)$ is a cumulative probability that an outcome of f is not less than x_i. By applying non-linear transformation w to $\sum_{j=1}^{i}p(x_j)$ one will have:

$$U(f)=\sum_{i=1}^{n} w\left(\sum_{j=1}^{i} p(x_j)\right)(u(x_i)-u(x_{i+1}))$$

which can be rewritten

$$U(f)=\sum_{i=1}^{n}\left(w\left(\sum_{j=1}^{i} p(x_j)\right)-w\left(\sum_{j=1}^{i-1} p(x_j)\right)\right)u(x_i)$$

As

$$\left(w\left(\sum_{j=1}^{i} p(x_j)\right)-w\left(\sum_{j=1}^{i-1} p(x_j)\right)\right)\in[0,1]$$

And

$$\sum_{i=1}^{n}\left(w\left(\sum_{j=1}^{i} p(x_j)\right)-w\left(\sum_{j=1}^{i-1} p(x_j)\right)\right)=w\left(\sum_{j=1}^{n} p(x_j)\right)=w(1)=1,\triangleright$$

one can consider $q_i=\left(w\left(\sum_{j=1}^{i} p(x_j)\right)-w\left(\sum_{j=1}^{i-1} p(x_j)\right)\right)$ as a probability. However, q_i depends on ranking of outcomes x_i – for various prospects value of q_i will be various. For this reason, such representation is called rank-dependent expected utility (RDEU). For the case of many outcomes, PT is based on the use of RDEU model with a value function:

$$U(f)=\sum_{i=1}^{n} w\left(\sum_{j=1}^{i} p(x_j)\right)(v(x_i)-v(x_{i+1}))$$

The representation of RDEU is a special case of that of CEU. Considering $w\left(\sum_{j=1}^{i} p(x_j)\right)$ as a value of non-additive measure, one arrives at CEU representation [24]. CEU is more general than RDEU: RDEU requires that the probabilities are known, whereas CEU – not.

The second problem with PT is that it does not in general satisfy first-order stochastic dominance.

Kahneman and Tversky, the authors of PT, suggested CPT as a more advanced theory which is free of the PT's above mentioned drawbacks. CPT can be applied, in contrast to PT, both for decisions under risk and uncertainty. In CPT, gains and losses (measured by a value function) aggregated separately by Choquet integrals with different capacities and the results of aggregations are summed. The representation of CPT is:

$$U(f)=\int_S v(f^+(s))d\eta^+ + \int_S v(f^-(s))d\eta^-,$$

where $f^{+}(s)=\max(f(s),0)$ and $f^{-}(s)=\min(f(s),0)$, $\int_S$ is Choquet integral, v is a value function (the same as in PT), and η^{+},η^{-} are capacities. For the case with risky prospects, i.e. with known probabilities of states of nature it uses two weighting functions. Consider, without loss of generality, a risky prospect $f=(f(s_1),\ p(s_1);...;f(s_n),p(s_n))$ with $f(s_1)\geq,...,\geq f(s_k)\geq 0>>f(s_{k+1})\geq,...,\geq f(s_n)$. The CPT representation for f is the following:

$$U(f)=\sum_{i=1}^{k}\left(w\left(\sum_{j=1}^{i}p(s_j)\right)-w\left(\sum_{j=1}^{i-1}p(s_j)\right)\right)v(f(s_i))+$$

$$+\sum_{i=k+1}^{n}\left(w\left(\sum_{j=i}^{n}p(s_j)\right)-w\left(\sum_{j=i+1}^{n}p(s_j)\right)\right)v(f(s_i))$$

The use of different weighting functions w^{+} and w^{-} for probabilities of gains and losses comes from an experimental observation that people differently weight the same probability p depending on whether it is associated with gain or loss; however the same experimental evidence showed that both w^{+} and w^{-} are *S*-shaped [47]. w^{+} and w^{-} obtained by Kahneman and Tversky from the experimental data have the same form with different curvatures:

$$w^{+}(p)=\frac{p^{\gamma}}{(p^{\gamma}+(1-p)^{\gamma})^{1/\gamma}},\quad w^{-}(p)=\frac{p^{\delta}}{(p^{\delta}+(1-p)^{\delta})^{1/\delta}}$$

$\gamma=0.61$, $\delta=0.69$. As a result, w^{+} is more curved.

The dependence of beliefs on the sign of outcomes is referred to as sign-dependence. Sign-dependence for the case of decisions under uncertainty is modeled by using two different capacities η^{+},η^{-}. When $w^{+}(p)=1-w^{-}(1-p)$ or $\eta^{+}(A)=1-\eta^{-}(S\setminus A), A\subset S$, the sign-dependence disappears and CPT is reduced to RDEU or CEU respectively.

CPT is one of the most successful theories – it encompasses reference-dependence, rank-dependence and sign-dependence. It combines advantages of both PT and CEU: describing asymmetry of gains and losses and modeling non-additive beliefs under uncertainty. As opposed to PT, CPT satisfies first-order stochastic dominance.

However, CPT suffers from a series of notable disadvantages. One main disadvantage of CPT is that gains and losses are aggregated separately, and in this sense the model is additive. Drawbacks and violations of such additivity were empirically and experimentally shown in [5,6,36,53].

The other main disadvantage is that CPT, as it is shown in [39], may be good for laboratory works but not sufficiently suitable for real applications. The research in [39] shows that this is related to the fact that CPT is highly conditioned by the combination of parameters' values used in value and weighting functions. Roughly speaking, the same combination works well in one choice problem but badly in another. For example, such problems with parameterization is assumed as a possible reason of the experiments reported in [2], where it was illustrated that CPT fails to capture the choice between mixed gambles with moderate and equal probabilities.

One of the underlying motivations of the CEU is that a DM's beliefs may not be probabilistic in choice problems under uncertainty, i.e. they may be incompatible with a unique probability measure. In CEU, this evidence is accounted for by using a non-additive probability. Another way to describe non-probabilistic beliefs is modeling them not by one probability distribution but by a set of probability distributions (priors), which led to development of a large class of various utility models. This class is referred to as multiple priors models. The use of multiple priors allows describing the fact that in situations with insufficient information or vague knowledge, a DM can not have a precise (single) probabilistic belief on an event's occurrence, but have to allow for a range of values of probabilities, i.e. to have an imprecise belief. Indeed, various priors from a considered set will in general assign various probabilities to an event. From the other side, imprecise probabilistic beliefs imply existence of a set of priors.

The first two known examples showing incompatibility of a single probability measure with human choices were suggested by Daniel Ellsberg in 1961 [16]. These examples uncovered choices under uncertainty in which human intuition does not follow the sure thing principle. One example is called Ellsberg two-urn paradox. In this experiment they present a DM two urns each with 100 balls. A DM is allowed to see that urn I contains 50 white and 50 black balls, whereas no information is provided on a ratio of white and black balls in urn II. A DM is suggested choosing from bets on a color of a ball drawn at random: on a white color and on a black color. Each bet produces a prize of $100. After bet is chosen, a DM needs to choose the urn to play. Majority of people were indifferent whether to bet on white or on black. However, whatever bet was chosen, most of people strictly preferred to choose urn I to play – they prefer betting on an outcome with known probability 0.5 to betting on an outcome with probability that may take any value from [0,1]. This choice is inconsistent with any probabilistic belief on a color of a ball taken at random from urn II. Indeed, betting on white and then choosing urn I means that a DM believes on the number of white balls in urn II being smaller than in urn I, whereas betting on black and then choosing urn I means that a DM believes on the number of black balls in urn II being smaller than in urn I. No single probabilistic belief on colors for urn II may simultaneously explain these two choices – the probabilities of a white and a black balls drawn at random from urn II cannot be simultaneously smaller than 0.5.

The second example is called Ellsberg single-urn experiment. In this example a DM is offered to choose among bets on colors of balls in one urn. This urn contains 90 balls among which 30 are red and the other 60 are blue and yellow in an unknown proportion. The following bets on a color of a ball taken at random are suggested (Table 2.1):

Table 2.1 Ellsberg single-urn decision problem

	Red	Blue	Yellow
f	\$100	0	0
g	0	\$100	0
f'	\$100	0	\$100
g'	0	\$100	\$100

For example, f yields \$100 if a ball drawn is red and 0 otherwise whereas f' yields \$100 whether a ball drawn is red or yellow and 0 otherwise. Majority of subjects prefer f to g (i.e. they prefer an outcome with known probability 1/3 to an outcome with unknown probability being somewhere between 0 and 2/3). At the same time, majority prefer g' to f' (i.e. they prefer an outcome with known probability 2/3 to an outcome with unknown probability being somewhere between 1/3 and 1). These two choices cannot be explained by beliefs described by a single probability distribution. Indeed, if to suppose that the beliefs are probabilistic, then the first choice implies that subjects think that a red ball is more probable to be drawn than a blue ball: $p(red) > p(blue)$. The second choice implies subjects thinking that a blue or yellow ball drawn is more probable than a red or yellow ball drawn – $p(blue) + p(yellow) > p(red) + p(yellow)$ which means $p(blue) > p(red)$. This contradicts the beliefs underlying the first choice. Indeed, information on occurrence of complementary events *blue* and *yellow* represented only by probability of their union cannot be uniquely separated into probabilities of *blue* and *yellow*.

The observed choices contradict the sure thing principle, according to which $f \succ g$ should imply $f' \succ g'$. Indeed, f' and g' can be obtained from f and g respectively by changing from 0 to \$100 the equal outcomes of f and g produced for yellow ball. However, the evidence shows: $f \succ g$ and $g' \succ f'$. Violation of the sure thing principle also takes place in the Ellsberg two-urns experiment. The intuition behind the real choices in these experiments is that people tend to prefer probabilistic outcomes to uncertain ones. This phenomenon is referred to as uncertainty aversion, or ambiguity aversion. Term ambiguity was suggested by Daniel Ellsberg and defined as follows: "a quality depending on the amount, type, reliability and 'unanimity' of information, and giving rise to one's 'degree of confidence' in an estimate of relative likelihoods" [16]. Mainly, ambiguity is understood as uncertainty with respect to probabilities [7].

The principle of uncertainty aversion was formalized in form of an axiom by Gilboa and Schmeidler [26]. To understand formal description of uncertainty aversion, let us recall the Ellsberg two-urn example. Suppose an act f yielding \$100 for a white ball drawn from an unknown urn and an act g yielding \$100 for a black ball drawn from an unknown urn. It is reasonable to consider these acts equivalent: $f \sim g$. Now mix these acts as $\frac{1}{2}f+\frac{1}{2}g$, obtaining an act which yields a lottery of getting \$100 with probability ½ and 0 with probability ½ no matter what ball is drawn. That is, by mixing two uncertain bets we got a risky bet – the hedging effect takes place. The obtained act is equivalent to a bet on any color, say white, for a known urn that provides the same lottery. But as this bet is preferred to f and g, we may state that $\frac{1}{2}f+\frac{1}{2}g \succsim f$. The uncertainty aversion axiom is a generalization of this and it states that for equivalent acts f and g, their mix is weakly preferred to each of them: $\alpha f+(1-\alpha)g \succsim f$. This axiom is one of the axioms underlying a famous utility model called Maximin Expected Utility (MMEU) [26]. According to the axiomatic basis of this model, there exist a unique closed and convex set C of priors (probability measures) P over states of nature, such that

$$f \succsim g \Leftrightarrow \min_{P\in C}\int_S u(f(s))dP \geq \min_{P\in C}\int_S u(g(s))dP,$$

where u is unique up to a positive linear transformation. In simple words, an overall utility of an act is a minimum among all its expected utilities each obtained for prior one $P \in C$. The consideration of C as convex does not affect generality. Applying MMEU to both Ellsberg paradoxes, one can easily arrive at the observed preferences of most people.

Ghirardato, Maccheroni and Marinacci suggested a generalization of MMEU [20] consisting in using all its underlying axioms except uncertainty aversion axiom. The obtained model is referred to as $\alpha-$MMEU and states that $f \succsim g$ iff

$$\alpha \min_{P\in C}\int_S u(f(s))dP+(1-\alpha)\max_{P\in C}\int_S u(f(s))dP \geq$$
$$\geq \alpha \min_{P\in C}\int_S u(g(s))dP+(1-\alpha)\max_{P\in C}\int_S u(g(s))dP$$

$\alpha \in [0,1]$ is referred to as a degree of ambiguity aversion, or an ambiguity attitude. The higher α is the more ambiguity averse a DM is, and when $\alpha = 1$ we get MMEU. When $\alpha = 0$, the model describes *ambiguity seeking*, i.e. a DM relies on ambiguity because it includes the best possible realization of priors. The values $\alpha \in (0,1)$ describe balance between ambiguity aversion and ambiguity seeking to reflect the fact that a person may not have extreme attitudes to ambiguity.

The important issue is the relation between CEU and MMEU. One can easily find that the Ellsberg single-urn paradox may be explained by CEU under the capacity v satisfying the following:

$$v(\{s_r\}) = v(\{s_r, s_b\}) = v(\{s_r, s_y\}) = \frac{1}{3},$$

$$v(\{s_b\}) = v(\{s_y\}) = 0,$$

$$v(\{s_b\}) = v(\{s_y\}) = 0,$$

$$v(\{s_b, s_y\}) = \frac{2}{3},$$

where s_r, s_b, s_y denote states of nature being extractions of a red, a blue and a yellow balls respectively. Schmeidler showed that an ambiguity aversion is modeled by CEU if and only if a capacity satisfies $v(A \cup B) + v(A \cap B) \geq v(A) + v(B)$. Such v is called a convex capacity. Schmeidler proved that under assumption of ambiguity aversion, CEU is the special case of the MMEU:

$$(Ch)\int_S u(f(s))dv = \min_{P \in C} \int_S u(f(s))dP$$

where $(Ch)\int_S$ denotes Choquet integral, v is a convex capacity, and C is a set of probability measures defined as $C = \{P \mid P(A) \geq v(A), \forall A \subset S\}$. Such C is called a core of a convex capacity v. The capacity given above that expresses Ellsberg paradox, is a capacity whose value for each event is equal to a minimum among all the possible probabilities for this event, and therefore, satisfies $v(A) \leq P(A), \forall A \subset S$. Given a set of priors C, a convex capacity v satisfying $v(A) = \min_{P \in C} P(A)$ is called a lower envelope of C.

However, MMEU is not always a generalization of CEU. If v is not convex, there is no C for which CEU and MMEU coincide. Indeed, CEU does not presuppose an ambiguity aversion. For example, if you use concave v, CEU will model ambiguity seeking behavior [24]. Also, a capacity may not have a core.

If to compare CEU and MMEU, the latter has important advantages. Choquet integration and a non-additive measure are concepts that are not well-known and use of them requires specific mathematical knowledge. From the other side, there exist difficulties related to interpretation and construction of a capacity. Capacity, as a belief, may be too subjective. In contrast, a set of priors is straightforward and clear as a possible 'range' for unknown probability distribution. The idea to compute minimal EU is very intuitive and accepted easily. At the same time, it often requires to solve well-known optimization problems like those of linear programming. However, determination of a set of priors as the problem of strict constraining a range of possible probabilities is influenced by insufficient knowledge on reality. The use of probability itself may be questionable. In such cases, the use of a non-additive belief obtained from experience-based knowledge may be a good alternative.

The main disadvantages of the MMEU are that in real problems it is difficult to strictly constrain the set of priors and various priors should not be considered equally relevant to a problem at hand. From the other side, in MMEU each act is evaluated on the base of only one prior. In order to cope with these problems Klibanoff et al. suggested a smooth ambiguity model as a more general way to formalize decision making under ambiguity than MMEU [34]. In this model an assessment of a DM's subjective probabilistic beliefs to various probability distributions is used to represent important additional knowledge which differentiates priors in terms of their relevance to a decision situation considered. In this approach the authors use the following representation:

$$U(f)=\int_C \phi\left(\int_S u(f)dP\right)dv$$

Here $P\in C$ is a possible prior from a set of priors C, v is a probabilistic belief over C, ϕ is a nonlinear function reflecting extent of ambiguity aversion. ϕ rules out reduction of a second-order probability model to a first-order probability model. An introduction of a subjective second-order probability measure defined over multiple priors was also suggested by Chew et al.[11], Segal [45], Seo [46] and others.

In [8] they suggested a model in which an overall utility for an action is obtained as $U(f)=\min_{\rho\in L_\beta}\frac{1}{\varphi(\rho)}\int_S u(f)d\rho$. Here φ is a confidence function whose value $\varphi(\rho)\in[0,1]$ is the relevance of the distribution ρ to the decision problem and $L_\beta=\{\rho:\varphi(\rho)\geq\beta\},\beta\in(0,1]$ is a set of distributions considered by a DM.

A wide class of multiple priors models is referred to as variational preferences models suggested in [37], robust control idea-based model suggested in [29], ε-contamination model suggested in [14]. The generalized representation for these models is the following:

$$U(f)=\min_{\rho\in\Delta(S)}\left[\int_S u(f)d\rho+c(\rho)\right]$$

where c is a "cost function" whose value $c(\rho)$ is higher for less relevant distribution ρ and $\Delta(S)$ is the set of all the distributions over S.

2.2 Analysis of the Existing Theories of Decision

Analyzing the above mentioned decision theories on the base of preference frameworks and types of decision relevant information on alternatives, states of nature, probabilities and outcomes, we arrived at conclusion that these theories are developed for a well-described environment of thought experiments and laboratory examples. Such environment is defined mainly three simplified directions of decision making theory: decision making under uncertainty, decision making under risk and decision making under ambiguity. We will try to discuss main advantages and disadvantages of the existing theories which belong to these directions.

The decision making methods developed for situations of uncertainty includes Laplace insufficient reason criterion, Savage minimax regret criterion, Hurwitz criterion, Wald maximin solution rule etc. Maximin solution rule models extreme pessimism in decision making, whereas its generalization, Hurwitz criterion uses linear combination of pessimistic and optimistic solutions. The main shortcoming of these methods is that most of them are not developed to deal with any information on probabilities of states of nature, whereas in real-life decision making DM's almost always have some amount of such information.

The main methods of decision making under risk are von Neumann and Morgenstern EU, subjective expected utility of Savage, Hodges-Lehmann criterion [30], prospect theory (PT) and others. EU theories have strong scientific foundations together with simplicity of the utility models. But, unfortunately, they are based on idealized preference and information frameworks. A decision maker is considered fully rational and relying on coarsely computational, inanimate reasoning. Decision relevant information is based on assumption that future may be perfectly described by means of states of nature like extraction of a white or a black ball. Likelihood of states of nature is described by objective probabilities or subjective probabilistic beliefs. However, due to absence of sufficient information, the first ones are commonly unknown. On the other hand, even if objective probabilities are known, beliefs of a DM do not coincide with them due to psychological aspects. Subjective probabilities are slightly more realistic, but are incapable to describe choices under uncertainty even in thought experiments.

Reconsideration of preferences framework underlying EU resulted in development of various advanced preferences frameworks as generalizations of the former. These generalizations can be divided into two types: rank-dependence generalization and sign-dependence generalization [9]. Advanced preference frameworks include various reconsiderations and weakening of independence axiom [18,21,26,44], human attitudes to risk and uncertainty (rank-dependent generalization), gains and losses [32,47] etc.

PT is suggested for decision making under risk and models such important psychological aspects as asymmetry of gains and losses and distorted evaluation of probabilities. However, this theory requires probabilities and outcomes to be exactly known which sufficiently restricts its application in real-world problems. From the other side, the psychological biases are precisely described in this theory, whereas extents of psychological distortion of decision-relevant information (probabilities and outcomes) are rather imprecisely known.

A large number of studies is devoted to decision making under ambiguity [3,7,10,15,17-20,22-24,28,31,33]. Ambiguity is commonly referred to the cases when probabilities are not known but are supposed to vary within some ranges. The terms 'uncertainty' and 'ambiguity' are not always clearly distinguished and defined but, in general, are related to non-probabilistic uncertainty. In turn, decision making under uncertainty often is considered as an extreme non-probabilistic case – when no information on probabilities is available. From the other side, this case is also termed as decision making under complete ignorance. At the same time, sometimes, this is considered as ambiguity represented by simultaneous consideration of all the probability distributions. The studies on decision making under ambiguity are conducted in two directions – a development of models based on multiple probability distributions, called multiple priors models [25,26], and a formation of approaches based on non-additive measures such as CEU [50-52]. Mainly, these models consider so-called ambiguity aversion as a property of human behavior to generally prefer outcomes related to non-ambiguous events to those related to ambiguous ones.

CEU is a good model describing common inconsistency of the independence assumption with human decision reasoning and non-additivity of beliefs. Non-additivity is an actual property of beliefs conditioned by scarce relevant information on likelihood of events or by psychological distortion of probabilities. This property allows CEU to describe the famous Ellsberg and Allais paradoxes. CEU is a one of the most successful utility models, it is used as a criterion of decision making under ambiguity and decision making under risk.

However, non-additive measure used in CEU is precise (numerical) the use of which is questionable for real-world problems where probabilities are not as perfectly (precisely) constrained as in Ellsberg paradoxes. From the other side, CEU is developed for precise utility-based measuring of outcomes that also is not adequate when the latter are imprecisely known and uncertain being related to future.

CPT encompasses PT and CEU basics, and as a result, is both rank-dependent and sign-dependent generalization of EU. Nowadays, CPT is one of the most successful theories and allows taking into account both asymmetry of gains and losses and ambiguity attitudes. However, no interaction of these behavioral factors is considered. It is naive to assume that these factors don't exhibit some mix in their influence on choices, as a human being can hardly consider them independently due to psychological issues and restricted computational abilities. From the other side, CPT, as well as PT and CEU, is developed for perfectly described information and well-defined preferences of a DM (this also restricts its ability to adequately model human choices affected by ensemble of behavioral factors).

Multiple priors models were suggested for situations when objective probabilities are not known and a DM cannot assign precise probabilistic beliefs to events but has imprecise beliefs. Indeed, in real world, due to incompleteness of information or knowledge rather a range of probabilities but not a single probability is used describing likelihood of an event. This requires using a set of probability distributions instead of a unique prior. Within such a set all distributions are considered equally relevant. In general, it is much more adequate but still poor formulation of probability-relevant information available for a DM – in real-life problems a DM usually has some information that allows determining which priors are more and which are less relevant. Without doubt, a DM has different degrees of belief to different relevant probability distributions. In real-world problems, the values of probabilities are imprecise but cannot be sharply constrained as in Ellsberg paradoxes, which are specially designed problems. For addressing this issue, models with second-order probabilities were suggested [3,11,34,45]. For example, in the smooth ambiguity model [34] Klibanoff et al. suggested using a subjective probability measure to reflect a DM's belief on whether a considered subset of multiple priors contains a 'true' prior. The use of this second-order probability allows to depart from extreme evaluation of acts by their minimal or/and maximal expected utilities, and to take into account influence of each relevant prior to acts' overall utilities. However, for a human being an assessment of a precise subjective probability to each prior becomes almost impossible when the number of priors is large. Such a hard procedure does not correspond to extremely limited computational capability of humans. If probabilistic beliefs over states of nature often fail, why they should often work over a more complicated structure – a set of priors?

Second-order precise probability model is a non-realistic description of human beliefs characterized by imprecision and associated with some psychological aspects that need to be considered as well. The other disadvantage of the belief representation suggested in [34] is that the problem of investigation of consistency of subjective probability-relevant information is not discussed – consistent multiple priors are supposed to be given in advance. However, a verification of consistency of beliefs becomes a very important problem. An extensive investigation of this issue is covered in [50].

The other existing multiple priors models like the model based on confidence function and the variational preferences models considered in [37] also use complicated techniques to account for a relevance of a prior to a considered problem. From the other side, in these models each decision is evaluated only on one prior and the pessimistic evaluation (min operator) is used.

We mentioned only main existing models for decision making under ambiguity in this chapter. Let us note that the mentioned and other existing utility models for decision under ambiguity are based on rather complicated techniques and may be too subtle to be applied under vagueness of real-life decision-relevant information. In real life, amount, type, reliability and 'unanimity' of information, considered in Ellsberg's formulation of ambiguity, are not perfectly known that presents some difficulties in application of the existing models and decreases trust to the obtained results.

Generalizing the above mentioned drawbacks of the existing theories, we may conclude that the existing decision models yielded good results, but nowadays there is a need in generation of more realistic decision models. The main reason for this is that the existing decision theories are in general developed for thought experiments characterized by precise, perfect decision relevant information. The paradigm of construction of 'elegant' models does not match imperfect nature of decision making and relevant information supported by perceptions. Indeed, even the most advanced utility models are motivated by behavioral phenomena that were observed in thought experiment with simplified conditions. However, as David Schmeidler states, "*Real life is not about balls and urns*". The existing decision theories cannot be sufficiently accurate because in real-world decision situations human preferences are vague and decision-relevant information on environment and a DM's behavior is imperfect as perception-based. In contrast, humans are able to make proper decisions in such imperfect conditions. Modeling of this outstanding capability of humans, even to some extent, is a difficult but a promising study which stands for a motivation of the research suggested in this book. Economy is a human-centric system and this means that the languages of new decision models should be not languages based on binary logic and probability theory, but human-centric computational schemes for dealing with perception-based information. In our opinion, such languages are natural language (NL), in particular, precisiated NL (PNL) [56], and a geometric visual language (GVL) [27] or a geometric description language (GDL) [27]. New theories should be based on a more general and adequate view on imperfect perception-based information about environment and a DM's behavior. The main purpose of such a generalization is to construct models that are sufficiently flexible to deal with imperfect nature of decision-relevant information. Such flexibility would allow taking into account more relevant information and could yield more realistic (not more precise!) results and conclusions.

The other main disadvantage of the existing decision theories relates to modeling of behavioral factors underlying decision making like risk attitude, ambiguity attitude, altruism etc. The most of the existing theories are based on precise parametrical modeling which is too coarse and "inanimate" approach to model human activity conditioned by emotions, perceptions, mental factors etc. The existing non-parametric approaches are more adequate, but, they also are based on perfect and precise description of human decision activity. There is a need for a fundamental approach to modeling behavioral decision making. In this book we suggest a theory of behavioral decision making in which a DM's behavior, i.e. subjective conditions, is taken into a consideration at the same level of abstraction as objective conditions. We model a DM's behavior by a set of his/her states each representing one principal behavior or, in other words, one principal subjective behavioral condition a DM may exhibit. In line with states of nature, states of a DM constitute in the suggested theory a common basis for decision analysis. Such framework is more general than behavioral basics of the existing theories like CPT and allows for a transparent analysis.

In real-world decision making imperfect information is supported by perceptions and is expressed in NL or in framework of visual images. In this book we suggest new approaches to decision making under imperfect information on decision environment and a DM's behavior. The suggested approaches utilize synthesis of the fuzzy sets theory[55,56] as a mathematical tool for description and reasoning with perception-based information and probability theory.

In the subsequent chapters we suggest new approaches to decision making under imperfect information on decision environment and a DM's behavior.

References

1. Akerlof, G.A., Shiller, R.J.: Animal Spirits. How Human Psychology Drives the Economy, and Why it Matters for Global Capitalism. Princeton University Press (2009)
2. Baltussen, G., Thierry, P., van Vliet, P.: Violations of Cumulative Prospect theory in Mixed Gambles with Moderate Probabilities. Management Science 52(8), 1288–1290 (2006)
3. Becker, J., Sarin, R.: Economics of Ambiguity in probability. Working paper, UCLA Graduate School of Management (1990)
4. Billot, A.: An existence theorem for fuzzy utility functions: a new elementary proof. Fuzzy Sets and Systems 74, 271–276 (1995)
5. Birnbaum, M.H.: New tests of cumulative prospect theory and the priority heuristic: Probability-outcome tradeoff with branch splitting. Judgment and Decision Making 3(4), 304–316 (2008)
6. Birnbaum, M.H., Johnson, K., Longbottom Lee, J.: Tests of Cumulative Prospect Theory with graphical displays of probability. Judgment and Decision Making 3(7), 528–546 (2008)
7. Camerer, C., Weber, M.: Recent Developments in Modeling Preferences. Journal of Risk and Uncertainty 5, 325–370 (1992)
8. Chateauneuf, A., Faro, J.: Ambiguity through confidence functions. Journal of Mathematical Economics 45(9-10), 535–558 (2009)
9. Chateauneuf, A., Wakker, P.: An Axiomatization of Cumulative Prospect Theory for Decision Under Risk. Journal of Risk and Uncertainty 18(2), 137–145 (1999)
10. Chen, Z., Epstein, L.G.: Ambiguity, Risk, and Asset Returns in Continuous Time. Econometrica 70, 1403–1443 (2002)
11. Chew, S.H., Karni, E., Safra, Z.: Risk aversion in the theory of expected utility with rank-dependent probabilities. Journal of Economic Theory 42, 370–381 (1987)
12. Choquet, G.: Theory of capacities. Annales de l'Institut Fourier 5, 131–295 (1953)
13. de Finetti, B.: Theory of Probability, vol. 1. John Wiley and Sons, New York (1974)
14. Eichberger, J., Kelsey, D.: E-Capacities and the Ellsberg Paradox. Theory and Decision 46, 107–138 (1999)
15. Einhorn, H., Hogarth, R.: Ambiquity and Uncertainty in Probabilistic inference. Psychology Review 92, 433–461 (1985)
16. Ellsberg, D.: Risk, Ambiquity and the Savage Axioms. Quarterly Journal of Economics 75, 643–669 (1961)
17. Epstein, L.G.: A Definition of Uncertainty Aversion. Review of Economic Studies 66, 579–608 (1999)

18. Epstein, L.G., Schneider, M.: Ambiguity, information quality and asset pricing. Journal of Finance 63(1), 197–228 (2008)
19. Franke, G.: Expected utility with ambiguous probabilities and "Irrational Parameters". Theory and Decision 9, 267–283 (1978)
20. Ghirardato, P., Maccheroni, F., Marinacci, M.: Differentiating Ambiguity and Ambiguity Attitude. Journal of Economic Theory 118, 133–173 (2004)
21. Ghirardato, P., Klibanoff, P., Marinacci, M.: Additivity with multiple priors. Journal of Mathematical Economics 30, 405–420 (1998)
22. Ghirardato, P., Marinacci, M.: Range Convexity and Ambiguity Averse Preferences. Economic Theory 17, 599–617 (2001)
23. Ghirardato, P., Marinacci, M.: Ambiguity Made Precise: A Comparative Foundation. Journal of Economic Theory 102, 251–289 (2002)
24. Gilboa, I.: Theory of Decision under Uncertainty. Cambridge University Press, Cambridge (2009)
25. Gilboa, I., Maccheroni, F., Marinacci, M., Schmeidler, D.: Objective and subjective rationality in a multiple prior model. Econometrica 78(2), 755–770 (2010)
26. Gilboa, I., Schmeidler, D.: Maximin Expected utility with a non-unique prior. Journal of Mathematical Economics 18, 141–153 (1989)
27. Guirimov, B.G., Gurbanov Ramiz, S., Aliev Rafik, A.: Application of fuzzy geometry in decision making. In: Proc. of the Sixth International Conference on Soft Computing and Computing with Words in System Analysis, Decision and Control (ICSCCW 2011), Antalya, pp. 308–316 (2011)
28. Guo, P.: One-Shot Decision Theory. IEEE Transactions on Systems, Man and Cybernetics – Part A: Systems and Humans 41(5), 917–926 (2011)
29. Hansen, L., Sargent, T.: Robust Control and Model Uncertainty. American Economic Review 91, 60–66 (2001)
30. Hodges, J.L., Lehmann, E.: The use of previous experience in reaching statistical decisions. The Annals of Mathematical Statistics 23, 396–407 (1952)
31. Huettel, S.A., Stowe, C.J., Gordon, E.M., Warner, B.T., Platt, M.L.: Neural signatures of economic preferences for risk and ambiguity. Neuron 49, 765–775 (2006)
32. Kahneman, D., Tversky, A.: Prospect theory: an analysis of decision under uncertainty. Econometrica 47, 263–291 (1979)
33. Karni, E.: Decision Making under Uncertainty: The case of state dependent preferences. Harvard University Press, Cambridge (1985)
34. Klibanoff, P., Marinacci, M., Mukerji, S.: A smooth model of decision making under ambiguity. Econometrica 73(6), 1849–1892 (2005)
35. Kojadinovic, I.: Multi-attribute utility theory based on the Choquet integral: A theoretical and practical overview. In: 7th Int. Conf. on Multi-Objective Programming and Goal Programming, Loire Valley, City of Tours, France (2006)
36. Labreuche, C., Grabisch, M.: Generalized Choquet-like aggregation functions for handling bipolar scales. European Journal of Operational Research 172, 931–955 (2006)
37. Maccheroni, F., Marinacci, M., Rustichini, A.: Ambiguity aversion, Robustness, and the Variational Representation of Preferences. Econometrica 74, 1447–1498 (2005)
38. Miyamoto, J.M., Wakker, P.: Multiattribute Utility Theory Without Expected Utility Foundations. Operations Research 44(2), 313–326 (1996)
39. Neilson, W., Stowe, J.: A Further Examination of Cumulative Prospect Theory Parameterizations. The Journal of Risk and Uncertainty 24(1), 31–46 (2002)

40. Paolo, P.: Cumulative prospect theory and second order stochastic dominance criteria: an application to mutual funds performance
41. Quiggin, J.: A theory of anticipated utility. Journal of Economic Behavior and Organization 3, 323–343 (1982)
42. Ramsey, F.P.: Truth and Probability. In: Braithwaite, R.B., Plumpton, F. (eds.) The Foundations of Mathematics and Other Logical Essays. K Paul, Trench, Truber and Co., London (1931)
43. Savage, L.J.: The Foundations of Statistics. Wiley, New York (1954)
44. Schmeidler, D.: Subjective probability and expected utility without additivity. Econometrita 57(3), 571–587 (1989)
45. Segal, U.: The Ellsberg paradox and risk aversion: An anticipated utility approach. International Economic Review 28, 175–202 (1987)
46. Seo, K.: Ambiguity and Second-Order Belief. Econometrica 77(5), 1575–1605 (2009)
47. Tversky, A., Kahneman, D.: Advances in Prospect theory: Cumulative Representation of Uncertainty. Journal of Risk and Uncertainty 5(4), 297–323 (1992)
48. von Neumann, J., Morgenstern, O.: Theory of games and economic behaviour. Princeton University Press (1944)
49. Wakker, P.P., Zank, H.: State Dependent Expected Utility for Savage's State Space. Mathematics of Operations Reseach 24(1), 8–34 (1999)
50. Walley, P.: Statistical Reasoning with Imprecise Probabilities. Chapman and Hall, London (1991)
51. Walley, P.: Measures of uncertainty in expert systems. Artificial Intelligence 83(1), 1–58 (1996)
52. Walley, P., de Cooman, G.: A behavioral model for linguistic uncertainty. Information Sciences 134(1-4), 1–37 (2001)
53. Wu, G., Markle, A.B.: An Empirical Test of Gain-Loss Separability in Prospect Theory. Management Science 54(7), 1322–1335 (2008)
54. Yaari, M.E.: The Dual Theory of Choice under Risk. Econometrica 55(1), 95–115 (1987)
55. Yager, R.R.: On global requirements for implication operators in fuzzy modus ponens. Fuzzy Sets and Systems 106, 3–10 (1999)
56. Zadeh, L.A.: Generalized theory of uncertainty (GTU) – principal concepts and ideas. Computational Statistics & Data Analysis 51, 15–46 (2006)

Chapter 3
Uncertain Preferences and Imperfect Information in Decision Making

3.1 Vague Preferences

One of the main aspects defining solution of a decision problem is a preferences framework. In its turn one of the approaches to formally describe preferences is the use of utility function. Utility function is a quantitative representation of a DM's preferences and any scientifically ground utility model has its underlying preference assumptions.

The first approach to modeling human preferences was suggested by von Neumann and Morgenstern [70] in their expected utility (EU) model. This approach is based on axioms of weak order, independence and continuity of human preferences over actions set A. As it was shown by many experiments and discussions conducted by economists and psychologists, the assumption of independence appeared non-realistic [4,9,20]. There were suggested a lot of preferences frameworks which departs from that of EU by modeling a series of key aspects of human behavior.

Reconsideration of preferences framework underlying EU resulted in development of various advanced preferences frameworks as generalizations of the former. These generalizations can be divided into two types: rank-dependence generalization and sign-dependence generalization [11]. Advanced preference frameworks include various reconsiderations and weakening of the independence axiom [24,27,31,63] , human attitudes to risk and uncertainty (rank-dependent generalization), gains and losses [42,68] (sign-dependent generalization) etc.

Various notions of uncertainty aversion (ambiguity aversion) in various formulations were included into many preference models starting from Schmeidler's Choquet expected utility (CEU) [63] preferences framework and Gilboa and Schmeidler's maximin expected utility preferences framework [31], to more advanced uncertainty aversion formulations of Ghirardato, Maccheroni, Marinacci, Epstein, Klibanoff [10,12,21,23,25,27,28,44,45].

In [25] they provide axiomatization for $\alpha-$MMEU – a convex combination of minimal and maximal expected utilities, where minimal expected utility is multiplied by a degree of ambiguity aversion (ambiguity attitude). This approach allows differentiating ambiguity and ambiguity attitude. However, the approach in

R.A. Aliev: *Fuzzy Logic-Based Generalized Theory of Decisions*, STUDFUZZ 293, pp. 89–125.
DOI: 10.1007/ 978-3-642-34895-2_3

[25] allows to evaluate overall utility only comparing ambiguity attitudes of DMs with the same risk attitudes. The analogous features of comparative ambiguity aversion are also presented in [28]. Ambiguity aversion as an extra risk aversion is considered in [12,22].

In [43] they suggested a smooth ambiguity model as a more general way to formalize decision making under ambiguity than MMEU. In this model probability-relevant information is described by assessment of DM's subjective probabilistic beliefs to various relevant probability distributions. In contrast to other approaches to decision making under ambiguity, the model provides a strong separation between ambiguity and ambiguity attitude. To describe whether a DM is an ambiguity averse, loving or neutral it is suggested to use well known technique of modeling risk attitudes. More concretely, to reflect a considered DM's reaction to ambiguity it is suggested to use a concave nonlinear function with a special parameter α as the degree of ambiguity aversion – the larger α correspond to a more ambiguity averse DM. In its turn ambiguity loving is modeled by a convex nonlinearity. As opposed to the models in [21,25,28], the model in [43] allows for comparison of ambiguity attitudes of DMs whose risk attitudes are different.

The other important property included into some modern preference frameworks is a tradeoff-consistency which reflects strength of preferences with respect to coordinates of probabilistic outcomes.

The preference framework of the Cumulative Prospect Theory (CPT), suggested by Kahneman and Tversky [68], as opposed to the other existing frameworks includes both rank-dependence and sign-dependence features [11].

But are the modern preferences frameworks sufficiently adequate to model human attitudes to alternatives? Unfortunately, the modern preferences frameworks miss very important feature of human preferences: human preferences are vague [58]. Humans compare even simple alternatives linguistically using certain evaluation techniques such as "much better", "much worse", "a little better", "almost equivalent" [81] etc. So, a preference is a matter of an imprecise degree and this issue should be taken into account in formulation of preferences framework. Let us consider an example.

Suppose that Robert wants to decide among two possible jobs a_1, a_2 based on the following criteria: salary, excitement and travel time. The information Robert has is that the job a_1 offers notably higher salary, slightly less travel time and is significantly less interesting as compared to the job a_2. What job to choose?

Without doubt, evaluations like these are subjective and context-dependent but are often faced. If to suppose that for Robert salary is "notably" more important than the time issues and "slightly" more important than excitement then it may be difficult to him to compare these alternatives. The relevant information is too vague for Robert to clearly give preference to any of the alternatives. Robert may feel that superiority of the a_1 on the first criterion is approximately compensated by the superiority of the a_2 in whole on the second and the third criteria. But, at the same time Robert may not consider these jobs equally good. As a result, in contrast to have unambiguous preferences, Robert has some "distribution" of his preferences among alternatives. In other words, he may think that to some degree the job a_1 is as good as the job a_2 and, at the same time, that to some degree job a_2 is as good as the job a_1.

In this example we see that vagueness of subjective comparison of alternatives on the base of some criteria naturally passes to the preferences among alternatives. For example, the term "notably higher" is not sharply defined but some vague term because various point estimates to various extents correspond to this term – for a given point estimate its correspondence to a "notably higher" term may not be true or false but partially true. This makes use of interval description of such estimates inadequate as no point may partially belong to an interval – it belongs or not. It is impossible to sharply differentiate "notably higher" and not "notably higher" points. As a result, vague estimates (in our case vague preferences) cannot be handled and described by classical logic and precise techniques. Fuzzy logic [1,81] is namely the tool to handle vague estimates and there is a solid number of works devoted to fuzzy and linguistic preference relations [57,74]. This is due to the fact that vagueness is more adequately measured by fuzziness. As a result, fuzzy degree-based preference axiomatization is more adequate representation from behavioral aspects point of view as it is closer to human thinking. In view of this, linguistic preference relations as a natural generalization of classical preference relations are an appropriate framework to underlie human-like utility model.

Fuzzy preferences or fuzzy preference relations (FPRs) are used to reflect the fact that in real-world problems, due to complexity of alternatives, lack of knowledge and information and some other reasons, a DM can not give a full preference to one alternative from a pair. Preferences remain "distributed" reflecting that one alternative is to some extent better than another. In contrast to classical preference relations, FPR shows whether an alternative a is more preferred to b than alternative c is preferred to d.

Given a set of alternatives A, any fuzzy preference relation on A is a mapping $R: A \times A \to T$ where T is a totally ordered set. Very often fuzzy preference relation is considered as $R: A \times A \to [0,1]$ which assigns to any pair of alternatives $a,b \in A$ a degree of preference $R(a,b) \in [0,1]$ to which a is preferred to b. The higher $R(a,b)$ is, the more a is preferred to b. In other words, FPR is characterized by membership function $\mu_R(a,b) = R(a,b)$ which returns a degree of membership of a pair (a,b) to R. FPR is a valued extension of classical preference relations. For example, a weak order is a special case of FPR when $R: A \times A \to \{0,1\}$ with $R(a,b) = 1$ if and only if $a \succeq b$ and $R(a,b) = 0$ otherwise.

Consider a general case of a classical preference relation (CPR) implying that a is either strictly preferred, or equivalent or incomparable to b. This means that CPR is decomposed into a strict preference relation P, indifference preference relation I and incomparability preference relation J, that is, $\forall a,b \in A$ either $(a,b) \in P$, or $(a,b) \in I$ or $(a,b) \in J$. An important extension of this case to FPR can be defined as follows:

$$P(a,b)+P(b,a)+I(a,b)+J(a,b)=1$$

where $P,I,J: A \times A \rightarrow [0,1]$ are fuzzy strict, fuzzy indifference and fuzzy incomparability preference relations respectively.

Another important type of FPR is described by a function $R: A \times A \rightarrow [0,1]$ where $R(a,b)=1$ means full strict preference of a over b, which is the same as $R(b,a)=0$ (full negative preference) and indifference between a over b is modeled as $R(a,b)=R(b,a)=1/2$. In general, R is an additive reciprocal, i.e. $R(a,b)+R(b,a)=1$. This is a degree-valued generalization of completeness property of classical relation, and $R(a,b)>1/2$ is a degree of a strict preference. However, such an R excludes incomparability.

Consider yet another important type of FPR within which indifference is modeled by $R(a,b)=R(b,a)=1$, incomparability – by $R(a,b)=R(b,a)=0$ and completeness – by $\max(R(a,b),R(b,a))=1$.

FPR are a useful tool to handle vague preferences. Linguistic preferences, or linguistic preference relations (LPRs), sometimes called fuzzy linguistic preferences, are generalization of FPR used to account for a situations when a DM or an expert cannot assign precise degree of preference of one alternative to another, but express this degree in a form of linguistic terms like "much better", "a little worse" etc. Indeed, under imperfect environment where relevant information is NL-based, there is no sufficient information to submit exact degrees, but is natural to express degrees in NL also corresponding to the kind of initial information.

To formalize LPR it is first necessary to define a set of linguistic terms as a set of verbal expressions of preference degrees which would be appropriate for a considered problem. As a rule, they consider a finite and totally ordered linguistic term set $T=\{t_i\}, i \in \{0,...,m\}$ with an odd cardinal ranging between 5 and 13. Each term is semantically represented by a fuzzy number, typically triangular or trapezoidal, placed over some predefined scale, e.g. [0,1]. For example: "no preference" – (0,0,0), "slightly better" – (0,0.3,0.5), "more or less better" – (0.3,0.5,0.7), "sufficiently better" – (0.5,0.7,1), "full preference" – (0.7,1,1). The cardinality of the term set is usually an odd.

Consider a finite set of alternatives $A=\{f_i, i=1,2,...,n\ (n \geq 2)\}$. Then an LPR is formally defined as follows:

Definition 3.1 [36]. Let $A=\{f_i, i=1,2,...,n\ (n\geq 2)\}$ be a finite set of alternatives, then a linguistic preference relation $\tilde{R}$ is a fuzzy set in A^2 characterized by a membership function

$$\mu_{\tilde{R}}: A^2 \to T$$

$$\mu(f_i, f_j) = \tilde{r}_{ij},\ \forall f_i, f_j \in A$$

indicating the linguistic preference degree of alternative f_i over f_j, , i.e. $\tilde{r}_{ij} \in T$.

So, LPR is represented by a membership function whose values are not precise degrees in [0,1] but fuzzy numbers in [0,1]. This means that LPR is a kind of FPR if to recall that the latter is in general defined by MF whose range is an ordered structure.

Traditional Fuzzy Linguistic Approach (TFLA). TFLA preserves fuzzy information about degrees of preference by direct computations over fuzzy numbers and, as a result, is of a high computational complexity. There are various other approaches to modeling LPR, some of which allow reducing computational complexity of the TFLA or suggesting some reasonable trade-off between preserving information and computational complexity. One of them is referred to as ordinal fuzzy linguistic modeling (OFLM). This approach is based on an idea of the adopting symbolic computations [35] over indices of terms in a term set instead of direct computations over the terms themselves as fuzzy numbers. This makes the approach sufficiently simpler in terms of computational complexity than TFLA. In OFLM, they consider a finite linguistic term set with an odd cardinality and the terms described by fuzzy numbers over the unit interval [0,1]. Also, a mid term is used to express approximate equivalence of alternatives by a fuzzy number with a mode equal to 0.5 and labeled like "almost equivalent". The other terms are distributed around the mid term expressing successively increasing preference degrees to the right and their symmetrical counterparts to the left. For example: "sufficiently worse", "more or less worse", "slightly worse", "almost equivalent", "slightly better", "more or less better", "sufficiently better".

There exist also approaches to model uncertainty of preferences, other than FPR. These approaches accounts for comparison of ill-known alternatives under crisp (non-fuzzy) preference basis. In one of these approaches, which is used for modeling valued tournament relations, $R(x,y)$ measures the *likelihood of a crisp weak preference* $x \succeq y$ [13,17]. Formally, $R(x,y)$ is defined as follows:

$$R(x,y) = P(x \succ y) + \frac{1}{2} P(x \sim y),$$

where $x \sim y \Leftrightarrow x \succeq y$ and $y \succeq x$, which implies $R(x,y)+R(y,x)=1$. Thus, uncertainty of preference is described by a probability distribution P over possible conventional preference relations, i.e., $T_i \subset A \times A$ with $x \succeq_i y \Leftrightarrow (x, y) \in T_i$ and $P(T_i) = p_i,\ i = 1,\ldots,N$. Then

$$R(x, y) = \sum_{i:x \succ_i y} p_i + \sum_{i:x \sim_i y} \frac{1}{2} p_i$$

For more details one can refer to [14]. This approach to modeling uncertain preferences was the first interpretation of fuzzy preference relations in the existing literature and was considered in the framework of the voting theory. It is needed to mention that such relations may be considered fuzzy because a degree of preference is used whereas the approach itself is probabilistic. However, this degree is a measured uncertainty about preferences which are themselves crisp but are not known with certainty. So, in its kernel, this approach does not support an idea underlying FPR – preference itself is a matter of a degree.

Another application of this approach may be implemented when there exists a utility function which quantifies preferences between alternatives a,b,c on some numerical scale and the latter is supported by additional information in form of a probability distribution. Then, $R(x, y) = P(u(x) > u(y))$, where $u : S \to R$ is a utility function.

Other measures of uncertainty can also be used to describe uncertainty of preference. One of them is the possibility measure, by using thereof the uncertain preference is defined as

$$R(x, y) = P(x \succeq y),$$

where $P(x \succeq y)$ is the degree of possibility of preference. The use of the possibility theory defines $\max(R(x, y), R(y, x)) = 1$. At the same time, in terms of the possibility theory, $1 - R(x, y) = N(x \succeq y)$ is the degree of certainty of a strict preference.

A large direction in the realm of modeling uncertain preferences is devoted to modeling *incomplete preferences*. In line with transitivity, completeness of preferences is often considered as a reasonable assumption. However, transitivity is used as a consistency requirement whereas completeness is used as a requirement which exclude indecisiveness. The reasonability and intuitiveness of these basics are not the same: for completeness they may loss their strength as compared to transitivity because in real choice problems lack of information, complexity of alternatives, psychological biases etc may hamper someone's choice up to indecisiveness. From the other side, indecisiveness may take place in group decision making when members' preferences disagree. The issue that completeness may be questionable was first addressed by Aumann [6]:

"Of all the axioms of utility theory, the completeness axiom is perhaps the most questionable. Like others of the axioms, it is inaccurate as a description of real life; but unlike them, we find it hard to accept even from the normative viewpoint.

For example, certain decisions that [an] individual is asked to make might involve highly hypothetical situations, which he will never face in real life; he might feel that he cannot reach an "honest" decision in such cases. Other decision problems might be extremely complex, too complex for intuitive "insight," and our individual might prefer to make no decision at all in these problems. Is it "rational" to force decisions in such cases?"

If to assume that preferences are not complete, one has to reject the use numerical utility functions and has to deal with more complex representations. As it is argued in [56], the use of a numerical utility is naturally leads to loss of information and then should not be dogmatic if one intends to model bounded rationality and imperfect nature of choice. In [56] they suggest to handle incomplete preferences by means of a vector-valued utility function as its range is naturally incompletely ordered. The other main argument for such an approach is that the use of a vector-valued utility is simpler than dealing with preferences themselves and in this case well-developed multi-objective optimization techniques may be applied. The idea of incomplete preferences underlying the approach in [56] is realized by the following assumption: given the set of alternatives A there exist at least one pair of alternatives $x, y \in A$ for which neither $x \succeq y$ nor $y \succeq x$ is assumed.

There exist also other approaches dealing with incomplete preferences by means of imprecise beliefs and/or imprecise utilities. The following classification of these approaches is given in [54]:

1) Probabilities alone are considered imprecise. For this setting preferences are represented by a convex set of probability distributions and a unique, utility function u(). Such models are widely used in robust Bayesian statistics [41, 61,73];
2) Utilities alone are considered imprecise. In this setting preferences are represented by a set of utility functions {u(c)} and a unique probability distribution p(s). Such representations were axiomatized and applied to economic models by Aumann [6] and Dubra, Maccheroni and Ok [19];
3) Both probabilities and utilities are considered imprecise. This is represented by sets of probability distributions {p(s)} and utility functions {u()}. These sets are considered separately from each other allowing for all arbitrary combinations of their elements. This is the traditional separation of imprecise information about beliefs and outcomes. Independence of two sets is practically justified and simplifies the decision analysis. However, this approach does not have axiomatic foundations. From the other side, the set of pairs may be non-convex and unconnected [40,41].

In order to compare adequacy of FPR and incomplete preferences models, we can emphasize the following classification of preference frameworks in terms of increasing uncertainty: complete orders, FPR, incomplete preferences. The first and the third one are idealized frameworks: the first implies that preference is absolutely clear, the third deals with the case when some alternatives are absolutely not comparable. Incomplete preference deals with lack of any information which can elucidate preferences. This is a very rare case in the sense that in most of real-world situations some such information does exist, though it requires to be obtained. In its turn FPR implies that preference itself is not "single-valued" and should reflect competition of alternatives even if the related information is precise.

3.2 Imperfect Information

In real-life decision making problems DM is almost never provided with perfect, that is, ideal decision-relevant information to determine states of nature, outcomes, probabilities, utilities etc and has to construct decision background structure based on his/her perception and envision. In contrast, relevant information almost always comes imperfect. Imperfect information is information which in one or more respects is imprecise, uncertain, incomplete, unreliable, vague or partially true [79]. We will discuss these properties of imperfect information and relations among them.

Two main concepts of imperfect information are imprecision and uncertainty. Imprecision is one of the widest concepts including variety of cases. For purposes of differentiation between imprecision and uncertainty, Prof. L.A. Zadeh suggested the following example: "*For purposes of differentiation it is convenient to use an example which involves ethnicity. Assume that Robert's father is German and his mother's parents were German and French. Thus, Robert is 3/4 German and 1/4 French. Suppose that someone asks me: What is Robert's ethnicity. If my answer is: Robert is German, my answer is imprecise or, equivalently, partially true. More specifically, the truth value of my answer is 3/4. No uncertainty is involved. Next, assume that Robert is either German or French, and that I am uncertain about his ethnicity. Based on whatever information I have, my perception of the likelihood that Robert is German is 3/4. In this case, 3/4 is my subjective probability that Robert is German. No partiality of truth is involved.*" In the first case imprecision is only represented by partial truth and no uncertainty is involved. As Prof. L.A. Zadeh defines, such imprecision is referred to as strict imprecision or s-imprecision for short. In the second case, imprecision is only represented by uncertainty and no partial truth is involved.

Information is partially true if it is neither absolutely true nor absolutely false but in an intermediate closeness to reality. For example, suppose you needed to write down ten pages of a text and have already written 8 pages. Certainly 'the work is done' is not absolutely true and is not absolutely false, and, if to assume that all pages are written equivalently difficult, 'the work is done' is true with degree 0.8. Form the other side, 'the work is not done' is not true and is not false from viewpoint of intuition because it is not informative and requires to be substituted by a more concrete evaluation.

Another example on imprecision and uncertainty is provided by P. Smets:

"To illustrate the difference between imprecision and uncertainty, consider the following two situations:

1. John has at least two children and I am sure about it.
2. John has three children but I am not sure about it.

In case 1, the number of children is imprecise but certain. In case 2, the number of children is precise but uncertain. Both aspects can coexist but are distinct. Often the more imprecise you are, the most certain you are, and the more precise, the less certain. There seems to be some Information Maximality Principle that requires that the 'product' of precision and certainty cannot be beyond a certain critical level. Any increase in one is balanced by a decrease in the other."

Imprecision is a property of the content under consideration: either more than one or no realization is compatible with the available information [65].

One realization of imprecise information is ambiguous information. Ambiguous information is information which may have at least to different meanings. For example, a statement 'you are aggressive' is ambiguous because aggressive may mean 'belligerent' or 'energetic'. For example, homonyms are typical carriers of ambiguity.

Ambiguous information may be approximate, e.g. 'the temperature of water in the glass is between 40 and 50°C is approximate if the temperature is 47°C. Ambiguous information like 'the temperature is close to 100C' is vague. Such vague information is fuzzy, because in this case the temperature is not sharply bounded. Both 99C and 103 corresponds to this, but the first corresponds stronger. Correspondence of a temperature value to 'the 'temperature is close to 100C' smoothly decreases as this value moves away from 100C. In general, vague information is information which is not well-defined; it is carried by a 'loose concept'. The worst case of vague information is unclear information. Ambiguous information may also be incomplete: "the vacation will be in a summer month" because a summer month may be either June, July or August.

Uncertain information is commonly defined as information which is not certain. P. Smets defines uncertainty as a property that results from a lack of information about the world for deciding if the statement is true or false. The question on whether uncertainty is objective or subjective property is still rhetoric.

Objective uncertainty may be probabilistic or non-probabilistic. Probabilistic uncertainty is uncertainty related to randomness – probability of an event is related to its tendency to occur. Main kinds of non-probabilistic uncertainty are possibilistic uncertainty and complete ignorance. Possibilistic uncertainty reflects an event's 'ability' to occur. To be probable, an event has to be possible. At the same time, very possible events may be a little probable. The dual concept of possibility is necessity. Necessity of an event is impossibility of the contrary event to occur. Complete ignorance is related to situations when no information on a variable of interest (e.g. probability) is available. For case of probability complete ignorance may be described by a set of all probability distributions.

Objective uncertainty relates to evidence on a likelihood of phenomena. Subjective uncertainty relates to DM's opinion on a likelihood of phenomena. More specifically, subjective uncertainty is a DM's belief on occurrence of an event. Classification of subjective uncertainty is very wide and its primitive forms are, analogously to that of objective uncertainty, subjective probability, subjective possibility and subjective necessity. The structures of these forms of subjective uncertainty are the same as those of objective probability, possibility and necessity. However, the sources of them differ: subjective uncertainty is a DM's opinion, whereas objective uncertainty is pure evidence. For example, mathematical structure of subjective probability is a probability measure but the values of this measure are assigned on the base of a DM's opinion under lack of evidence. Analogously, subjective possibility and necessity are a DM's opinions on possibility and necessity of an event.

Unreliable information is information to which an individual does not trust or trusts weakly due to the source of this information. As a result, an individual does not rely on this information. For example, you may not trust to the meteorological forecast if it is done by using old technology and equipment.

Imperfect information is impossible to be completely caught in terms of understanding what this concept means (e.g. uncertainty concept), and thus, cannot be perfectly classified. Any classification may have contradictions, flows and changes of concepts.

In real-world, imperfect information is commonly present in all the components of the decision making problem. States of nature reflects possible future conditions which are commonly ill-known whereas the existing theories are based on perfect construction – on partition of the future objective conditions into mutually exclusive and exhaustive states. Possible realizations of future are not completely known. The future may result in a situation which was not thought and unforeseen contingencies commonly take place [26]. From the other side, those states of nature which are supposed as possible, are themselves vaguely defined and it is not always realistic to strictly differentiate among them. The outcomes and probabilities are also not well known, especially taking into account that they are related to ill-known states of nature. However, the existing theories do not pay significant attention to these issues. The most of the theories, including the famous and advanced theories, take into account only imperfect information related to probabilities. Moreover, this is handled by coarse description of ambiguity – either by exact constraints on probabilities (a set of priors) or by using subtle techniques like probabilistic constraints or specific non-linear functions. These are, however, approaches rather for frameworks of the designed experiments but not for real-world decision problems when information is not sufficiently good to apply such techniques. In the Table 3.1. below we tried to classify decision situations on the

Table 3.1 Classification of decision-relevant information

		Probabilities			
Outcomes	Utilities	Precise	Complete Ignorance	Ambiguous	Imperfect
Precise	Precise	Situation 1	Situation 2	Situation 3	Situation 4
	Fuzzy	Situation 5	Situation 6	Situation 7	Situation 8
Complete Ignorance	Precise	Situation 9	Situation 10	Situation 11	Situation 12
	Fuzzy	Situation 13	Situation 14	Situation 15	Situation 16
Ambiguous	Precise	Situation 17	Situation 18	Situation 19	Situation 20
	Fuzzy	Situation 21	Situation 22	Situation 23	Situation 24
Imperfect	Precise	Situation 25	Situation 26	Situation 27	Situation 28
	Fuzzy	Situation 29	Situation 30	Situation 31	Situation 32

base of different types of decision relevant information that one can be faced with and the utility models that can be applied. In this table, we identify three important coordinates (dimensions). The first one concerns information available for probabilities, the second captures information about outcomes, while the third looks at the nature of utilities and their description. The first two dimensions include precise information (risk), complete ignorance (absence of information), ambiguous information, and imperfect information. Two main types of utilities are considered, namely precise and fuzzy. Decision-relevant information setups are represented at the crossing of these coordinates; those are cells containing Situations from 1 to 32. They capture combinations of various types of probabilities, outcomes, and utilities.

The most developed scenarios are those positioned in entries numbered from 1 to 4 (precise utility models). A limited attention has paid to situations 5-8 with fuzzy utilities, which are considered in [5,7,29,51]. For the situations 9-12 with complete ignorance with respect to outcomes and with precise utilities a few works related to interactive obtaining of information were suggested. For situations 13-16, to our knowledge, no works were suggested. Few studies are devoted to the situations with ambiguous outcomes (situations 17-20) [37,38,39] and precise utilities and no works to ambiguous outcomes with fuzzy utilities are available (situations 21-24). For situations 25-32 a very few studies were reported including the existing fuzzy utility models [5,7,29,51]. The case with imperfect probabilities, imperfect outcomes, and fuzzy utilities (situation 32) generalize all the other situations. An adequate utility model for this situation is suggested in [3] and is expressed in Chapter 4 of the present book.

The probability theory has a large spectrum of successful applications. However, the use of a single probability measure for quantification of uncertainty has severe limitations main of which are the following [3]: 1) precise probability is unable to describe complete ignorance (total lack of information); 2) one can determine probabilities of some subsets of a set of possible outcomes but cannot always determine probabilities for all the subsets; 3) one can determine probabilities of all the subsets of a set of possible outcomes but it will require laborious computations.

Indeed, classical probability imposes too strong assumptions that significantly limit its use even in simple real-world or laboratory problems. Famous Ellsberg experiments and Schmeidler's coin example are good illustrative cases when available information appears insufficient to determine actual probabilities. Good discussion of real-world tasks which are incapable to be handled within probabilistic framework is given in [30]. In real problems, quality of decision-relevant information does not require the use of a single probability measure. As a result, probabilities cannot be precisely determined and are imprecise. For such cases, they use constraints on a probability of an event A in form of lower and upper probabilities denoted $\underline{P}(A)$ and $\overline{P}(A)$ respectively. That is, a probability $P(A)$ of an event A is not known precisely but supposed to be somewhere between $\underline{P}(A)$ and $\overline{P}(A)$: $P(A) \in [\underline{P}(A), \overline{P}(A)]$ where $0 \leq \underline{P}(A) \leq \overline{P}(A) \leq 1$; in more general formulation, constraints in form of lower and upper expectations for a random variable are used. In special case when $\underline{P}(A) = \overline{P}(A)$ a framework of lower and upper probabilities degenerates to a single probability $P(A)$. Complete lack of knowledge about likelihood of A is modeled by $\underline{P}(A) = 0$ and $\overline{P}(A) = 1$. This means that when likelihood of an event is absolutely unknown, they suppose that probability of this event may take any value from [0,1] (from impossibility to occur up to certain occurrence).

Constraints on probabilities imply existence of a set of probability distributions, that is, multiple priors, which are an alternative approach to handle incomplete information on probabilities. Under the certain consistency requirements the use of multiple priors is equivalent to the use of lower and upper probabilities. Approaches in which imprecise probabilities are handled in form of intervals $[p_1, p_2]$. Such representation is termed as *interval probabilities*.

An alternative way to handle incomplete information on probabilities is the use of non-additive probabilities, typical cases of which are lower probabilities and upper probabilities and their convex combinations. However, multiple priors are more general and intuitive approach to handle incomplete probability information than non-additive probabilities.

The most fundamental axiomatization of imprecise probabilities was suggested by Peter Walley who suggested the term *imprecise probabilities.* The behavioral interpretation of Walley's axiomatization is based on buying and selling prices for gambles. Walley's axiomatization is more general than Kolmogorov's axiomatization of the standard probability theory. The central concept in Walley's theory is the lower prevision concept which generalizes standard (additive) probability, lower and upper probabilities and non-additive measures. However, in terms of generality, the concept of lower prevision is inferior to multiple priors. Another disadvantage of lower prevision theory is its high complexity that limits its practical use.

Alternative axiomatizations of imprecise probabilities were suggested by Kuznetsov [47] and Weichselbergern [75] for the framework of interval probabilities. Weichselberger generalizes Kolmogorov's axioms to the case of interval probabilities but, as compared to Walley, does not suggest a behavioral interpretation. However, his theory of interval probability is more tractable in practical sense.

What is the main common disadvantage of the existing imprecise probability theories? This disadvantage is missing the intrinsic feature of probability-related information which was pointed out by L. Savage even before emergence of the existing imprecise probability theories:

Savage wrote [62]: "…there seem to be some probability relations about which we feel relatively 'sure' as compared with others…. The notion of 'sure' and 'unsure' introduced here is vague, and my complaint is precisely that neither the theory of personal probability as it is developed in this book, nor any other device known to me renders the notion less vague". Indeed, in real-world situations we don't have sufficient information to be definitely sure or unsure in whether that or another value of probability is true. Very often, our sureness stays at some level and does not become complete being hampered by a lack of knowledge and information. That is, sureness is a matter of a strength, or in other words, of a degree. Therefore, 'sure' is a loose concept, a vague concept. In our opinion, the issue is that in most real-world decision-making problems, relevant information perceived by DMs involves possibilistic uncertainty. Fuzzy probabilities are the tools for resolving this issue to a large extent because they represent a degree of truth of a considered numeric probability.

Fuzzy probabilities are superior from the point of view of human reasoning and available information in real-world problems than interval probabilities which are rather the first departs from precise probabilities frameworks. Indeed, interval probabilities only show that probabilities are imprecise and no more. In real-world, the bounds of an interval for probability are subjectively 'estimated' but not calculated or actually known as they are in Ellsberg experiment. Subjective assignments of probability bounds will likely inconsistent with human choices in real-world problems as well as subjective probabilities do in Ellsberg experiment. Reflecting imperfect nature of real-world information, probabilities are naturally soft-constrained.

As opposed to second-order probabilities which are also used to differentiate probability values in terms of their relevance to available information, fuzzy probabilities are more relaxed constructs. Second-order probabilities are too exigent to available information and more suitable for designed experiments.

Fuzzy probability is formally a fuzzy number defined over [0,1] scale to represent an imprecise linguistic evaluation of a probability value. Representing likelihoods of mutually exclusive and exhaustive events, fuzzy probabilities are tied together by their basic elements summing up to one. Fuzzy probabilities define a fuzzy set $\tilde{P}^{\rho}$ of probability distributions ρ which is an adequate representation of imprecise probabilistic information related to objective conditions especially when the latter are vague. As compared to the use of second-order probabilities, the use of possibility distribution over probability distributions [2,3] is appropriate and easier for describing DM's (or experts') confidence. This approach does not require from DM to assign beliefs over priors directly. Possibility distribution can be constructed computationally from fuzzy probabilities assigned to states of nature [7,80]. This means that a DM or experts only need to assign linguistic evaluations of probabilities to states of nature as they usually do. For each linguistic evaluation a fuzzy probability can then be defined by construction of a membership function. After this possibility distribution can be obtained computationally [7,80] without involving a DM.

We can conclude that fuzzy probabilities [8,50,51,66,76] are a successful interpretation of imprecise probabilities which come from human expertise and perceptions being linguistically described. For example, in comparison to multiple priors consideration, for majority of cases, a DM has some linguistic additional information coming from his experience or even naturally present which reflects unequal levels of belief or possibility for one probability distribution or another. This means, that it is more adequate to consider sets of probability distributions as fuzzy sets which allow taking into account various degrees of belief or possibility for various probability distributions. Really, for many cases, some probability distributions are more relevant, some probability distributions are less relevant to the considered situation and also it is difficult to sharply differentiate probabilities that are relevant from those that are irrelevant. This type of consideration involves second-order uncertainty, namely, probability-possibility modeling of decision-relevant information.

The existing utility theories are based on Savage's formulation of states of nature as "*a space of* ***mutually exclusive*** *and* ***exhaustive*** *states*" [62]. This is a perfect consideration of environment structure. However, in real-world problems it is naïve to suppose that we can so perfectly partition future into mutually exclusive objective conditions and predict all possible objective conditions. Future is hidden from our eyes and only some indistinct, approximate trends can be seen. From the other side, unforeseen contingencies are commonly met which makes impossible to determine exhaustive states and also rules out sharp differentiation to exclusive objective conditions. This requires tolerance in describing each objective condition to allow for mistakes, misperceptions, flaws, that are due to imperfect nature of information about future. From the other side, tolerance may also allow for dynamic aspects due to which a state of nature may deviate from its initial condition.

In order to see difficulties with determination of states of nature let us consider a problem of differentiating future economic conditions into states of economy. Commonly, states of economy can be considered as "strong growth", "moderate growth", "stable economy", "recession". These are not 'single-valued' and cannot be considered as 'mutually exclusive' (as it is defined in Savage's formulation of state space): for example, moderate growth and stable economy don't have sharp boundaries and as a result, may not be "exclusive" – they may overlap. The same concerns 'moderate growth' and 'strong growth' states. For instance, when analyzing the values of the certain indicators that determine a state of economy it is not always possible to definitely label it as moderate growth or strong growth. Observing some actual situation an expert may conclude that it is "somewhere between" 'strong growth' and 'moderate growth', but "closer" to the latter. This means that to a larger extent the actual situation concerns the moderate growth and to a smaller extent to the strong growth. It is not adequate to sharply differentiate the values related to 'moderate growth' from those related to the "strong growth". In other words, various conditions labeled as "strong growth" with various extents concerns it, not equally. How to take into account the inherent vagueness of states of nature and the fact that they are intrinsically not exclusive but overlapping? Savage's definition is an idealized view on formalization of objective conditions for such cases. Without doubt, in real-life decision making it is often impossible to construct such an ideal formalization, due to uncertainty of relevant information. In general, a DM cannot exhaustively determine each objective condition that may be faced and cannot precisely differentiate them. Each state of nature is, essentially, some area under consideration which collects in some sense similar objective conditions one can face, that is some set of "elementary" states or quantities [26]. Unfortunately, in the existing decision making theories a small attention is paid to the essence and structure of states of nature, consideration of them is very abstract (formal) and is unclear from human perception point of view.

Formally speaking, a state of nature should be considered as a granule - not some single point or some object with abstract content. This will result to some kind of information granulation of objective conditions. Construction of states of nature on the base of similarity, proximity etc of objective conditions may be adequately modeled by using fuzzy sets and fuzzy information granulation concepts [78]. This will help to model vague and overlapping states of nature. For example, in the considered problem economic conditions may be partitioned into overlapping fuzzy sets defined over some relative scale representing levels of economic welfare. Such formalization will be more realistic for vagueness, ambiguity, partial truth, impreciseness and other imperfectness of future-related information.

In real-life decision making a DM almost always cannot precisely determine future possible outcomes and have to use imprecise quantities like, for example, *high profit*, *medium cost* etc. Such quantities can be adequately represented by ranges of numerical values with possibility distribution among them. From the other side, very often outcomes and utilities are considered in monetary sense, whereas a significantly smaller attention is paid to other types of outcomes and utilities. Indeed, utilities are usually subjectively assigned and, as a result, are heuristic evaluations. In extensive experiments conducted by Kahneman and Tversky, which uncovered very important aspects of human behavior only monetary outcomes are used. Without doubt, monetary consideration is very important, but it is worth to investigate also other types of outcomes which are naturally present in real-life decision activity. In this situation it is not suitable to use precise quantities because subjective evaluations are conditioned also by non-monetary issues such as health, time, reputation, quality etc. The latter are usually described by linguistic evaluations.

In order to illustrate impreciseness of outcomes in real-world problems, let us consider a case of an evaluation of a return from investment into bonds of an enterprise which will produce new products the next year. Outcomes (returns) of investment will depend on future possible economic conditions. Let us suppose these conditions of economy to be partitioned into states of nature labeled as "strong growth", "moderate growth", "stable economy", and "recession" which we considered above. It is impossible to precisely know values of outcomes of the investment under these states of nature. For example, the outcome of the investment obtained under "strong growth" may be evaluated "high" (off course with underlying range of numerical values). The vagueness of outcomes evaluations are resulted from uncertainty about future: impreciseness of a demand for the products produced by the enterprise in the next year, future unforeseen contingencies, vagueness of future economic conditions, political processes etc. Indeed, the return is tightly connected to the demand the next year which cannot be precisely known in advance. The investor does not really know what will take place the next year, but still approximately evaluate possible gains and losses by means of linguistic terms. In other words, the investor is not completely sure in some precise value of the outcome – the future is too uncertain for precise estimation to be reasonably used. The investor sureness is 'distributed' among various possible values of the perceived outcome. One way to model is the use of a probabilistic outcome, i.e. to use probability distribution (if discrete set of numerical outcomes is considered) or probability density function (for continuous set) over possible basic outcomes [30] to encode the related objective probabilities or subjective probabilities. However, this approach has serious disadvantages. Using objective probabilities requires good representative data which don't exist as a demand for a new product is considered. Even for the case of a common product, a good statistics does not exist because demands for various years take place in various environmental conditions. The use of subjective probabilities is also not suitable as they commonly fail to describe human behavior and perception under ambiguity of information.

The use of probabilistic outcomes does not also match human perceptions which are expressed in form of linguistic evaluations of outcomes. Humans don't think within the probabilistic framework as this is too strong for computational abilities of a human brain; thus, a more flexible formalization is needed to use. Fuzzy set theory provides more adequate representation of linguistic evaluations. By using this theory a linguistic evaluation of an outcome may be formalized by a membership function (a fuzzy set) representing a soft constraint on possible basic outcomes. In contrast to probabilistic constraint, a membership function is not based on strong assumptions and does not require good data. A membership function is directly assigned by a DM to reflect his/her experience, perception, envision etc. which cannot be described by classical mathematics but may act well under imperfect information. Fuzzy sets theory helps to describe future results as imprecise and overlapping, especially under imprecise essence of states of nature. Also, a membership function may reflect various basic outcomes' possibilities, which are much easier to determine than probabilities.

From the other side, the use of fuzzy sets allows to adequately describe non-monetary outcomes like health, reputation, quality which are often difficult to be defined in terms of precise quantities.

3.3 Measures of Uncertainty

Uncertainty is intrinsic to decision making environment. No matter whether we deal with numerical or non-numerical events, we are not completely sure in their occurrence. Numerical events are commonly regarded as values of a random variable. Non-numerical events can be encoded by, for example, natural numbers and then treated as values of a random variable. To formally take into account uncertainty in decision analysis, we need to use some mathematical constructs which will measure quantitatively an extent to what that or another event is likely to occur. Such constructs are called measures of uncertainty. The most famous measure is the probability measure. Probability measure assigns its values to events to reflect degrees to which events are likely to occur. These values are called probabilities. Probability is a real number from $[0,1]$, and the more likely an event to occur the higher is its probability. Probability equal to 0 implies that it is impossible for an event to occur or we are completely sure that it cannot occur, and probability equal to 1 means that an event will necessary occur or we are completely sure in its occurrence. The axiomatization of the standard probability measure was suggested by Kolmogorov [46]. Prior to proceeding to the Kolmogorov's axiomatization, let us introduce the necessary concepts. The first concept is the space of elementary events. Elementary event, also called an atomic event, is the minimal event that may occur, that is, an event that cannot be divided into smaller events. Denote S the space of elementary events and denote $s \in S$ an elementary event. A subset $H \subseteq S$ of the space of elementary events $s \in S$ is called an event. An event H occurs if any $s \in H$ occurs. The next concept is a σ-algebra of subsets denoted $\mathcal{F}$.

Definition 3.2. σ-algebra [46]**.** A set $\mathcal{F}$, elements of which are subsets of S (not necessarily all) is called σ-algebra if the following hold:

(1) $S \in \mathcal{F}$

(2) if $H \in \mathcal{F}$ then $H^c \in \mathcal{F}$

(3) if $H_1, H_2, ... \in \mathcal{F}$ then $H_1 \cup H_2 \cup ... \in \mathcal{F}$

Now let us proceed to the Kolmogorov's axiomatization of a probability measure.

Definition 3.3. Probability Measure [46]. Let S be a space of elementary events and $\mathcal{F}$ is a σ-algebra of its subsets. The probability measure is a function $P:\mathcal{F}\rightarrow[0,1]$ satisfying:

(1) $P(H)\geq 0$ for any $H\in\mathcal{F}$.

(2) For any set $H_1,H_2,...\in\mathcal{F}$ with $H_i\cap H_j,...\varnothing$: $P(\bigcup_{i=1}^{\infty}H_i)=\sum_{i=1}^{\infty}P(H_i)$

(3) $P(S)=1$

The first condition is referred to as non-negativity. The second condition is referred to as additivity condition. The third condition implies that the event S will necessary occur. Conditions (1)-(3) are called probability axioms. From (1)-(3) it follows $P(\varnothing)=0$ which means that it is impossible that no elementary event $s\in S$ will occur. Let us mention that probability of a union $H\cup G$ of two arbitrary events is $P(H\cup G)=P(H)+P(G)-P(H\cap G)$. When $H\cap G=\varnothing$ one has $P(H\cup G)=P(H)+P(G)$.

Definition 3.3 provides mathematical structure of a probability measure. Consider now natural interpretations of a probability measure. There exist two main types of probabilities: objective probabilities and subjective probabilities. Objective probabilities, also called empirical probabilities, are quantities which are calculated on the base of real evidence: experimentations, observations, logical conclusions. They also can be obtained by using statistical formulas. Objective probabilities are of two types: experimental probabilities and logical probabilities. Experimental probability of an event is a frequency of its occurrence. For example, a probability that a color of a car taken at random in a city is white is equal to the number of white cars divided by the whole number of the cars in the city. Logical probability is based on a reasonable conclusion on a likelihood of an event. For instance, if a box contains 70 white and 30 black balls, a probability that a ball drawn at random is white is 70/100=0.7.

The use of objective probabilities requires very restrictive assumptions. For experimental probabilities the main assumptions are as follows:

(1) Experimentation (or observations) must take place under the same conditions and it must be assumed that the future events will also take place under these conditions. Alternatively, there need to be present clear dynamics of conditions in future;
(2) Observations of the past behavior must include representative data (e.g., observations must be sufficiently large).

As to logical probabilities, their use must be based on quite reasonable conclusions. For example, if to consider the box with balls mentioned above, an assumption must be made that the balls are well mixed inside the box (not a layer of white balls is placed under the layer of black balls) to calculate probability of a white ball drawn as 70/100=0.7.

From the other side, as Kahneman, Tversky and others showed [42] , that even if objective probabilities are known, beliefs of a DM don't coincide with them. As being perceived by humans, objective probabilities are affected by some kind of distortion – they are transformed into so-called decision weights and mostly small probabilities are overweighted, whereas high probabilities are underweighted. The overweighting and underweighting of probabilities also are different for positive and negative outcomes [68].

Subjective probability is a degree of belief of an individual in the likelihood of an event. Formally, subjective probabilities are values of a probability measure. From interpretation point of view, subjective probability reflects an individual's experience, perceptions and is not based on countable and, sometimes, detailed facts like objective probability. Subjective probabilities are more appropriate and 'smart' approach for measuring likelihood of events in real-life problems because in such problems the imperfect relevant information conflicts with the very strong assumptions underlying the use of objective probabilities. Real-life relevant information is better handled by experience and knowledge that motivates the use of a subjective basis.

Subjective probability has a series of disadvantages. One of the main disadvantages is that different people would assign different subjective probabilities. It is difficult to reason uniquely accurate subjective probabilities among those assigned by different people. Indeed, given a lack of information, people have to guess subjective probabilities as they suppose. As it is mentioned in [52], using subjective probabilities is a "symptom of the problem, not a solution". Subjective probability is based not only on experience but also on feelings, emotions, psychological and other factors which can distort its accuracy. The other main disadvantage is that subjective probability, due to its preciseness and additivity, fails to describe human behavior under ambiguity.

The use of the additive probability measure is unsuitable to model human behavior conditioned by uncertainty of the real-world, psychological, mental and other factors. In presence of uncertainty, when true probabilities are not exactly known, people often tend to consider each alternative in terms of the worst case within the uncertainty and don't rely on good cases. In other words, most of people prefer those decisions for which more information is available. This behavior is referred to as ambiguity aversion – people don't like ambiguity and wish certainty. Even when true probabilities are known, most people exhibit non-linear attitude to probabilities – change of likelihood of an event from impossibility to some chance or from a very good chance to certainty are treated much more strongly than the same change somewhere in the range of medium probabilities. Therefore, attitude to values of probabilities is qualitative.

Due to its additivity property, the classical (standard) probability measure cannot reflect the above mentioned evidence. Axiomatizations of such evidence required to highly weaken assumptions on a DM's belief which was considered as the probability measure. The resulted axiomatizations are based either on non-uniqueness of probability measure or on non-additivity of a measure of uncertainty reflecting humans' beliefs. The first axiomatization of choices based on a non-additive measure was suggested by Schmeidler [63]. This is a significant generalization of additive measures-based decisions because the uncovered non-additive measure inherited only normalization and monotonicity properties from the standard probability measure.

Nowadays non-additive measures compose a rather wide class of measures of uncertainty. Below we list non-additive measures used in decision making under ambiguity. For these measures a unifying term *non-additive probability* is used.

We will express the non-additive probabilities in the framework of decision making under ambiguity. Let S be a non-empty set of states of nature and $\mathcal{F}$ be a family of subsets of S. We will consider w.l.o.g. $\mathcal{F}=2^S$.

The definition of a non-additive probability is as follows [63].

Definition 3.4 [63]. **Non-additive Probability.** A set function $v:\mathcal{F}\to[0,1]$ is called a non-additive probability if it satisfies the following:

1. $v(\varnothing)=0$
2. $\forall H,G\in\mathcal{F}$, $H\subset G$ implies $v(H)\le v(G)$
3. $v(S)=1$

The non-additive probability is also referred to as *Choquet capacity*. Condition (2) is called monotonicity with respect to set inclusion and conditions (1) and (3) are called normalization conditions. Thus, a non-additive probability does not have to satisfy $v(H\cup G)=v(H)+v(G)$. A non-additive probability is called *super-additive* if $v(H\cup G)\ge v(H)+v(G)$ and *sub-additive* if $v(H\cup G)\le v(H)+v(G)$, provided $H\cap G=\varnothing$.

There exist various kinds of non-additive probability many of which are constructed on the base of a set C of probability measures P over S. The one of the well known non-additive probabilities is the *lower envelope* $v_*:\mathcal{F}\to[0,1]$ which is defined as follows:

$$v_*(H)=\min_{P\in C}P(G) \tag{3.1}$$

The dual concept of the lower envelope is the *upper envelope* $v^*:\mathcal{F}\to[0,1]$ which is defined by replacing $\min$ operator in (3.1) by $\max$ operator. Lower and upper envelopes are respectively minimal and maximal probabilities of an event $H\subset S$. Therefore, $v_*(H)\le P(H)\le v^*(H),\forall H\subset S,P\in C$. Lower envelope is super-additive, whereas upper envelope is sub-additive. A non-additive probability can also be defined as a convex combination of $v_*(H)$ and

$v^*(H)$: $v(H)=\alpha v_*(H)+(1-\alpha)v^*(H)$, $\alpha\in[0,1]$. The parameter α is referred to as a degree of ambiguity aversion. Indeed, α is an extent to which belief $v(H)$ is based on the smallest possible probability of an event H ; $1-\alpha$ is referred to as a degree of ambiguity seeking.

The generalizations of lower and upper envelopes are *lower* and *upper probabilities* which are respectively super-additive and sub-additive probabilities. Lower and upper probabilities, denoted respectively $\underline{v}$ and $\overline{v}$, satisfy $\underline{v}(H)=1-\overline{v}(H^c)\ \forall H\in S$, where $H^c=S\setminus H$.

The special case of lower envelopes and, therefore, of lower probabilities are *2-monotone Choquet capacities*, also referred to as *convex capacities.* A non-additive probability is called 2-monotone Choquet capacity if it satisfies

$$v(H\cup G)\geq v(H)+v(G)-v(H\cap G),\ \forall H,G\subset S$$

A generalization of *2-monotone* capacity is an *n-monotone* capacity. A capacity is an *n-monotone*, if for any sequence $H_1,...,H_n$ of subsets of S the following holds:

$$v(H_i\cup...\cup H_n)\geq\sum_{\substack{I\subset\{1,...,n\}\\ I\neq\varnothing}}(-1)^{|I|-1}v\left(\bigcap_{i\subset I}H_i\right)$$

A capacity which is *n-monotone* for all n is called *infinite monotone capacity* or a *belief function.*

The belief function theory, also known as Dempster-Shafer theory, or mathematical theory of evidence, or theory of random sets, was suggested by Dempster in [15], and developed by Shafer in [64]. Belief functions are aimed to be used for describing subjective degrees of belief to an event, phenomena, or object of interest. We will not explain this theory but just mention that it was not directly related to decision making. As it was shown in [33,34], axiomatization of this theory is a generalization of the Kolmogorov's axioms of the standard probability theory. Due to this fact, a value of a belief function denoted $Bel()$ for an event H can be considered as a lower probability, that is, as a lower bound on a probability of an event H . An upper probability in the belief function theory is termed as a *plausibility function* and is denoted Pl . So, in the belief functions theory probability $P(H)$ of an event H is evaluated as $Bel(H)\leq P(H)\leq Pl(H)$.

The motivation of using non-additive probabilities in decision making problems is the fact that information on probabilities is imperfect, which can be incomplete, imprecise, distorted by psychological factors etc. Non-additive measure can be determined from imprecise objective or subjective probabilities of states of nature. Impreciseness of objective probabilities can be conditioned by the lack of information ruling out determination of actual exact probabilities (as in Ellsberg experiments). Impreciseness of subjective probabilities can be conditioned by natural impreciseness of human beliefs. Let us consider the case when imprecise information is represented in form of interval probabilities. Given a set of states of nature $S=\{s_1,s_2,...,s_n\}$, interval probabilities are defined as follows [32].

Definition 3.5 [32]. **Interval Probability.** The intervals $\overline{P}(s_i)=[p_{*i},p_i^*]$ are called the interval probabilities of S if for $p_i\in[p_{*i},p_i^*]$ there exist $p_1\in[p_{*1},p_1^*],...,p_{i-1}\in[p_{*i-1},p_{i-1}^*]$, $p_{i+1}\in[p_{*i+1},p_{i+1}^*]$,..., $p_n\in[p_{*n},p_n^*]$ such that

$$\sum_{i=1}^{n} p_i = 1$$

From this definition it follows, in particular, that interval probabilities cannot be directly assigned as numerical probabilities. The issue is that in the case of interval probabilities, the requirement to numerical probabilities to sum up to one must be satisfied throughout all the probability ranges. Sometimes, interval probabilities $\overline{P}(s_i)=\overline{P}_i$ can be directly assigned consistently to $n-1$ states of nature $s_1,s_2,...,s_{j-1},s_{j+1},...,s_n$, and on the base of these probabilities, an interval probability $\overline{P}(s_j)=\overline{P}_j$ for the rest one state of nature s_j will be calculated. For example, consider a set of states of nature with three states $S=\{s_1,s_2,s_3\}$. Let interval probabilities for s_2 and s_3 be assigned as follows:

$$\overline{P}_2=[0.2,0.3],\ \overline{P}_3=[0.5,0.6]$$

Then, according to the conditions in Definition 3.5, $\overline{P}_1$ will be determined as follows:

$$\overline{P}_1=[1-0.3-0.6,\ 1-0.2-0.5]=[0.1,0.3].$$

Given interval probabilities $\overline{P}_i=[p_{*i},p_i^*]$ of states of nature $s_i, i=1,...,n$ a value $v_*(A)$ of a lower probability for an event A can be determined as follows:

$$\begin{aligned}
& v_*(A)=\min\sum_{s_i\in A} p_i \\
& s.t. \\
& \sum_{i=1}^{n} p_i = 1 \\
& p_i \le p_i^* \\
& p_i \ge p_{*i}
\end{aligned} \qquad (3.2)$$

A value $v^*(A)$ of an upper probability for an event A can be determined by replacing *min* operator by *max* operator in the above mentioned problem. Consider an example. Given interval probabilities $\overline{P}_1=[0.1,0.3]$, $\overline{P}_2=[0.2,0.3]$, $\overline{P}_3=[0.5,0.6]$, the values of the lower and upper probabilities v_* and v^* for $A=\{s_1,s_3\}$, obtained as solutions of the problem (3.2) are $v_*(A)=0.7$ and $v^*(A)=0.8$.

The above mentioned measures of uncertainty can be listed in terms of the increasing generality between the probability measure and the Choquet capacity as follows:

probability measure $\Rightarrow$ belief function $\Rightarrow$ convex capacity $\Rightarrow$ lower envelope $\Rightarrow$ lower probability $\Rightarrow$ Choquet capacity

In [48], it is suggested a decision model based on a new kind of measure called bi-capacity. Bi-capacity is used to model interaction between 'good' and 'bad' performances with respect to criteria. As compared to capacity, bi-capacity is a two-place set function. The values the bi-capacity takes are from [-1,1]. More formally, the bi-capacity is defined as a set function

$$\eta: W \to [-1,1], \text{ where } W=\{(H,G): H,G \subset I, H \cap G=\varnothing\}$$

satisfying

$$H \subset H' \Rightarrow \eta(H,G) \le \eta(H',G),\ G \subset G' \Rightarrow \eta(H,G') \le \eta(H,G)$$

and

$$\eta(I,\varnothing)=1,\ \eta(\varnothing,I)=-1,\ \eta(\varnothing,\varnothing)=0.$$

I is the set of indexes of criteria. The attributes in H are satisfied attributes whereas the attributes in G are dissatisfied ones. The integral with respect to bi-capacity as a representation of an overall utility of an alternative $f: I \to R$ is defined as follows:

$$U(f)=\sum_{l=1}^{n}(u(f_{(l)})-u(f_{(l+1)}))\eta(\{(1),...,(l)\}\cap I^+,\{(1),...,(l)\}\cap I^-), \qquad (3.3)$$

provided $u(f_{(l)}) \ge u(f_{(l+1)})$; $I^+=\{i \in I: u(f_i) \ge 0\}$, $I^-=I \setminus I^+$ where $u(f_{(l)})$ is a utility of a value of *(l)*-th criterion for f, $\eta(\cdot,\cdot)$ is a bi-capacity.

In special case, when η is equal to the difference of two capacities η_1 and η_2 as $\eta(H,G)=\eta_1(H)-\eta_2(G)$, (3.3) reduces to the CPT model. In general case, as compared to CPT, (3.3) is not an additive representation of separately aggregated satisfied and dissatisfied criteria that provides more smart way for decision making.

The disadvantage of a bi-capacity relates to difficulties of its determination, in particular, to computational complexity. In details the issues are discussed in [48].

The bi-capacity-based aggregation which was axiomatized for multicriteria decision making [48] can also be applied for decisions under uncertainty due to formal parallelism between these two problems [53]. Indeed, states of nature are criteria on base of which alternatives are evaluated.

The non-additive measures provide a considerable success in modeling of decision making. However, the non-additive measures only reflect the fact that human choices are non-additive and monotone, which may be due to attitudes to uncertainty, distortion of probabilities etc, but nothing more. However, in real-world it is impossible to accurately determine precisely the 'shape' of a non-additive measure due to imperfect relevant information. Indeed, real-world probabilities of subsets and subsets themselves, outcomes, interaction of criteria, etc are imprecisely and vaguely defined. From the other side, attitudes to uncertainty, extent of probabilities distortion and other behavioral issues violating additivity are also imperfectly known. These aspects rule out exact determination of a uniquely accurate non-additive measure.

Above we considered non-additive measures which are used in the existing decision theories to model non-additivity of DM's behavior. Main shortcoming of using non-additive measures is the difficulty of the underlying interpretation. One approach to overcome this difficulty is to use a lower envelope of a set of priors as a non-additive probability and then to use it in CEU model. However, in real-world problems determination of the set of priors itself meets difficulty of imposing precise constraint determining what prior should be included and what should not be included into this set. In other words, due to lack of information, it is impossible to sharply constraint a range for a probability of a state of nature, that is, to assign accurate interval probability. From the other side, if the set of priors is defined, why to construct lower envelope and use it in the CEU? It is computationally simpler to use the equivalent model – MMEU. Let us consider very important a class of non-additive measures called *fuzzy measures*. Fuzzy measures have their own interpretations that do not require using a set of priors to define them and makes construction of these measures computationally simple. Finally, we will consider an effective extension of non-additive measures called *fuzzy-valued* fuzzy measures which have a good suitability for measuring vague real-world information.

The first fuzzy measure we consider is a *possibility measure*. Possibility means an ability of an event to occur. It was recently mentioned that probability of an event can hardly be determined due to a series of reason, whereas possibility of occurrence of an event is easier to be evaluated. Possibility measure has also its interpretation in terms of multiple priors.

Possibility measure [77] is a non-additive set function $\Pi : \mathcal{F}(S) \to [0,1]$ defined over a σ-algebra $\mathcal{F}(S)$ of subsets of S and satisfying the following conditions:

(1) $\Pi(\varnothing) = 0$

(2) $\Pi(S) = 1$

(3) For any collection of subsets $H_i \in \mathcal{F}(S)$ and any set of indexes I the following holds:

$$\Pi(\bigcup_{i \in I} H_i) = \sup_{i \in I} \Pi(H_i)$$

Possibility measure Π can be represented by *possibility distribution function,* or possibility distribution, for short. Possibility distribution is a function $\pi : S \to [0,1]$ and by means of π possibility measure Π is determined as follows:

$$\Pi(H) = \sup_{s \in H} \pi(s)$$

Condition (2) predetermines normalization condition $\sup_{s \in S} \pi(s) = 1$. Given S as a set of states of nature, possibility measure provides information on possibility of occurrence of an event $H \subset S$. A possibility distribution π_1 is more informative than π_2 if $\pi_1(s) \le \pi_2(s), \forall s \in S$.

The dual concept of the possibility is the concept of necessity. Necessity measure is a set function $N : P(S) \to [0,1]$ that is defined as $N(H) = 1 - \Pi(H^c)$, $H^c = S \setminus H$. This means, for example, that if an event H is necessary (will necessary happen), then the opposite event H^c is impossible.

From the definitions of possibility and necessity measures one can find that the following hold:

1) $N(H) \le \Pi(H)$
2) if $\Pi(H) < 1$ then $N(H) = 0$
3) if $N(H) > 0$ then $\Pi(H) = 1$
4) $\max(\Pi(H), \Pi(H^c)) = 1$
5) $\min(N(H), N(H^c)) = 0$

The possibility differs from probability in various aspects. First, possibility of two sets H and G provided $H \cup G = \varnothing$ is equal to the maximum possibility among those of H and $G : \Pi(H \cup G) = \max(\Pi(H), \Pi(G))$. In its turn probability $H \cup G$ is equal to the sum of those of H and G: $P(H \cup G) = P(H) + P(G)$.

Another difference between possibility measure and probability measure is that the first is compositional that make it more convenient from computational point of view. For example, given $P(H)$ and $P(G)$, we cannot determine precisely $P(H \cup G)$, but can only determine its lower bound which is equal to $\max(P(H), P(G))$ and an upper bound which is equal to $\min(P(H)+P(G),1)$. At the same time possibility of $H \cup G$ is exactly determined based on $\Pi(H)$ and $\Pi(G): \Pi(H \cup G) = \max(\Pi(H), \Pi(G))$. However, the possibility of an intersection is not exactly defined: it is only known that $\Pi(A \cap B) \leq \min(\Pi(A), \Pi(B))$. As to necessity measure, it is exactly defined only for an intersection of sets: $N(H \cap G) = \min(N(H), N(G))$.

Yet another difference is that as compared to probability, possibility is able to model complete ignorance, that is, absence of any information. Absence of any information about H is modeled in the possibility theory as $\Pi(H) = \Pi(H^c) = 1$ and $N(H) = N(H^c) = 0$. From this it follows $\max(\Pi(H), \Pi(H^c)) = 1$ and $\min(N(H), N(H^c)) = 0$.

The essence of the possibility is that it models rather qualitative information about events than quantitative one. Possibility measure only provides ranking of events in terms of their comparative possibilities. For example, $\pi(s_1) \leq \pi(s_2)$ implies that s_1 is more possible than s_2. $\pi(s) = 0$ implies that occurrence of s is impossible whereas $\pi(s) = 1$ implies that s is one of the most possible realizations. The fact that possibility measure may be used only for analysis at qualitative, comparative level [69], was proven by Pytyev in [60], and referred to as the principle of relativity in the possibility theory. This principle implies that possibility measure cannot be used to measure actual possibility of an event but can only be used to determine whether the possibility of one event is higher, equal to, or lower than the possibility of another event. Due to this feature, possibility theory is less self-descriptive than probability theory but requires much less information for analysis of events than the latter.

One of the interpretations of possibility measure is an upper bound of a set of probability measures [18,62,72]. Let us consider the following set of probability measures coherent with possibility measure Π:

$$\mathcal{P}(\Pi) = \{P : P(H) \leq \Pi(H), \forall H \subseteq S\}$$

Then the upper bound of probability for an event H is

$$\overline{P}(H) = \sup_{P \in \mathcal{P}(\Pi)} P(H)$$

and is equal to possibility $\Pi(H)$. The possibility distribution is then defined as

$$\pi(s) = \overline{P}(\{s\}), \forall s \in S$$

Due to normalization condition $\sup_{s\in S}\pi(s)=1$, the set $\mathcal{P}(\Pi)$ is always not empty. In [18,72] they show when one can determine a set of probability measures given possibility measure.

Analogous interpretation of possibility is its representation on the base of lower and upper bounds of a set of distribution functions. Let information about unknown distribution function F for a random variable X is described by means of a lower $\underline{F}$ and an upper $\overline{F}$ distribution functions: $\underline{F}(x)\le F(x)\le \overline{F}(x), \forall x\in X$. The possibility distribution π then may be defined as

$$\pi(x)=\min(\overline{F}(x),1-\underline{F}(x)).$$

Baudrit and Dubois showed that a set of probabilities generated by possibility distribution π is more informative than a set of probabilities generated by equivalent distribution functions.

In order to better explain what possibility and necessity measures are, consider an example with a tossed coin. If to suppose that heads and tails are equiprobable, then the probabilities of heads and tails will be equal to 0.5. As to possibilities, we can accept that both heads and tails are very possible. Then, we can assign the same high value of possibility to both events, say 0.8. At the same time, as the result of tossing the coin is not intentionally designed, we can state that the necessity of both events is very small. It also follows from $\mathrm{N}(\{heads\})=1-\Pi(\{tails\}), \mathrm{N}(\{tails\})=1-\Pi(\{heads\})$. As this example suggests, we can state that possibility measure may model ambiguity seeking (hope for a good realization of uncertainty), where as necessity measure may model ambiguity aversion.

One of the most practically efficient and convenient fuzzy measures are *Sugeno* measures. Sugeno measure is a fuzzy measure $g:P(S)\to[0,1]$ that satisfies

(1) $g(\varnothing)=0$,

(2) $g(S)=1$;

(3) $H\subset G\Rightarrow g(H)\le g(G)$;

(4) $H_i\uparrow H$ or $H_i\downarrow H\Rightarrow \lim_{i\to+\infty} g(H_i)=g(H)$

From these conditions it follows $g(H\cup G)\ge \max(g(H),g(G))$ and $g(H\cap G)\le \min(g(H),g(G))$. In special case, when $g(H\cup G)==\max(g(H),g(G))$, Sugeno measure g is the possibility measure. When $g(H\cap G)=\min(g(H),g(G))$, Sugeno measure g is the necessity measure.

The class of Sugeno measures that became very widespread due to its practical usefulness are g_λ measures. g_λ measure is defined by the following condition referred to as the λ-rule:

$$g_\lambda(H \cup G) = g_\lambda(H) + g_\lambda(G) + \lambda g_\lambda(H) g_\lambda(G),\ \lambda \in [-1, +\infty)$$

For the case of $H = S$, this condition is called normalization rule. λ is called normalization parameter of g_λ measure. For $\lambda > 0$ g_λ measure satisfy $g_\lambda(H \cup G) > g_\lambda(H) + g_\lambda(G)$ that generates a class of superadditive measures. For $\lambda > 0$ one gets a class of subadditive measures: $g_\lambda(H \cup G) <$ $< g_\lambda(H) + g_\lambda(G)$. The class of additive measures is obtained for $\lambda = 0$.

One type of fuzzy measures is defined as a linear combination of possibility measure and probability measure. This type is referred to as g_v measure. g_v measure is a fuzzy measure that satisfies the following:

(1) $g_v(\varnothing) = 0$

(2) $g_v(S) = 1$

(3) $\forall i \in N, H_i \in \mathcal{F}(S), \forall i \neq j$

(4) $H_i \cap H_j = \varnothing \Rightarrow g_v\left(\bigcup_{i \in N} H_i\right) = (1 - v) \bigvee_{i \in N} g_v(H_i) + v \sum_{i \in N} g_v(H_i),\ v \geq 0$

(5) $\forall H, G \in \mathcal{F}(S): H \subseteq G \Rightarrow g_v(H) \leq g_v(G)$

g_v is an extension of a measure suggested by Tsukamoto which is a special case obtained when $v \in [0,1]$ [67]. For $v \in [0,1]$ one has a convex combination of possibility and probability measures. For purposes of decision making this can be used to model behavior which is inspired by a mix of probabilistic judgement and an extreme non-additive reasoning, for instance, ambiguity aversion. Such modeling may be good as reflecting that a person is not only an uncertainty averse but also thinks about some ‘average’, i.e. approximate precise probabilities of events. This may be justified by understanding that, from one side, in real-world situations we don’t know exactly the boundaries for a probability of an event. From the other, we don’t always exhibit pure ambiguity aversion by try to guess some reasonable probabilities in situations of ambiguity.

When $v = 0$, g_v is the possibility measure and when $v = 1$, g_v is the probability measure. For $v = 1$, g_v describes uncertainty that differs from both probability and possibility [59].

Fuzzy measures are advantageous type of non-additive measures as compared to non-additive probabilities because they mainly have clear interpretation and some of them are “self-contained”. The latter means that some fuzzy measures, like possibility measure, don’t require a set of priors for their construction. Moreover, a fuzzy measure can be more informative than a set of priors or a set of priors can be obtained from it. Despite of these advantages, fuzzy measures are also not sufficiently adequate for solving real-world decision problems. The issue is that fuzzy measures suffer from the disadvantage of all the widespread additive and non-additive measures: fuzzy measures are numerical representation of uncertainty. In contrast, real-world uncertainty cannot be precisely described – it is not to be caught by a numerical function. This aspect is, in our opinion, one of the most essential properties of real-world uncertainty.

The precise non-additive measures match well the backgrounds of decision problems of the existing theories which are characterized by perfect relevant information: mutually exclusive and exhaustive states of nature, sharply constrained probabilities. However, as we discussed above chapter, real-world decision background is much more 'ill-defined'. Essence of information about states of nature makes them rather blurred and overlapping but not perfectly separated. For example, evaluations like 'moderate growth' and 'strong growth' of economy cannot be precisely bounded and may overlap to that or another extent. This requires to use fuzzy sets as more adequate descriptions of real objective conditions. Probabilities of states of nature are also fuzzy as they cannot be sharply constrained. This is conditioned by lack of specific information, by the fact that human sureness in occurrence of events stays in form of linguistic estimations like "very likely", "probability is medium", "probability is small" etc which are fuzzy. From the other side, this is conditioned by fuzziness of states of nature themselves. When the "strong growth" and "moderate growth" and their likelihoods are vague and, therefore, relations between them are vague – how to obtain precise measure? Natural impreciseness, fuzziness related to states of nature must be kept as the useful data medium in passing from probabilities to a measure – the use of precise measure cannot be sufficiently reasonable and leads to loss of information. From the other side, shape of non-additivity of a DM's behavior cannot be precisely determined, whereas some linguistic, approximate, but still ground relevant information can be obtained. Fuzziness of the measure in this case serves as a good interpretation.

Thus, a measure which models human behavior under real-world imperfect information should be considered not only as non-additive, but also as fuzzy imprecise quantity that will reflect human evaluation technique based on, in general, linguistic assessments. In this sense a more adequate construction that better matches imperfect real-world information is *a fuzzy number-valued fuzzy measure.* Prior to formally express what is a fuzzy number-valued measure, let us introduce some formal concepts. The first concept is a set of fuzzy states of nature $\mathcal{S}=\{\tilde{S}_1,...,\tilde{S}_n\}$, where $\tilde{S}_i, i=1,...,n$ is a fuzzy set defined over a universal set U in terms of membership function $\mu_{\tilde{S}_i}: U \to [0,1]$. The second concept relates to comparison of fuzzy numbers:

Definition 3.6. [82]. For $\tilde{A},\tilde{B}\in \mathcal{E}^1$, we say that $\tilde{A}\le\tilde{B}$, if for every $\alpha\in(0,1]$, $A_1^\alpha \le B_1^\alpha$ and $A_2^\alpha \le B_2^\alpha$.

We consider that $\tilde{A}<\tilde{B}$, if $\tilde{A}\le\tilde{B}$, and there exists an $\alpha_0\in(0,1]$ such that $A_1^{\alpha_0} < B_1^{\alpha_0}$, or $A_2^{\alpha_0} < B_2^{\alpha_0}$.

We consider that $\tilde{A}=\tilde{B}$ if $\tilde{A}\le\tilde{B}$, and $\tilde{B}\le\tilde{A}$

The third concept is a fuzzy infinity:

Definition 3.7. Fuzzy Infinity [82]. Let $\tilde{A}$ be a fuzzy number. For every positive real number M, there exists a $\alpha_0\in(0,1]$ such that $M<A_2^{\alpha_0}$ or $A_1^{\alpha_0}<-M$. Then $\tilde{A}$ is called fuzzy infinity, denoted by $\tilde{\infty}$.

Now denote $\mathcal{E}_+^1=\left\{\tilde{A}\in\mathcal{E}\middle|\tilde{A}\ge 0\right\}$. Thus, $\mathcal{E}_+^1$ is a space of fuzzy numbers defined over $\mathcal{R}_+$. Let Ω be a nonempty finite set and $\mathcal{F}$ be σ-algebra of subsets of Ω. A definition of a fuzzy number-valued fuzzy measure as a monotone fuzzy number-valued set function suggested by Zhang [82] and referred to as a (*z*)-fuzzy measure, is as follows:

Definition 3.8. Fuzzy Number-Valued Fuzzy Measure ((z)-Fuzzy Measure) [82]. A (*z*) fuzzy measure on $\mathcal{F}$ is a fuzzy number-valued fuzzy set function $\tilde{\eta}:\mathcal{F}\to\mathcal{E}_+^1$ with the properties:

(1) $\tilde{\eta}(\varnothing)=0$;

(2) if $\mathcal{H}\subset\mathcal{G}$ then $\tilde{\eta}(\mathcal{H})\le\tilde{\eta}(\mathcal{G})$;

(3) if $\mathcal{H}_1\subset\mathcal{H}_2\subset...,\mathcal{H}_n\subset...\in\mathcal{F}$, then $\tilde{\eta}(\bigcup_{n=1}^{\infty}\mathcal{H}_n)=\lim_{n\to\infty}\tilde{\eta}(\mathcal{H}_n)$;

(4) if $\mathcal{H}_1\supset\mathcal{H}_2\supset...,\mathcal{H}_n\in\mathcal{F}$, and there exists n_0 such that $\tilde{\eta}(\mathcal{H}_{n_0})\ne\tilde{\infty}$, then

$$\tilde{\eta}(\bigcap_{n=1}^{\infty}\mathcal{H}_n)=\lim_{n\to\infty}\tilde{\eta}(\mathcal{H}_n).$$

Here limits are taken in terms of supremum metric d [16,49]. A pair $(\Omega,\tilde{\mathcal{F}}(\Omega))$ is called a fuzzy measurable space and a triple $(\Omega,\tilde{\mathcal{F}}(\Omega),\tilde{\eta})$ is called a (*z*) fuzzy measure space.

So, a fuzzy number-valued fuzzy measure $\tilde{\eta}:\mathcal{F}\to\mathcal{E}_+^1$ assigns to every subset of Ω a fuzzy number defined over [0,1]. Condition (2) of Definition 3.8 is called monotonicity condition. $\tilde{\eta}:\mathcal{F}\to\mathcal{E}_+^1$ is monotone and is free of additivity requirement. Consider an example. Let $\Omega=\{\omega_1,\omega_2,\omega_3\}$. The values of the fuzzy number-valued set function $\tilde{\eta}$ for the subsets of Ω can be as the triangular fuzzy numbers given in Table 3.2:

Table 3.2 The values of the fuzzy number-valued set function $\tilde{\eta}$

$\mathcal{H}\subset\Omega$	$\{\omega_1\}$	$\{\omega_2\}$	$\{\omega_3\}$	$\{\omega_1,\omega_2\}$	$\{\omega_1,\omega_3\}$	$\{\omega_2,\omega_3\}$
$\tilde{\eta}(\mathcal{H})$	(0.3,0.4,0.4)	(0,0.1,0.1)	(0.3,0.5,0.5)	(0.3,0.5,0.5)	(0.6,0.9,0.9)	(0.3,0.6,0.6)

Fuzzy number-valued set function $\tilde{\eta}$ is a fuzzy number-valued fuzzy measure. For instance, one can verify that condition 2 of Definition 3.8 for $\tilde{\eta}$ is satisfied.

Let us consider $\tilde{\eta}$ as a fuzzy number-valued lower probability constructed from linguistic probability distribution $\tilde{P}^l$:

$$\tilde{P}^l = \tilde{P}_1/\tilde{S}_1 + \tilde{P}_2/\tilde{S}_2 + ... + \tilde{P}_n/\tilde{S}_n$$

Linguistic probability distribution $\tilde{P}^l$ implies that a state $\tilde{S}_i \in \mathcal{S}$ is assigned a linguistic probability $\tilde{P}_i$ that can be described by a fuzzy number defined over [0,1]. Let us shortly mention that the requirement for numeric probabilities to sum up to one is extended for linguistic probability distribution $\tilde{P}^l$ to a wider requirement which includes degrees of consistency, completeness and redundancy that will be described in details in Chapter 4. Given $\tilde{P}^l$, we can obtain from it a fuzzy set $\tilde{P}^\rho$ of possible probability distributions $\rho(s)$. We can construct a fuzzy-valued fuzzy measure from $\tilde{P}^\rho$ as its lower probability function [55] by taking into account a degree of correspondence of $\rho(s)$ to $\tilde{P}^l$. A degree of membership of an arbitrary probability distribution $\rho(s)$ to $\tilde{P}^\rho$ (a degree of correspondence of $\rho(s)$ to $\tilde{P}^l$) can be obtained by the formula

$$\pi_{\tilde{P}}(\rho(s)) = \min_{i=\overline{1,n}}(\pi_{\tilde{P}_i}(p_i)) \quad ,$$

where $p_i = \int_S \rho(s)\mu_{\tilde{S}_i}(s)ds$ is numeric probability of fuzzy state $\tilde{S}_i$ defined by $\rho(s)$. $\pi_{\tilde{P}_i}(p_i) = \mu_{\tilde{P}_i}\left(\int_S \rho(s)\mu_{\tilde{S}_i}(s)ds\right)$ is the membership degree of p_i to $\tilde{P}_i$.

To derive a fuzzy-number-valued fuzzy measure $\tilde{\eta}_{\tilde{P}^l}$ we suggest to use the following formulas [3]:

$$\eta(\mathcal{H}) = \bigcup_{\alpha\in(0,1]} \alpha\left[\eta_1^\alpha(\mathcal{H}), \eta_2^\alpha(\mathcal{H})\right] \tag{3.4}$$

where

$$\eta_1^\alpha(\mathcal{H}) = \inf\left\{\int_S \rho(s)\max_{\mathcal{S}\in\mathcal{H}}\mu_{\mathcal{S}}(s)ds \,\middle|\, \rho(s)\in P^{\rho^\alpha}\right\},$$

$$\eta_2^\alpha(\mathcal{H}) = \inf\left\{\int_S \rho(s)\max_{\mathcal{S}\in\mathcal{H}}\mu_{\mathcal{S}}(s)ds \,\middle|\, \rho(s)\in core(\tilde{P}^\rho)\right\}, \tag{3.5}$$

$$P^{\rho^\alpha} = \left\{\rho(s) \,\middle|\, \min_{i=\overline{1,n}}(\pi_{\tilde{P}_i}(p_i)) \geq \alpha\right\},\; core(\tilde{P}^\rho) = P^{\rho^{\alpha=1}},\; \mathcal{H}\subset\mathcal{S}$$

The support of $\tilde{\eta}$ is defined as $\operatorname{supp}\tilde{\eta} = cl\left(\bigcup_{\alpha\in(0,1]} \eta^{\alpha}\right)$. So, $\tilde{\eta}_{\tilde{P}^l}$ is constructed by using $\mu_{\tilde{S}}(s)$ which implies that in construction of the non-additive measure $\tilde{\eta}_{\tilde{P}^l}$ we take into account impreciseness of the information on states of nature themselves. Detailed examples on construction of a fuzzy number-valued measure are considered in the upcoming chapters.

In this section we will discuss features of various existing precise additive and non-additive measures and fuzzy-valued fuzzy measures. The discussion will be conducted in terms of a series of criteria suggested in [72]: interpretation, calculus, consistency, imprecision, assessment, computation. The emphasis will be given to situations in which all the relevant information is described in NL.

Interpretation, Calculus and Consistency. Linguistic probabilities-based fuzzy-valued lower and upper probabilities and their convex combinations have clear behavioral interpretation: they represent ambiguity aversion, ambiguity seeking and their various mixes when decision-relevant information is described in NL. Updating these measures is to be conducted as updating the underlying fuzzy probabilities according to fuzzy Bayes' rule and new construction of these measures from the updated fuzzy probabilities.

Formal validity of the considered fuzzy-valued measures is defined from verification of degrees of consistency, completeness and redundancy of the underlying fuzzy probabilities as initial judgments.

Among the traditional measures, Bayesian probability and coherent lower previsions suggested by Walley [72] (these measures are crisp, non-fuzzy) are only measures which satisfy the considered criteria. Bayesian probability has primitive behavioral interpretation, on base of which the well-defined rules of combining and updating are constructed. Coherent lower previsions have clear and more realistic behavioral interpretation. The rules for updating, combining and verification of consistency of lower previsions are based on the natural extension principle [71,72] which is a general method. However, it is very complex both from analytical and computational points of view.

Possibility theory and the Dempster-Shafer theory, as it is mentioned in [71,72], suffer from lack of the methods to verify consistency of initial judgments and conclusions.

Imprecision. Fuzzy-valued lower and upper previsions and their convex combinations are able to transfer additional information in form of possibilistic uncertainty from states of nature and associated probabilities to the end up measuring of events. As a result, these measures are able to represent vague predicates in NL and partial and complete ignorance as degenerated cases of linguistic ambiguity.

Dempster-Shafer theory is a powerful tool for modeling imprecision and allows to model complete ignorance. However, this theory suffers from series of significant disadvantages [69]. Determination of basic probabilities in this theory may lead to contradictory results. From the other side, under lack of information on some elements of universe of discourse, values of belief and plausibility functions for these elements become equal to zero which means that occurrence of them will not take place. However, this is not justified if the number of observations is small.

Possibility theory is able to model complete ignorance and requires much less information for modeling than probability theory. Possibility measure, as opposed to probability measure, is compositional, which makes it computationally more convenient. However, possibility measure has a serious disadvantage as compared with the probability measure. This theory allows only for qualitative comparative analysis of events – it allows determining whether one event is more or less possible than another, but does not allow determining actual possibilities of events.

Dempster-Shafer theory, lower prevision theory and possibility theory can be considered as special cases of multiple priors representations [69]. In this sense, belief and plausibility functions can be considered as an upper and lower bounds of probability respectively. Possibility theory also can be used for representation of bounds of multiple priors and is used in worst cases of statistical information.

Possibility theory, Dempster-Shafer theory and coherent lower previsions as opposed to Bayesian probabilities are able to model ignorance, impreciseness and NL-based vague evaluations. However, as these theories are based on precise modeling of uncertainty, use of them lead to significant roughening of NL-based information. For example, linguistic description of information on states of nature and their probabilities creates a too high vagueness for these precise measures to be believable or reliable in real-life problems.

Assessment. Fuzzy-valued lower probability is obtained from the linguistic probability assessments which are practical and human-like estimations for real-world problems. Coherent lower prevision can also be obtained from the same sources, but, as a precise quantity, it will be not reliable as very much deviated from vague and imprecise information on states of nature and probabilities. From the other side, fuzzy-valued lower or upper probabilities are computed from fuzzy probabilities.

The other main advantage of fuzzy-valued lower probabilities and fuzzy probabilities constructed for NL-based information is that they, as opposed to all the other measures, don't require independence or non-interaction assumptions on the measured events, which are not accurate when we deal with overlapping and similar objective conditions.

Computation. The construction of unknown fuzzy probability, the use of fuzzy Bayes' formula and construction of a fuzzy-valued lower prevision are quite complicated variational or nonlinear programming problems. However, the complexity here is the price we should pay if we want to adequately formalize and compute from linguistic descriptions. However, as opposed to the natural extension-based complex computations of coherent lower previsions which involves linear programming, the computation of fuzzy-valued previsions is more intuitive, although arising the well known problems of nonlinear optimization.

Computations of coherent lower previsions (non-fuzzy) can be reduced to simpler computations of possibility measures and belief functions as their special cases, but it will lead to the loss of information.

Adequacy of the use of a fuzzy-valued lower (upper) probability consists in its ability to represent linguistic information as the only adequate relevant information on dependence between states of nature in real-life problems. The existing non-additive measures, being numerical-valued, cannot adequately represent such information. To some extent it can be done by lower previsions, but in this case one deals with averaging of linguistic information to precise values which leads to loss of information.

References

1. Aliev, R.A., Aliev, R.R.: Soft Computing and its Application. World Scientific, New Jersey (2001)
2. Aliev, R.A., Alizadeh, A.V., Guirimov, B.G., Huseynov, O.H.: Precisiated information-based approach to decision making with imperfect information. In: Proc. of the Ninth International Conference on Application of Fuzzy Systems and Soft Computing (ICAFS 2010), Prague, Czech Republic, pp. 91–103 (2010)
3. Aliev, R.A., Pedrycz, W., Fazlollahi, B., Huseynov, O.H., Alizadeh, A.V., Guirimov, B.G.: Fuzzy logic-based generalized decision theory with imperfect information. Information Sciences 189, 18–42 (2012)
4. Allais, M., Hagen, O. (eds.): Expected Utility Hypotheses and the Allais Paradox: Contemporary Discussions of the Decisions Under Uncertainty with Allais' Rejoinder. D. Reidel Publishing Co., Dordrecht (1979)
5. Alo, R., de Korvin, A., Modave, F.: Using Fuzzy functions to select an optimal action in decision theory. In: Proc. of the North American Fuzzy Information Processing Society (NAFIPS), pp. 348-353 (2002)
6. Aumann, R.: Utility Theory without the Completeness Axiom. Econometrica 30, 445–462 (1962)
7. Borisov, A.N., Alekseyev, A.V., Merkuryeva, G.V., Slyadz, N.N., Gluschkov, V.I.: Fuzzy information processing in decision making systems. Radio i Svyaz, Moscow (1989) (in Russian)
8. Buckley, J.J.: Fuzzy Probability and Statistics. Studies in Fuzziness and Soft Computing, vol. 270. Springer, Heidelberg (2003)
9. Camerer, C., Weber, M.: Recent Developments in Modeling Preferences. Journal of Risk and Uncertainty 5, 325–370 (1992)
10. Casadesus-Masanell, R., Klibanoff, P., Ozdenoren, E.: Maxmin expected utility through statewise combinations. Economics Letters 66, 49–54 (2000)
11. Chateauneuf, A., Wakker, P.: An Axiomatization of Cumulative Prospect Theory for Decision Under Risk. Journal of Risk and Uncertainty 18(2), 137–145 (1999)
12. Chen, Z., Epstein, L.G.: Ambiguity, Risk and Asset Returns in Continuous Time. Econometrica 70, 1403–1443 (2002)

13. de Baets, B., de Meyer, H.: Transitivity frameworks for reciprocal relations: cycle transitivity versus FG-transitivity. Fuzzy Sets Syst. 152, 249–270 (2005)
14. de Baets, B., de Meyer, H., De Schuymer, B., Jenei, S.: Cyclic evaluation of transitivity of reciprocal relations. Soc. Choice Welfare 26, 217–238 (2006)
15. Dempster, A.P.: Upper and lower probabilities induced by a multivalued mapping. Annals of Mathematical Statistics 38, 325–339 (1967)
16. Diamond, P., Kloeden, P.: Metric spaces of fuzzy sets. Theory and applications. World Scientific, Singapoure (1994)
17. Dubois, D.: The role of fuzzy sets in decision sciences: Old techniques and new directions. Fuzzy Sets and Systems 184, 3–28 (2011)
18. Dubois, D., Prade, H.: When upper probabilities are possibility measures. Fuzzy Sets and Systems 49, 65–74 (1992)
19. Dubra, J., Maccheroni, F., Ok, E.: Expected utility theory without the completeness axiom. J. Econom. Theory 115, 118–133 (2004)
20. Ellsberg, D.: Risk, Ambiquity and the Savage Axioms. Quarterly Journal of Economics 75, 643–669 (1961)
21. Epstein, L.G.: A Definition of Uncertainty Aversion. Review of Economic Studies 66, 579–608 (1999)
22. Epstein, L.G., Wang, T.: Intertemporal Asset Pricing under Knightian Uncertainty. Econometrica 62, 283–322 (1994)
23. Epstein, L.G., Zhang, J.: Subjective Probabilities on Subjectively Unambiguous Events. Econometrica 69, 265–306 (2001)
24. Epstein, L.G., Schneider, M.: Ambiguity, information quality and asset pricing. Journal of Finance 63(1), 197–228 (2008)
25. Ghirardato, P., Maccheroni, F., Marinacci, M.: Differentiating Ambiguity and Ambiguity Attitude. Journal of Economic Theory 118, 133–173 (2004)
26. Ghirardato, P.: Coping with ignorance: unforeseen contingencies and non-additive uncertainty. Economic Theory 17, 247–276 (2001)
27. Ghirardato, P., Klibanoff, P., Marinacci, M.: Additivity with multiple priors. Journal of Mathematical Economics 30, 405–420 (1998)
28. Ghirardato, P., Marinacci, M.: Ambiguity Made Precise: A Comparative Foundation. Journal of Economic Theory 102, 251–289 (2002)
29. Gil, M.A., Jain, P.: Comparison of Experiments in Statistical Decision Problems with Fuzzy Utilities. IEEE Transactions on Systems, Man and Cybernetícts 22(4), 662–670 (1992)
30. Gilboa, I.: Theory of Decision under Uncertainty. Cambridge University Press, Cambridge (2009)
31. Gilboa, I., Schmeidler, D.: Maximin Expected utility with a non-unique prior. Journal of Mathematical Economics 18, 141–153 (1989)
32. Guo, P., Tanaka, H.: Decision making with interval probabilities. European Journal of Operational Research 203, 444–454 (2010)
33. Halpern, J.Y.: Reasoning about uncertainty, p. 483. The MIT Press, Massachusetts (2003)
34. Halpern, J.Y., Fagin, R.: Two views of belief: Belief as generalized probability and belief as evidence. Artificial Intelligence 54, 275–317 (1992)
35. Herrera, F., Herrera-Viedma, E.: Linguistic decision analysis: steps for solving decision problems under linguistic information. Fuzzy Sets and Systems 115, 67–82 (2000)
36. Herrera, F., Herrera-Viedma, E.: Choice functions and mechanisms for linguistic preference relations. European Journal of Operational Research 120, 144–161 (2000)

37. Ho, J.L.Y., Keller, P.L., Keltyka, P.: Effects of Outcome and Probabilistic Ambiguity on Managerial Choices. Journal of Risk and Uncertainty 24(1), 47–74 (2002)
38. Hu, Q., Yu, D., Guo, M.: Fuzzy preference based rough sets. Information Sciences, 180: Special Issue on Intelligent Distributed Information Systems 15, 2003–2022 (2010)
39. Huettel, S.A., Stowe, C.J., Gordon, E.M., Warner, B.T., Platt, M.L.: Neural signatures of economic preferences for risk and ambiguity. Neuron 49, 765–775 (2006)
40. Insua, D.R.: Sensitivity Analysis in Multiobjective Decision Making. Springer, New York (1990)
41. Insua, D.R.: On the foundations of decision making under partial information. Theory and Decision 33, 83–100 (1992)
42. Kahneman, D., Tversky, A.: Prospect theory: an analysis of decision under uncertainty. Econometrica 47, 263–291 (1979)
43. Klibanoff, P., Marinacci, M., Mukerji, S.: A smooth model of decision making under ambiguity. Econometrica 73(6), 1849–1892 (2005)
44. Klibanoff, P.: Maxmin Expected Utility over Savage Acts with a Set of Priors. Journal of Economic Theory 92, 35–65 (2000)
45. Klibanoff, P.: Characterizing uncertainty aversion through preference for mixtures. Social Choice and Welfare 18, 289–301 (2001)
46. Kolmogorov, A.N.: Foundations of the theory of probability. Chelsea Publishing Company, New York (1936)
47. Kuznetsov, V.P.: Interval Statistical Models. Radio i Svyaz Publ., Moscow (1991) (in Russian)
48. Labreuche, C., Grabisch, M.: Generalized Choquet-like aggregation functions for handling bipolar scales. European Journal of Operational Research 172, 931–955 (2006)
49. Lakshmikantham, V., Mohapatra, R.: Theory of fuzzy differential equations and inclusions. Taylor & Francis, London (2003)
50. Lü, E.-L., Zhong, Y.M.: Random variable with fuzzy probability. Applied Mathematics and Mechanics 24(4), 491–498 (2003), doi:0.1007/BF02439629
51. Mathieu-Nicot, B.: Fuzzy Expected Utility. Fuzzy Sets and Systems 20(2), 163–173 (1986)
52. Meredith, J., Shafer, S., Turban, E.: Quantitative business modeling. South-Western, Thomson Learning, USA (2002)
53. Modave, F., Grabisch, M., Dubois, D., Prade, H.: A Choquet Integral Representation in Multicriteria Decision Making. Technical Report of the Fall Symposium of Association for the Advancement of Artificial Intelligence (AAAI), pp. 22–29. AAAI Press, Boston (1997)
54. Nau, R.: The shape of incomplete preferences. Ann. Statist. 34, 2430–2448 (2006)
55. Nguyen, H.T., Walker, E.A.: A first Course in Fuzzy logic. CRC Press, Boca Raton (1996)
56. Ok, E.A.: Utility representation of an incomplete preference relation. J. Econ. Theory 104, 429–449 (2002)
57. Ovchinnikov, S.: On fuzzy preference relations. International Journal of Intelligent Systems 6, 225–234 (1991)
58. Pena, J.P.P., Piggins, A.: Strategy-proof fuzzy aggregation rules. Journal of Mathematical Economics 43, 564–580 (2007)
59. Pospelov, D.A.: Fuzzy Sets in Models of Control and Artificial Intelligence. Nauka, Moscow (1986) (in Russian)
60. Pytyev, Y.P.: Possibility. Elements of theory and practice. Editorial URSS, Moscow (2000) (in Russian)

61. Rigotti, R., Shannon, C.: Uncertainty and risk in financial markets. Econometrica 73, 203–243 (2005)
62. Savage, L.J.: The Foundations of Statistics. Wiley, New York (1954)
63. Schmeidler, D.: Subjective probability and expected utility without additivity. Econometrita 57(3), 571–587 (1989)
64. Shafer, G.A.: Mathematical theory of evidence. Princeton University Press (1976)
65. Smets P Imperfect information: Imprecision – Uncertainty, `http://sites.poli.usp.br/d/pmr5406/Download/papers/Imperfect_Data.pdf`
66. Talasova, J., Pavlacka, O.: Fuzzy Probability Spaces and Their Applications in Decision Making. Austrian Journal of Statistics 35(2, 3), 347–356 (2006)
67. Tsukamoto, Y.: Identification of preference measure by means of fuzzy integrals. In: Ann. Conf. of JORS, pp. 131–135 (1972)
68. Tversky, A., Kahneman, D.: Advances in Prospect theory: Cumulative Representation of Uncertainty. Journal of Risk and Uncertainty 5(4), 297–323 (1992)
69. Utkin, L.V.: Risk Analysis and decision making under incomplete information. Nauka, St. Petersburg (2007) (in Russian)
70. von Neumann, J., Morgenstern, O.: Theory of games and economic behaviour. Princeton University Press (1944)
71. Walley, P.: Statistical Reasoning with Imprecise Probabilities. Chapman and Hall, London (1991)
72. Walley, P.: Measures of uncertainty in expert systems. Artificial Intelligence 83(1), 1–58 (1996)
73. Walley, P.: Statistical inferences based on a second-order possibility distribution. International Journal of General Systems 9, 337–383 (1997)
74. Wang, T.C., Chen, Y.H.: Applying fuzzy linguistic preference relations to the improvement of consistency of fuzzy. AHP Information Sciences 78(19), 3755–3765 (2008)
75. Weichselberger, K.: The theory of interval probability as a unifying concept for uncertainty. International Journal of Approximate Reasoning 24, 149–170 (2000)
76. Yager, R.R.: Decision Making with Fuzzy Probability Assessments. IEEE Transactions on Fuzzy Systems 7(4), 462–467 (1999)
77. Zadeh, L.A.: Fuzzy sets as a basis for a theory of possibility. Fuzzy Sets and Systems 1, 3–28 (1978)
78. Zadeh, L.A.: Toward a theory of fuzzy Information Granulation and its centrality in human reasoning and fuzzy logic. Fuzzy Sets and Systems 90(2), 111–127 (1997)
79. Zadeh, L.A.: Computing with words and perceptions—a paradigm shift. In: Proceedings of the IEEE International Conference on Information Reuse and Integration, Las Vegas, pp. 450–452. IEEE Press, Nevada (2009)
80. Zadeh, L.A., Aliev, R.A., Fazlollahi, B., Alizadeh, A.V., Guirimov, B.G., Huseynov, O.H.: Decision Theory with Imprecise Probabilities, Contract on Application of Fuzzy Logic and Soft Computing to communications, planning and management of uncertainty, Berkeley, Baku, vol. 95 (2009)
81. Zadeh, L.A., Klir, G., Yuan, B.: Fuzzy sets, fuzzy logic, and fuzzy systems: selected papers By Lotfi Asker Zadeh. World Scientific Publishing Co. (1996)
82. Zhang, G.Q.: Fuzzy number-valued fuzzy measure and fuzzy number-valued fuzzy integral on the fuzzy set. Fuzzy Sets and Systems 49, 357–376 (1992)

Chapter 4
A Generalized Fuzzy Logic-Based Decision Theory

4.1 Decision Model

In real-world decision making problems usually we are not provided with precise and credible information. In contrast, available information is vague, imprecise, partial true and, as a result, is described in natural language (NL) [2,3]. NL-based information creates fuzzy environment in which decisions are commonly made. The existing decision theories are not developed for applications in fuzzy environment and consequently require more deterministic information. There exist a series of fuzzy approaches to decision making like fuzzy AHP [28,49], fuzzy TOPSIS [49,74], fuzzy Expected Utility [21,34,50]. However, they are mainly fuzzy generalizations of the mathematical structures of the existing theories used with intent to account for vagueness, impreciseness and partial truth. Direct fuzzification of the existing theories often leads to inconsistency and loss of properties of the latter (for example, loss of consistency and transitivity of preference matrices in AHP method when replacing their numerical elements by fuzzy numbers). Let us consider the existing works devoted to the fuzzy utility models and decisions under fuzzy uncertainty [1,17,20,21,24,34-36,50,63]. In [21] they presented axioms for LPR in terms of linguistic probability distributions over fuzzy outcomes and defined fuzzy expected utility on this basis. But, unfortunately, an existence of a fuzzy utility function has not been proved. [20] is an extensive work devoted to the representation of FPR. In this paper, an existence of a utility function representing a fuzzy preorder is proved. However, in this work a utility function itself is considered as a non-fuzzy real-valued function. In [50] it is formulated conditions for existence and continuity of a numerical and fuzzy-valued expected utility under some standard conditions of a FPR (viz. reflexivity, transitivity, continuity, etc.). The author proves theorems on existence of a fuzzy expected utility for the cases of probabilistic and possibilistic information on states of nature. The possibilistic case, as it is correctly identified by the author, appears to be more adequate to deal with real-world problems. However, in this model, probabilities and outcomes are considered as numerical entities. This notably limits the use of the suggested model for real-life decision problems where almost all the information is described in NL. A new approach for decision

R.A. Aliev: *Fuzzy Logic-Based Generalized Theory of Decisions*, STUDFUZZ 293, pp. 127–189.
DOI: 10.1007/ 978-3-642-34895-2_4

making under possibilistic information on states of nature when probabilistic information is absent is considered in [35]. In [34] they suggest representation of a fuzzy preference relation by fuzzy number-valued expected utility on the basis of fuzzy random variables. However, an existence of a fuzzy utility function has not been shown. In [17] they consider a fuzzy utility as a fuzzy-valued Choquet integral with respect to a real-valued fuzzy measure obtained based on a set of possible probability distributions and with a fuzzy integrand. Unfortunately, the existence of the suggested fuzzy utility is not proved.

We can conclude that the existing fuzzy approaches to decision making have significant disadvantages. Fuzzy approaches in which an existence of a utility function is proved, uses fuzzy sets to describe only a part of the components of decision problems. Those approaches that are based on fuzzy description of the most part of a decision problem are lack of mathematical proof of an existence of a utility function. From the other side, many of the existing fuzzy approaches follow too simple models like EU model.

It is needed to develop original and mathematically grounded fuzzy decision theories which are based on initial fuzzy information on all the components of a decision problem. Such fuzzy theories should take into account initial information stemming from fuzzy environment in end-up comparison of alternatives.

In the present chapter we present a fuzzy-logic-based decision theory with imperfect information. This theory is developed for the framework of mix of fuzzy information and probabilistic information and is based on a fuzzy utility function represented as a fuzzy-valued Choquet integral. Being developed for imperfect information framework, the suggested theory differs from the CEU theory as follows:

1) Spaces of fuzzy sets [2,3,25,44] instead of a classical framework are used for modeling states of nature and outcomes 2) Fuzzy probabilities are considered instead of numerical probability distributions[2,3] 3) Linguistic preference relation [21,81] (LPR) is used instead of binary logic-based preference relations 4) Fuzzy number-valued utility functions [15,25,44] are used instead of real-valued utility functions 5) Fuzzy number-valued fuzzy measure [2,3,86] is used instead of a real-valued non-additive probability.

These aspects form fundamentally a new statement of the problem – the problem of decision making with imperfect information. This problem is characterized by second-order uncertainty, namely by fuzzy probabilities. In this framework, we prove representation theorems for a fuzzy-valued utility function. Fuzzy-valued utility function will be described as a fuzzy-valued Choquet integral [6,15,79] with a fuzzy number-valued integrand and a fuzzy number-valued fuzzy measure. Fuzzy number-valued integrand will be used to model imprecise linguistic utility evaluations. It is contemplated that fuzzy number-valued fuzzy measure that can be generated by fuzzy probabilities will better reflect the features of impreciseness and non-additivity related to human behavior. The fuzzy utility model we consider is more suitable for human evaluations and vision of decision problem and related information.

Prior to formally state a problem of decision making with imperfect information as the problem of decision making under mix of fuzzy and probabilistic uncertainties, we will provide the necessary mathematical background. This is the background of spaces of fuzzy numbers and fuzzy functions and the related operations described below.

The first concept we consider is the space of all fuzzy subsets of $\mathcal{R}^n$ denoted $\mathcal{E}^n$ [25,44] which satisfy the conditions of normality, convexity, and are upper semicontinuous with compact support. It is obvious that $\mathcal{E}^1$ is the set of fuzzy numbers defined over $\mathcal{R}$. Then let us denote by $\mathcal{E}^1_{[0,1]}$ the corresponding space of fuzzy numbers defined over the unit interval $[0,1]$. Once the space of fuzzy sets is chosen, a metrics on it must be chosen to define other concepts such as limits, closures and continuity. We suggest to use fuzzy-valued metrics (the use of which is more adequate to measure distances between fuzzy objects) definition of which is given in Chapter 1.

Let Ω be a nonempty finite set, $\mathcal{F}$ be σ-algebra of subsets of Ω and $\tilde{\eta}:\mathcal{F}\to\mathcal{E}^1_+$ be a fuzzy number valued fuzzy measure (see Definition 3.8 in Chapter 3). Let us provide a definition of a fuzzy-valued Choquet integral as a Choquet integral of a fuzzy-valued function with respect to $\tilde{\eta}:\mathcal{F}\to\mathcal{E}^1_+$.

Definition 4.1. Fuzzy-Valued Choquet Integral [79]. Let $\tilde{\varphi}:\Omega\to\mathcal{E}^1$ be a measurable fuzzy-valued function on Ω and $\tilde{\eta}$ be a fuzzy-number-valued fuzzy measure on $\mathcal{F}$. The Choquet integral of $\tilde{\varphi}$ with respect to $\tilde{\eta}$ is defined as

$$\int_\Omega \tilde{\varphi}d\tilde{\eta}=\sum_{i=1}^{n}\left(\tilde{\varphi}(\omega_{(i)}))-_h\tilde{\varphi}(\omega_{(i+1)})\right)\cdot\tilde{\eta}(\mathcal{H}_{(i)})$$

where index (i) implies that elements $\omega_i\in\Omega, i=1,...,n$ are permuted such that $\tilde{\varphi}(\omega_{(i)}))\geq\tilde{\varphi}(\omega_{(i+1)})$, $\tilde{\varphi}(\omega_{(n+1)})=0$ by convention and $\mathcal{H}_{(i)}\subseteq\Omega$.

In the suggested theory $\tilde{\eta}$ will be constructed on the base of the linguistic information on distribution of probabilities over Ω. This requires using the following concepts:

Definition 4.2. Linguistic Probabilities of This Random Variable [21]. The set of linguistic probabilities $\tilde{P}^l=\{\tilde{P}_1,...,\tilde{P}_i,...,\tilde{P}_n\}$ and corresponding values $\{X_1,...,X_i,...,X_n\}$ of a random variable X are called a distribution of linguistic probabilities of this random variable.

Definition 4.3. Fuzzy Set-Valued Random Variable [21]. Let a discrete variable $\tilde{X}$ takes a value from the set $\{\tilde{X}_1,...,\tilde{X}_n\}$ of possible linguistic values, each of which is a fuzzy variable $\langle x_i, U_x, \tilde{X}_i \rangle$ described by a fuzzy set $\tilde{X}_i = \int_{U_x} \mu_{\tilde{X}_i}(x)/x$. Let the probability that $\tilde{X}$ takes a linguistic value $\tilde{X}_i$ be characterized by a linguistic probability $\tilde{P}_i \in \tilde{P}^l$, $\tilde{P}^l = \left\{ \tilde{P} \middle| \tilde{P} \in \mathcal{E}^1_{[0,1]} \right\}$. The variable $\tilde{X}$ is then called a fuzzy set-valued random variable.

Definition 4.4. Linguistic Lottery [21]. Linguistic lottery is a fuzzy set-valued random variable with known linguistic probability distribution. Linguistic lottery is represented by a vector:

$$\tilde{L} = \left(\tilde{P}_1, \tilde{X}_1; ...; \tilde{P}_i, \tilde{X}_i; ...; \tilde{P}_n, \tilde{X}_n \right)$$

Let us consider an example. Let us have the linguistic lottery $\tilde{L} = \left(\tilde{P}_1, \tilde{X}_1; \tilde{P}_2, \tilde{X}_2; \tilde{P}_3, \tilde{X}_3 \right)$, where $\tilde{P}_i$ and $\tilde{X}_i$ are described by triangular and trapezoidal fuzzy numbers defined over [0,1]: $\tilde{X}_1 = (0.1, 0.3, 0.5)$ ('small'), $\tilde{X}_2 = (0.3, 0.5, 0.7)$ ('medium'), $\tilde{X}_3 = (0.5, 0.7, 0.9)$ ('large'), $\tilde{P}_1 = (0.5, 0.7, 0.9)$ ('high'), $\tilde{P}_2 = (0.0, 0.2, 0.4)$ ('low'), $\tilde{P}_3 = (0.0, 0.0, 0.1, 0.4)$ ('very low'). Then the considered linguistic lottery is:

$$\tilde{L} = \begin{pmatrix} (0.5, 0.7, 0.9), (0.1, 0.3, 0.5); \\ (0.0, 0.2, 0.4), (0.3, 0.5, 0.7); \\ (0.0, 0.0, 0.1, 0.4), (0.5, 0.7, 0.9) \end{pmatrix}$$

On the base of the above mentioned concepts we can proceed to the formal statement of problem of decision making with imperfect information.

Let $\mathcal{S} = \left\{ \tilde{S}_1, ..., \tilde{S}_m \right\} \subset \mathcal{E}^n$ be a set of fuzzy states of nature, $\mathcal{X} = \left\{ \tilde{X}_1, ..., \tilde{X}_l \right\} \subset \mathcal{E}^n$ be a set of fuzzy outcomes, $\mathcal{Y}$ be a set of *distributions of linguistic probabilities* over $\mathcal{X}$, i.e. $\mathcal{Y}$ is a set of *fuzzy number-valued functions* [5,8,10]: $\mathcal{Y} = \left\{ \tilde{y} \middle| \tilde{y} : \tilde{\mathcal{X}} \to \mathcal{E}^1_{[0,1]} \right\}$. For notational simplicity we identify $\mathcal{X}$ with the subset $\left\{ \tilde{y} \in \mathcal{Y} \middle| \tilde{y}(\tilde{X}) = 1 \, \textit{for some} \; \tilde{X} \in \mathcal{X} \right\}$ of $\mathcal{Y}$. Denote by $\mathcal{F}_\mathcal{S}$ a σ-algebra of subsets of $\mathcal{S}$. Denote by $\mathcal{A}_0$ the set of all $\mathcal{F}_\mathcal{S}$-measurable [54,79] fuzzy finite valued step functions [75] from $\mathcal{S}$ to $\mathcal{Y}$ and denote by $\mathcal{A}_c$ the constant fuzzy functions in $\mathcal{A}_0$. We call a function $\tilde{f} : \mathcal{S} \to \mathcal{Y}$ a fuzzy finite valued step function if there is a finite partition of $\mathcal{S}$ to $\mathcal{H}_i \subset \mathcal{S}, i = 1, 2, ..., m$, $\mathcal{H}_j \cap \mathcal{H}_k = \varnothing$, for $j \neq k$, such that $\tilde{f}(\tilde{S}) = \tilde{y}_i$ for all $\tilde{S}\in\mathcal{H}_i$. In this case $\tilde{g}:\mathcal{S}\to\mathcal{Y}$ is called a constant fuzzy function if for some $\tilde{y}\in\mathcal{Y}$ one has $\tilde{g}(\tilde{S})=\tilde{y}$ for all $\tilde{S}\in\mathcal{S}$. Thus the constant fuzzy function is a special case of a fuzzy finite valued step function.

Let $\mathcal{A}$ be a convex subset [56] of $\mathcal{Y}^{\mathcal{S}}$ which includes $\mathcal{A}_c$. $\mathcal{Y}$ can be considered as a subset of some linear space, and $\mathcal{Y}^{\mathcal{S}}$ can then be considered as a subspace of the linear space of all fuzzy functions from $\mathcal{S}$ to the first linear space. Let us now define convex combinations in $\mathcal{Y}$ pointwise [56]: for $\tilde{y}$ and $\tilde{z}$ in $\mathcal{Y}$, and $\lambda\in(0,1)$, $\lambda\tilde{y}+(1-\lambda)\tilde{z}=\tilde{r}$, where $\tilde{r}(\tilde{X})=\lambda\tilde{y}(\tilde{X})+(1-\lambda)\tilde{z}(\tilde{X})$, $\tilde{y}(\tilde{X}),\tilde{z}(\tilde{X})\in\mathcal{E}^1_{[0,1]}$. The latter expression is defined based on the Zadeh's extension principle. Let $\mu_{\tilde{r}(\tilde{X})},\mu_{\tilde{y}(\tilde{X})},\mu_{\tilde{z}(\tilde{X})}:[0,1]\to[0,1]$ denote the membership functions of fuzzy numbers $\tilde{r}(\tilde{X}),\tilde{y}(\tilde{X}),\tilde{z}(\tilde{X})$, respectively. Then for $\mu_{\tilde{r}(\tilde{X})}:[0,1]\to[0,1]$ we have:

$$\mu_{\tilde{r}(\tilde{X})}(r(\tilde{X}))=\sup\min_{\substack{r(\tilde{X})=\lambda y(\tilde{X})+(1-\lambda)z(\tilde{X})\\ y(\tilde{X})+z(\tilde{X})\le 1}}(\mu_{\tilde{y}(\tilde{X})}(y(\tilde{X})),\mu_{\tilde{z}(\tilde{X})}(z(\tilde{X}))),$$

$$r(\tilde{X}),y(\tilde{X}),z(\tilde{X})\in[0,1]$$

Convex combinations in $\mathcal{A}$ are also defined pointwise, i.e., for $\tilde{f}$ and $\tilde{g}$ in $\mathcal{A}$ $\lambda\tilde{f}+(1-\lambda)\tilde{g}=\tilde{h}$, where $\lambda\tilde{f}(\tilde{S})+(1-\lambda)\tilde{g}(\tilde{S})=\tilde{h}(\tilde{S})$ on $\mathcal{S}$.

To model LPR, let's introduce a linguistic variable "degree of preference" with term-set $\mathcal{T}=(T_1,\ldots,T_K)$. Terms can be labeled, for example, as "equivalence", "little preference", "high preference", and can each be described by a fuzzy number defined over some scale, for example, [0,1]. The fact that preference of $\tilde{f}$ against $\tilde{g}$ is described by some $T_i\in T$ is expressed as $\tilde{f}\,T_i\,\tilde{g}$. We denote LPR as $\succsim_l$ and below we sometimes, for simplicity, write $\tilde{f}\succsim_l^i\tilde{g}$ or $\tilde{f}\succ_l^i\tilde{g}$ instead of $\tilde{f}\,T_i\,\tilde{g}$.

In the suggested framework, we extend a classical neo-Bayesian nomenclature as follows: elements of $\mathcal{X}$ are fuzzy outcomes; elements of $\mathcal{Y}$ are linguistic lotteries; elements of $\mathcal{A}$ are fuzzy acts; elements of $\mathcal{S}$ are fuzzy states of nature; and elements of $\tilde{\mathcal{F}}_S$ are fuzzy events.

Definition 4.5. Comonotonic Fuzzy Acts [15]. Two fuzzy acts $\tilde{f}$ and $\tilde{g}$ in $\mathcal{Y}^{\mathcal{S}}$ are said to be comonotonic if there are no $\tilde{S}_i$ and $\tilde{S}_j$ in $\mathcal{S}$ for which $\tilde{f}(\tilde{S}_i)\succ_l\tilde{f}(\tilde{S}_j)$ and $\tilde{g}(\tilde{S}_j)\succ_l\tilde{g}(\tilde{S}_i)$ hold.

Two real-valued functions $a: S \to \mathcal{R}$ and $b: S \to \mathcal{R}$ are comonotonic iff $(a(\tilde{S}_i)-a(\tilde{S}_j))(b(\tilde{S}_i)-b(\tilde{S}_j)) \geq 0$ for all $\tilde{S}_i$ and $\tilde{S}_j$ in $\mathcal{S}$.

For a fuzzy number-valued function $\tilde{a}: \mathcal{S} \to \mathcal{E}^1$ denote by a^α, $\alpha \in (0,1]$ its α-cut and note that $a^\alpha = \left[a_1^\alpha, a_2^\alpha\right]$, where $a_1^\alpha, a_2^\alpha : \mathcal{S} \to \mathcal{R}$.

Two fuzzy functions $\tilde{a}, \tilde{b}: \mathcal{S} \to \mathcal{E}^1$ are said to be comonotonic iff the real-valued functions $a_1^\alpha, b_1^\alpha : \mathcal{S} \to \mathcal{R}$ and also $a_2^\alpha, b_2^\alpha : \mathcal{S} \to \mathcal{R}$, $\alpha \in (0,1]$ are comonotonic.

A constant act $\tilde{f} = \tilde{y}^S$ for some $\tilde{y}$ in $\mathcal{Y}$, and any act $\tilde{g}$ are comonotonic. An act $\tilde{f}$ whose statewise lotteries $\{\tilde{f}(\tilde{S})\}$ are mutually indifferent, i.e., $\tilde{f}(\tilde{S}) \sim_l \tilde{y}$ for all $\tilde{S}$ in $\mathcal{S}$, and any act $\tilde{g}$ are comonotonic.

It is common knowledge that under degrees of uncertainty humans evaluate alternatives or choices linguistically using certain evaluation techniques such as "much worse", "a little better", "much better", "almost equivalent" etc. In contrast to the classical preference relation, imposed on choices made by humans, LPR consistently expresses "degree of preference" allowing the analysis of preferences under uncertainty.

Below we give a series of axioms of the LPR $\succsim_l$ over $\mathcal{A}$ [7,15].

(i) ***Weak-Order***

(a) *Completeness.* Any two alternatives are comparable with respect to LPR: for all $\tilde{f}$ and $\tilde{g}$ in $\mathcal{A}$: $\tilde{f} \succsim_l \tilde{g}$ or $\tilde{g} \succsim_l \tilde{f}$. This means that for all $\tilde{f}$ and $\tilde{g}$ there exists such $T_i \in \mathcal{T}$ that $\tilde{f} \succsim_l^i \tilde{g}$ or $\tilde{g} \succsim_l^i \tilde{f}$.

(b) *Transitivity.* For all $\tilde{f}, \tilde{g}$ and $\tilde{h}$ in $\mathcal{A}$: If $\tilde{f} \succsim_l \tilde{g}$ and $\tilde{g} \succsim_l \tilde{h}$ then $\tilde{f} \succsim_l \tilde{h}$. This means that if there exist such $T_i \in \mathcal{T}$ and $T_j \in \mathcal{T}$ that $\tilde{f} \succsim_l^i \tilde{g}$ and $\tilde{g} \succsim_l^j \tilde{h}$, then there exists such $T_k \in \mathcal{T}$ that $\tilde{f} \succsim_l^k \tilde{h}$. Transitivity of LPR is defined on the base of the extension principle and fuzzy preference relation [18]. This axiom states that any two alternatives are comparable and assumes one of the fundamental properties of preferences (transitivity) for the case of fuzzy information;

(ii) ***Comonotonic Independence*:** For all pairwise comonotonic acts $\tilde{f}, \tilde{g}$ and $\tilde{h}$ in $\mathcal{A}$ if $\tilde{f} \succsim_l \tilde{g}$, then $\sigma\tilde{f} + (1-\sigma)\tilde{h} \succsim_l \sigma\tilde{g} + (1-\sigma)\tilde{h}$ for all $\sigma \in (0,1)$. This means that if there exist such $T_i \in \mathcal{T}$ that $\tilde{f} \succsim_l^i \tilde{g}$ then there exists such $T_k \in \mathcal{T}$ that $\sigma\tilde{f} + (1-\sigma)\tilde{h} \succsim_l^k \sigma\tilde{g} + (1-\sigma)\tilde{h}$, with $\tilde{f}, \tilde{g}$ and $\tilde{h}$ pairwise comonotonic. The axiom extends the independency property for comonotonic actions for the case of fuzzy information;

(iii) ***Continuity***: For all $\tilde{f},\tilde{g}$ and $\tilde{h}$ in $\mathcal{A}$: if $\tilde{f} \succ_l \tilde{g}$ and $\tilde{g} \succ_l \tilde{h}$ then there are σ and β in $(0,1)$ such that $\sigma\tilde{f}+(1-\sigma)\tilde{h} \succ_l \tilde{g} \succ_l \beta\tilde{f}+(1-\beta)\tilde{h}$. This means that if there exist such $T_i \in \mathcal{T}$ and $T_j \in \mathcal{T}$ that $\tilde{f} \succsim_l^i \tilde{g}$ and $\tilde{g} \succsim_l^j \tilde{h}$ then there exist such $T_k \in \mathcal{T}$ and $T_m \in \mathcal{T}$ that define preference of $\sigma\tilde{f}+$ $+(1-\sigma)\tilde{h} \succsim_l^k \tilde{g} \succsim_l^m \beta\tilde{g} +(1-\beta)\tilde{h}$;

(iv) ***Monotonicity***: For all $\tilde{f}$ and $\tilde{g}$ in $\mathcal{A}$: If $\tilde{f}(\tilde{S}) \succsim_l \tilde{g}(\tilde{S})$ on $\mathcal{S}$ then $\tilde{f} \succsim_l \tilde{g}$. This means that if for any $\tilde{S} \in \mathcal{S}$ there exists such $T \in \mathcal{T}$ that $\tilde{f}(\tilde{S}) \succsim_l \tilde{g}(\tilde{S})$, then there exists $T_i \in \mathcal{T}$ such that $\tilde{f} \succsim_l^i \tilde{g}$;

(v) ***Nondegeneracy***: Not for all $\tilde{f},\tilde{g} \in \mathcal{A}$, $\tilde{f} \succsim_l \tilde{g}$.

LPR $\succsim_l$ on $\mathcal{A}$ induces LPR denoted also by $\succsim_l$ on $\mathcal{Y}$: $\tilde{y} \succsim_l \tilde{z}$ iff $\tilde{y}^{\mathcal{S}} \succsim_l \tilde{z}^{\mathcal{S}}$, where $\tilde{y}^{\mathcal{S}}$ and $\tilde{z}^{\mathcal{S}}$ denotes the constant functions $\tilde{y}$ and $\tilde{z}$ on $\mathcal{S}$.

The presented axioms are formulated to reflect human preferences under a mixture of fuzzy and probabilistic information. Such formulation requires the use of a fuzzy-valued utility function. Formally, it is required to use a fuzzy-valued utility function $\tilde{U}$ such that

$$\forall \tilde{f},\tilde{g} \in \mathcal{A}, \tilde{f} \succsim_l \tilde{g} \Leftrightarrow \tilde{U}(\tilde{f}) \ge \tilde{U}(\tilde{g})$$

The problem of decision making with imperfect information is formalized as a 4-tuple $D_{DMII} = (\mathcal{S},\mathcal{Y},\mathcal{A},\succsim_l)$ and consists in determination of an optimal $\tilde{f}^* \in \mathcal{A}$, that is, $\tilde{f}^* \in \mathcal{A}$ for which $\tilde{U}(\tilde{f}^*) = \max_{\tilde{f}\in\mathcal{A}} \tilde{U}(\tilde{f})$.

Fuzzy utility function $\tilde{U}$ we adopt will be described as a fuzzy number-valued Choquet integral with respect to a fuzzy number-valued fuzzy measure. In its turn fuzzy number-valued fuzzy measure can be obtained from NL-described knowledge about probability distribution over $\mathcal{S}$. NL-described knowledge about probability distribution over $\mathcal{S}$ is expressed as $\tilde{P}^l = \tilde{P}_1/\tilde{S}_1 + \tilde{P}_2/\tilde{S}_2 + \tilde{P}_3/\tilde{S}_3 =$ *small/small + high/medium + small/large*, with the understanding that a term such as *high/medium* means that the probability, that $\tilde{S}_2 \in \mathcal{S}$ is medium, is high. So, $\tilde{P}^l$ is a *linguistic (fuzzy) probability distribution.*

Fuzzy Utility Function

In the discussions above, we have mentioned the necessity of the use of a fuzzy utility function as a suitable quantifying representation of vague preferences. Below we present a definition of a fuzzy number-valued utility function representing LPR (i)-(v) over an arbitrary set $\mathcal{Z}$ of alternatives.

Definition 4.6. Fuzzy Utility Function [7,10,15,16]. Fuzzy number-valued function $\tilde{U}(\cdot):\mathcal{Z}\to\mathcal{E}^1$ is a fuzzy utility function if it represents linguistic preferences $\succsim_l$ such that for any pair of alternatives $\tilde{Z}_1,\tilde{Z}_2\in\mathcal{Z}$, $\tilde{Z}_1 \succsim_l^i \tilde{Z}_2$ holds if and only if $\tilde{U}(\tilde{Z}_1)\geq\tilde{U}(\tilde{Z}_2)$, where T_i is determined on the base of $\tilde{d}_{fH}\left(\tilde{U}(\tilde{Z}_1),\tilde{U}(\tilde{Z}_2)\right)$.

Here we consider a set $\mathcal{Z}$ of alternatives as if they are a set $\mathcal{A}$ of actions $\tilde{f}:\mathcal{S}\to\mathcal{Y}$.

Below we present representation theorems showing the existence of a fuzzy number-valued Choquet-integral-based fuzzy utility function [6,7,15] that represents LPR defined over the set $\mathcal{A}$ of alternatives.

Theorem 4.1 [15,16]. *Assume that LPR $\succsim_l$ on $\mathcal{A}=\mathcal{A}_0$ satisfies (i) weak order, (ii) continuity, (iii) comonotonic independence, (iv) monotonicity, and (v) nondegeneracy. Then there exists a unique fuzzy number-valued fuzzy measure $\tilde{\eta}$ on $\tilde{\mathcal{F}}_S$ and an affine fuzzy number-valued function $\tilde{u}$ on $\mathcal{Y}$ such that for all $\tilde{f}$ and $\tilde{g}$ in $\mathcal{A}$:*

$$\tilde{f}\succsim_l\tilde{g} \text{ iff } \int_S \tilde{u}(\tilde{f}(\tilde{S}))d\tilde{\eta}\geq\int_S \tilde{u}(\tilde{g}(\tilde{S}))d\tilde{\eta}$$

where $\tilde{u}$ is unique up to positive linear transformations.

Theorem 4.2 [15,16]. *For a nonconstant affine fuzzy number-valued function $\tilde{u}$ on $\mathcal{Y}$ and a fuzzy number-valued fuzzy measure $\tilde{\eta}$ on $\tilde{\mathcal{F}}_S$ a fuzzy number-valued Choquet integral $\tilde{U}(\tilde{f})=\int_S \tilde{u}(\tilde{f}(\tilde{S}))d\tilde{\eta}$ induces such LPR on $\mathcal{A}_0$ that satisfies conditions (i)-(v). Additionally, $\tilde{u}$ is unique up to positive linear transformations.*

The direct theorem (Theorem 4.1) provides conditions for existence of the suggested fuzzy utility function representing LPR defined over a set of fuzzy actions under conditions of fuzzy probabilities. LPR formulated by using a series of axioms reflects the essence of human-like preferences under conditions of imperfect information. The converse theorem (Theorem 4.2) provides conditions under which a fuzzy utility function described as a fuzzy number-valued Choquet integral with a fuzzy number-valued integrand and a fuzzy number-valued fuzzy measure induces the formulated LPR.

In order to prove these theorems we need to use a series of mathematical results [15,16] we present below. These results are obtained for a fuzzy-valued Choquet integral [79] of a fuzzy number-valued function with respect to a fuzzy number-valued fuzzy measure. The general expression of the considered fuzzy-valued Choquet integral is

$$\tilde{I}(\tilde{a}) = \int_{\mathcal{S}} \tilde{a} d\tilde{\eta} = \int_0^{\infty} \tilde{\eta}(\{\tilde{S} \in \mathcal{S} | \tilde{a}(\tilde{S}) \geq \tilde{\delta}\}) d\tilde{\delta}, \tag{4.1}$$

where $\tilde{a}: \mathcal{S} \to \mathcal{E}^1$. In α-cuts we will have

$$\left[\tilde{I}(\tilde{a})\right]^{\alpha} = \left[I_1^{\alpha}(a_1^{\alpha}), I_2^{\alpha}(a_2^{\alpha})\right], \text{ where}$$

$$I_1^{\alpha}(a_1^{\alpha}) = \int_{\mathcal{S}} a_1^{\alpha} d\eta_1^{\alpha} = \int_0^{\infty} \eta_1^{\alpha}(\{\tilde{S} \in \mathcal{S} | a_1^{\alpha}(\tilde{S}) \geq \delta_1^{\alpha}\}) d\delta_1^{\alpha},$$

$$I_2^{\alpha}(a_2^{\alpha}) = \int_{\mathcal{S}} a_2^{\alpha} d\eta_2^{\alpha} = \int_0^{\infty} \eta_2^{\alpha}(\{\tilde{S} \in \mathcal{S} | a_2^{\alpha}(\tilde{S}) \geq \delta_2^{\alpha}\}) d\delta_2^{\alpha}.$$

Here $I_1^{\alpha}, I_2^{\alpha}: \mathcal{C} \to \mathcal{R}$ are monotonic and homogenous functions, where $\mathcal{C}$ is the space of bounded, $\tilde{\mathcal{F}}_{\mathcal{S}}$-measurable, and real-valued functions on $\mathcal{S}$. $\left[\tilde{I}(\tilde{a})\right]^{\alpha} = \left[I_1^{\alpha}(a_1^{\alpha}), I_2^{\alpha}(a_2^{\alpha})\right]$ is an α-cut of a fuzzy number because $I_1^{\alpha}(a_1^{\alpha}) \leq I_1^{\bar{\alpha}}(a_1^{\bar{\alpha}})$, $I_2^{\alpha}(a_2^{\alpha}) \geq I_2^{\bar{\alpha}}(a_2^{\bar{\alpha}})$ for $\alpha \leq \bar{\alpha}$ and $I_1^{\alpha}(a_1^{\alpha}) \leq I_2^{\alpha}(a_2^{\alpha})$ due to monotonicity property of $I_1^{\alpha}, I_2^{\alpha}: \mathcal{C} \to \mathcal{R}$ and the fact that $\eta_1^{\alpha}(a_1^{\alpha}) \leq \eta_2^{\alpha}(a_2^{\alpha})$.

Denote $\mathcal{S}^*$ the indicator function of $\mathcal{S}$. Consider the following result related to a fuzzy number-valued fuzzy measure $\tilde{\eta}$.

Theorem 4.3. *Let $\tilde{I}: \mathcal{B} \to \mathcal{E}^1$, where $\mathcal{B}$ is the space of bounded, $\tilde{\mathcal{F}}_{\mathcal{S}}$-measurable, fuzzy number-valued functions on $\mathcal{S}$, satisfying $\tilde{I}(\mathcal{S}^*) = 1$, be given. Assume also that the functional $\tilde{I}$ satisfies:*

(i) Comonotonic additivity: for comonotonic $\tilde{a}, \tilde{b} \in \mathcal{B}$, $\tilde{I}(\tilde{a} + \tilde{b}) = \tilde{I}(\tilde{a}) + \tilde{I}(\tilde{b})$ holds;

(ii) Monotonicity: if $\tilde{a}(\tilde{S}) \geq \tilde{b}(\tilde{S})$ for all $\tilde{S} \in \mathcal{S}$ then $\tilde{I}(\tilde{a}) \geq \tilde{I}(\tilde{b})$.

Under these conditions, defining $\tilde{\eta}(\mathcal{H}) = \tilde{I}(\mathcal{H}^*)$ for all $\mathcal{H} \in \tilde{\mathcal{F}}_{\mathcal{S}}$, where $\mathcal{H}^*$ denotes the indicator function of $\mathcal{H}$, we have

$$\tilde{I}(\tilde{a}) = \int_0^{\infty} \tilde{\eta}(\tilde{a} \geq \tilde{\delta}) d\tilde{\delta} + \int_{-\infty}^{0} \left(\tilde{\eta}(\tilde{a} \geq \tilde{\delta}) - 1\right) d\tilde{\delta}, \ \forall \tilde{a} \in \mathcal{B}, \tag{4.2}$$

such that

$$I_1^\alpha(a_1^\alpha) = \int_0^\infty \eta_1^\alpha(a_1^\alpha \geq \delta_1^\alpha) d\delta_1^\alpha + \int_{-\infty}^0 \left(\eta_1^\alpha(a_1^\alpha \geq \delta_1^\alpha) - 1\right) d\delta_1^\alpha$$

$$I_2^\alpha(a_2^\alpha) = \int_0^\infty \eta_2^\alpha(a_2^\alpha \geq \delta_2^\alpha) d\delta_2^\alpha + \int_{-\infty}^0 \left(\eta_2^\alpha(a_2^\alpha \geq \delta_2^\alpha) - 1\right) d\delta_2^\alpha .$$

Note that comonotonically additive and monotonic $\tilde{I}$ on $\mathcal{B}$ satisfies $\tilde{I}(\lambda\tilde{a}) = \lambda\tilde{I}(\tilde{a})$ for $\lambda > 0$. Indeed, α-cut of $\tilde{I}(\lambda\tilde{a})$ is defined as $\left[\tilde{I}(\lambda\tilde{a})\right]^\alpha = \left[I_1^\alpha([\lambda\tilde{a}]_1^\alpha), I_2^\alpha([\lambda\tilde{a}]_2^\alpha)\right]$, where $[\lambda\tilde{a}]_1^\alpha = \lambda a_1^\alpha$, $[\lambda\tilde{a}]_2^\alpha = \lambda a_2^\alpha$ because $\lambda > 0$. So, $I_1^\alpha(\lambda a_1^\alpha) = \lambda I_1^\alpha(a_1^\alpha), I_2^\alpha(\lambda a_2^\alpha) = \lambda I_2^\alpha(a_2^\alpha)$.

Thus, we will have:

$$\left[\tilde{I}(\lambda\tilde{a})\right]^\alpha = \left[I_1^\alpha(\lambda a_1^\alpha), I_2^\alpha(\lambda a_2^\alpha)\right] = \left[\lambda I_1^\alpha(a_1^\alpha), \lambda I_2^\alpha(a_2^\alpha)\right] =$$
$$= \lambda\left[I_1^\alpha(a_1^\alpha), I_2^\alpha(a_2^\alpha)\right] = \lambda\left[\tilde{I}(\tilde{a})\right]^\alpha$$

So, $\tilde{I}(\lambda\tilde{a}) = \lambda\tilde{I}(\tilde{a}), \lambda > 0$.

In order to prove Theorem 4.3 we need to use the following remark.

Remark 4.1. The integrand in (4.2) can be compactly expressed as follows:

$$\tilde{a}^*(\tilde{\delta}) = \begin{cases} \tilde{\eta}(\tilde{a} \geq \tilde{\delta}), \tilde{\delta} \geq 0 \\ \tilde{\eta}(\tilde{a} \geq \tilde{\delta}) - \tilde{\eta}(S), \tilde{\delta} < 0, \end{cases}$$

in sense that

$$a^{*\alpha}_1(\tilde{\delta}) = a^{*\alpha}_1(\delta_1^\alpha) = \begin{cases} \eta_1^\alpha(a_1^\alpha \geq \delta_1^\alpha), \delta_1^\alpha \geq 0 \\ \eta_1^\alpha(a_1^\alpha \geq \delta_1^\alpha) - \eta_1^\alpha(S), \delta_1^\alpha < 0, \end{cases}$$

$$a^{*\alpha}_2(\tilde{\delta}) = a^{*\alpha}_2(\delta_2^\alpha) = \begin{cases} \eta_2^\alpha(a_2^\alpha \geq \delta_2^\alpha), \delta_2^\alpha \geq 0 \\ \eta_2^\alpha(a_2^\alpha \geq \delta_2^\alpha) - \eta_2^\alpha(S), \delta_2^\alpha < 0. \end{cases}$$

If a_1^α, a_2^α are nonnegative, then $a^{*\alpha}_1, a^{*\alpha}_2 = 0$ for $\delta_1^\alpha, \delta_2^\alpha < 0$. If θ_1^α, θ_2^α are negative lower bounds of a_1^α, a_2^α respectively then $a^{*\alpha}_1(\delta_1^\alpha) = 0$ and $a^{*\alpha}_2(\delta_2^\alpha) = 0$ for $\delta_1^\alpha \leq \theta_1^\alpha$ and $\delta_2^\alpha \leq \theta_2^\alpha$ respectively. If $\vartheta_1^\alpha, \vartheta_2^\alpha$ are upper bound of a_1^α, a_2^α respectively, then α-cuts of (4.2) are equivalent to

$$I_1^\alpha(a_1^\alpha) = \int_{\theta_1^\alpha}^{\vartheta_1^\alpha} a^{*\alpha}_1(\delta_1^\alpha)d\delta_1^\alpha ,$$

$$I_2^\alpha(a_2^\alpha) = \int_{\theta_2^\alpha}^{\vartheta_2^\alpha} a^{*\alpha}_2(\delta_2^\alpha)d\delta_2^\alpha .$$

So, (4.2) is equivalent to $\tilde{I}(\tilde{a}) = \int_{\tilde{\theta}}^{\tilde{\vartheta}} \tilde{a}^*(\tilde{\delta})d\tilde{\delta}$ (the fact that $\vartheta_1^\alpha, \vartheta_2^\alpha$ and θ_1^α, θ_2^α can be considered as endpoints of α-cut of a fuzzy number is obvious). As $I_1^\alpha, I_2^\alpha : \mathcal{C} \to \mathcal{R}$ are comonotonically additive and monotonic, then, based on the results in [68], we can claim that α-cuts of (4.2) are implied by α-cuts of (4.1), and hence, (4.2) is implied by (4.1).

Proof of Theorem 4.3. *Remark 4.1 allows for proof of (4.1) for nonnegative fuzzy number-valued functions only. Assuming that (4.1) holds for any fuzzy finite step function, we will prove it for an arbitrary nonnegative* $\tilde{\mathcal{F}}_S$*-measurable fuzzy number-valued function [79]* $\tilde{a}$ *bounded by some* $\tilde{\lambda} \in \mathcal{E}^1$ *(that is,* $0 \le \tilde{a}(\tilde{S}) \le \tilde{\lambda}$*, hold for all* $\tilde{S} \in \mathcal{S}$ *). For* $n = 1,2,...$ *and* $1 \le k \le 2^n$ *we define*

$$\mathcal{H}_n^{k,1,\alpha} = \left\{ \tilde{S} \in \mathcal{S} \middle| \lambda_1^\alpha (k-1)/2^n < \left[\tilde{a}(\tilde{S})\right]_1^\alpha \le \lambda_1^\alpha k/2^n \right\} \text{ and}$$

$$\mathcal{H}_n^{k,2,\alpha} = \left\{ \tilde{S} \in \mathcal{S} \middle| \lambda_2^\alpha (k-1)/2^n < \left[\tilde{a}(\tilde{S})\right]_2^\alpha \le \lambda_2^\alpha k/2^n \right\}.$$

Define also $\left[\tilde{a}_n(\tilde{S})\right]_1^\alpha = \lambda_1^\alpha (k-1)/2^n$, $\left[\tilde{a}_n(\tilde{S})\right]_2^\alpha = \lambda_2^\alpha (k-1)/2^n$, $\left[\tilde{b}_n(\tilde{S})\right]_1^\alpha = = \lambda_1^\alpha k/2^n$, $\left[\tilde{b}_n(\tilde{S})\right]_2^\alpha = \lambda_2^\alpha k/2^n$.

Thus, for all $\tilde{S}$ and n:

$$\left[\tilde{a}_n(\tilde{S})\right]_1^\alpha \le \left[\tilde{a}_{n+1}(\tilde{S})\right]_1^\alpha \le \left[\tilde{a}(\tilde{S})\right]_1^\alpha \le \left[\tilde{b}_{n+1}(\tilde{S})\right]_1^\alpha \le \left[\tilde{b}_n(\tilde{S})\right]_1^\alpha \text{ and}$$

$$\left[\tilde{a}_n(\tilde{S})\right]_2^\alpha \le \left[\tilde{a}_{n+1}(\tilde{S})\right]_2^\alpha \le \left[\tilde{a}(\tilde{S})\right]_2^\alpha \le \left[\tilde{b}_{n+1}(\tilde{S})\right]_2^\alpha \le \left[\tilde{b}_n(\tilde{S})\right]_2^\alpha$$

hold. So, for all $\tilde{S}$ and n $\tilde{a}_n(\tilde{S}) \le \tilde{a}_{n+1}(\tilde{S}) \le \tilde{a}(\tilde{S}) \le \tilde{b}_{n+1}(\tilde{S}) \le \tilde{b}_n(\tilde{S})$ hold. Monotonicity of $I_1^\alpha, I_2^\alpha : \mathcal{C} \to \mathcal{R}$ implies $I_1^\alpha(a_{n1}^{\ \alpha}) \le I_1^\alpha(a_1^\alpha) \le I_1^\alpha(b_{n1}^{\ \alpha})$, $I_2^\alpha(a_{n2}^{\ \alpha}) \le I_2^\alpha(a_2^\alpha) \le I_2^\alpha(b_{n2}^{\ \alpha})$, in turn comonotonic additivity of $I_1^\alpha, I_2^\alpha : \mathcal{C} \to \mathcal{R}$ implies

$$0 \le I_1^\alpha(b_{n1}^\alpha) - I_1^\alpha(a_{n1}^\alpha) = \lambda_1^\alpha / 2^n \to 0 \text{ and}$$

$$0 \le I_2^\alpha(b_{n2}^\alpha) - I_2^\alpha(a_{n2}^\alpha) = \lambda_2^\alpha / 2^n \to 0, n \to \infty.$$

Based on the assumption about fuzzy finite step functions, it follows that

$$I_1^\alpha(a_{n1}^\alpha) = \int_0^{\lambda_1^\alpha} \eta_1^\alpha(a_{n1}^\alpha \ge \delta_1^\alpha) d\delta_1^\alpha,\ I_2^\alpha(a_{n2}^\alpha) = \int_0^{\lambda_2^\alpha} \eta_2^\alpha(a_{n2}^\alpha \ge \delta_2^\alpha) d\delta_2^\alpha, \text{ and}$$

$$I_1^\alpha(b_{n1}^\alpha) = \int_0^{\lambda_1^\alpha} \eta_1^\alpha(b_{n1}^\alpha \ge \delta_1^\alpha) d\delta_1^\alpha, I_2^\alpha(b_{n2}^\alpha) = \int_0^{\lambda_2^\alpha} \eta_2^\alpha(b_{n2}^\alpha \ge \delta_2^\alpha) d\delta_2^\alpha.$$

The monotonicity of $\eta_1^\alpha, \eta_2^\alpha$ and the definitions of $a_{n1}^\alpha, b_{n1}^\alpha, a_{n2}^\alpha, b_{n2}^\alpha$, $n = 1, 2, \dots$, imply $\eta_1^\alpha(a_{n1}^\alpha \ge \delta_1^\alpha) \le \eta_1^\alpha(a_1^\alpha \ge \delta_1^\alpha) \le \eta_1^\alpha(b_{n1}^\alpha \ge \delta_1^\alpha)$, $\eta_2^\alpha(a_{n2}^\alpha \ge \delta_2^\alpha) \le$ $\le \eta_2^\alpha(a_2^\alpha \ge \delta_2^\alpha) \le \eta_2^\alpha(b_{n2}^\alpha \ge \delta_2^\alpha)$. From these inequalities it follows that

$$\int_0^{\lambda_1^\alpha} \eta_1^\alpha(a_{n1}^\alpha \ge \delta_1^\alpha) d\delta_1^\alpha \le \int_0^{\lambda_1^\alpha} \eta_1^\alpha(a_1^\alpha \ge \delta_1^\alpha) d\delta_1^\alpha \le \int_0^{\lambda_1^\alpha} \eta_1^\alpha(b_{n1}^\alpha \ge \delta_1^\alpha) d\delta_1^\alpha \text{ a}$$

$$\int_0^{\lambda_2^\alpha} \eta_2^\alpha(a_{n2}^\alpha \ge \delta_2^\alpha) d\delta_2^\alpha \le \int_0^{\lambda_2^\alpha} \eta_2^\alpha(a_2^\alpha \ge \delta_2^\alpha) d\delta_2^\alpha \le \int_0^{\lambda_2^\alpha} \eta_2^\alpha(b_{n2}^\alpha \ge \delta_2^\alpha) d\delta_2^\alpha$$

hold.

So, $I_1^\alpha(a_1^\alpha) = \int_0^{\lambda_1^\alpha} \eta_1^\alpha(a_1^\alpha \ge \delta_1^\alpha) d\delta_1^\alpha$, $I_2^\alpha(a_2^\alpha) = \int_0^{\lambda_2^\alpha} \eta_2^\alpha(a_2^\alpha \ge \delta_2^\alpha) d\delta_2^\alpha$, that is,

$$\tilde{I}(\tilde{a}) = \int_0^{\tilde{\lambda}} \tilde{\eta}(\tilde{a} \ge \tilde{\delta}) d\tilde{\delta}.$$

Let us now prove that (4.1) holds for fuzzy finite step functions. Any nonnegative fuzzy step function $\tilde{a} \in \mathcal{B}$ has a unique α-cut representation $a_1^\alpha = \sum_{i=1}^k \delta_{i1}^\alpha \mathcal{H}_i^*$, $a_2^\alpha = \sum_{i=1}^k \delta_{i2}^\alpha \mathcal{H}_i^*$ for some k, where $\delta_{11}^\alpha > \delta_{21}^\alpha > \dots > \delta_{k1}^\alpha$, $\delta_{12}^\alpha > \delta_{22}^\alpha > \dots > \delta_{k2}^\alpha$ and the sets $\mathcal{H}_i$, $i = 1, \dots, k$ are pairwise disjoint. Defining $\delta_{k+11}^\alpha = 0$, $\delta_{k+12}^\alpha = 0$ we have:

$$\int_0^{\delta_{11}^\alpha} \eta_1^\alpha(a_1^\alpha \ge \delta_1^\alpha) d\delta_1^\alpha = \sum_{i=1}^k \left(\delta_{i1}^\alpha - \delta_{i+11}^\alpha\right) \eta_1^\alpha\left(\bigcup_{j=1}^i \mathcal{H}_j\right),$$

$$\int_0^{\delta_{12}^\alpha} \eta_2^\alpha (a_2^\alpha \geq \delta_2^\alpha) d\delta_2^\alpha = \sum_{i=1}^{k} \left(\delta_{i2}^\alpha - \delta_{i+12}^{\alpha} \right) \eta_2^\alpha \left(\bigcup_{j=1}^{i} \mathcal{H}_j \right).$$

So,

$$\int_0^{\tilde{\delta}_1} \tilde{\eta}(\tilde{a} \geq \tilde{\delta}) d\tilde{\delta} = \sum_{i=1}^{k} \left(\tilde{\delta}_i -_h \tilde{\delta}_{i+1} \right) \tilde{\eta} \left(\bigcup_{j=1}^{i} \mathcal{H}_j \right), \tag{4.3}$$

Note that throughout the study we use the Hukuhara difference. The induction hypothesis implies that for $k < n$

$$\tilde{I}(\tilde{a}) = \sum_{i=1}^{k} \left(\tilde{\delta}_i -_h \tilde{\delta}_{i+1} \right) \tilde{\eta} \left(\bigcup_{j=1}^{i} \mathcal{H}_j \right) \tag{4.4}$$

We need to prove it for $k = n$. Note that for $k = 1$ $I_1^\alpha(\delta_1^\alpha \mathcal{H}^*) = \delta_1^\alpha \eta_1^\alpha(\mathcal{H})$, and $I_2^\alpha(\delta_2^\alpha \mathcal{H}^*) = \delta_2^\alpha \eta_2^\alpha(\mathcal{H})$ hold, i.e, $\tilde{I}(\tilde{\delta}\mathcal{H}^*) = \tilde{\delta}\tilde{\eta}(\mathcal{H})$ holds.

Given endpoints of α-cut of $\tilde{a}$ as $a_1^\alpha = \sum_{i=1}^{k} \delta_{i1}^\alpha \mathcal{H}_i^*$, $a_2^\alpha = \sum_{i=1}^{k} \delta_{i2}^\alpha \mathcal{H}_i^*$, $\tilde{a} = \tilde{b} + \tilde{c}$, where $b_1^\alpha = \sum_{i=1}^{k-1} (\delta_{i1}^\alpha - \delta_{i+11}^{\alpha}) \mathcal{H}_i^*$, $b_2^\alpha = \sum_{i=1}^{k-1} (\delta_{i2}^\alpha - \delta_{i+12}^{\alpha}) \mathcal{H}_i^*$, $c_1^\alpha = \delta_{k1}^{\alpha} \left(\sum_{i=1}^{k} \mathcal{H}_j^* \right)$, $c_2^\alpha = \delta_{k2}^{\alpha} \left(\sum_{i=1}^{k} \mathcal{H}_j^* \right)$. From the induction hypothesis ($k - 1 < n$),

$$\tilde{I}(\tilde{b}) = \sum_{i=1}^{k-1} \left(\left(\tilde{\delta}_i -_h \tilde{\delta}_k \right) -_h \left(\tilde{\delta}_{i+1} -_h \tilde{\delta}_k \right) \right) \tilde{\eta} \left(\bigcup_{j=1}^{i} \mathcal{H}_j \right) = \sum_{i=1}^{k-1} \left(\tilde{\delta}_i -_h \tilde{\delta}_{i+1} \right) \tilde{\eta} \left(\bigcup_{j=1}^{i} \mathcal{H}_j \right)$$

and

$$\tilde{I}(\tilde{c}) = \tilde{\delta}_k \tilde{\eta} \left(\bigcup_{j=1}^{i} \mathcal{H}_j \right).$$

Thus, $\tilde{I}(\tilde{b}) + \tilde{I}(\tilde{c}) = \sum_{i=1}^{k} \left(\tilde{\delta}_i -_h \tilde{\delta}_{i+1} \right) \tilde{\eta} \left(\bigcup_{j=1}^{i} \mathcal{H}_j \right)$. From the other side, as $\tilde{b}$ and $\tilde{c}$ are comonotonic, $\tilde{I}(\tilde{a}) = \tilde{I}(\tilde{b}) + \tilde{I}(\tilde{c})$ and (4.4) for $k = n$ has been proved. The proof is completed.

Remark 4.2. From the opposite direction of Theorem 4.3 it follows that if a fuzzy functional $\tilde{I}$ is defined by (4.2) with respect to some fuzzy number-valued fuzzy measure, then it satisfies comonotonic additivity and monotonicity. One can easily obtain the proof by reversing the proof of Theorem 4.3 as follows. For a functional $\tilde{I}$ defined by (4.2) with respect to some fuzzy number-valued fuzzy measure $\tilde{\eta}$, it is needed to prove that it is comonotonically additive and monotonic. Monotonicity of $\tilde{I}$ follows from the fact that $\tilde{a} \geq \tilde{b}$ on $\mathcal{S}$ implies $\tilde{a}^* \geq \tilde{b}^*$ on $\mathcal{E}^1$.

So, at first it is needed to show comonotonic additivity for fuzzy finite step functions in $\mathcal{B}$. To this end the following two claims are given.

Claim 4.1. Two fuzzy finite step functions $\tilde{b}, \tilde{c} \in \mathcal{B}$ are comonotonic iff there exists an integer k, a partition of $\mathcal{S}$ into k pairwise disjoint elements $(\mathcal{H}_i)_{i=1}^k$ of $\tilde{\mathcal{F}}_{\mathcal{S}}$, and two k-lists of fuzzy numbers $\tilde{\beta}_1 \geq \tilde{\beta}_2 \geq \ldots \geq \tilde{\beta}_k$ and $\tilde{\gamma}_1 \geq \tilde{\gamma}_2 \geq \ldots \geq \tilde{\gamma}_k$ such that $\tilde{b} = \sum_{i=1}^{k} \tilde{\beta}_i \mathcal{H}_i^*$ and $\tilde{c} = \sum_{i=1}^{k} \tilde{\gamma}_i \mathcal{H}_i^*$. The proof is obvious.

Claim 4.2. Let $(\mathcal{H}_i)_{i=1}^k$ be $\tilde{\mathcal{F}}_{\mathcal{S}}$-measurable finite partition of $\mathcal{S}$ (if $i \neq j$, then $\mathcal{H}_i \cap \mathcal{H}_j = \varnothing$) and let $\tilde{a} = \sum_{i=1}^{k} \tilde{\delta}_i \mathcal{H}_i^*$ with $\tilde{\alpha}_1 \geq \tilde{\alpha}_2 \geq \ldots \geq \tilde{\alpha}_k$. Then for any fuzzy number-valued fuzzy measure $\tilde{\eta} : \tilde{\mathcal{F}}_{\mathcal{S}} \to \mathcal{E}^1$ we have:

$$\int_{-\infty}^{\infty} \tilde{a}^*(\tilde{\delta}) d\tilde{\delta} = \sum_{i=1}^{k} (\tilde{\delta}_i -_h \tilde{\delta}_{i+1}) \tilde{\eta}\left(\bigcup_{j=1}^{i} \mathcal{H}_j \right) \tag{4.5}$$

with $\tilde{\delta}_{k=1} = 0$.

For $\tilde{I}(\tilde{a})$ defined by the left side of (4.5) for fuzzy finite step functions, the formula (4.5) and Claim 4.1 imply additivity for comonotonic fuzzy finite step functions. Extension of this result to any comonotonic functions in $\mathcal{B}$ is obtained by computing appropriate limits in metrics $\tilde{d}_{fH}$.

It can be easily shown that Theorem 4.3 and its converse hold if $\mathcal{B}$ is substituted by $\mathcal{B}_0$, the set of all fuzzy finite step functions in $\mathcal{B}$. Also, for comonotonically additive and monotonic $\tilde{I} : \mathcal{B}_0 \to \mathcal{E}^1$ there exists a unique extension to all of $\mathcal{B}$, which satisfies comonotonic additivity and monotonicity. To prove this it is needed to pass to α-cuts of $\tilde{I}$ and then easily apply the facts that $\mathcal{B}$ is the (sup) norm closure of $\mathcal{B}_0$ in $(\mathcal{E}^1)^{\mathcal{S}}$ in metrics $\tilde{d}_{fH}$ and that monotonicity implies norm continuity.

Now let $\mathcal{B}(\mathcal{K})$ denote the set of functions in $\mathcal{B}$ with values in $\mathcal{K}$, and suppose that $\mathcal{K} \supset \{\tilde{v} \in \mathcal{E}^1 | -\tilde{\gamma} \leq \tilde{v} \leq \tilde{\gamma}\}$, where $\tilde{\gamma} \geq 0, -\tilde{\gamma} = -1\tilde{\gamma}$.

Corollary 4.1. Let $\tilde{I} : \mathcal{B}(\mathcal{K}) \to \mathcal{E}^1$ be given such that

1) for all $\tilde{\lambda} \in \mathcal{K}$ $\tilde{I}(\tilde{\lambda}\mathcal{S}^*) = \tilde{\lambda}$
2) if $\tilde{a}, \tilde{b}$ and $\tilde{c}$ are pairwise comonotonic, and $\tilde{I}(\tilde{a}) > \tilde{I}(\tilde{b})$, then $\tilde{I}(\sigma\tilde{a} + (1-\sigma)\tilde{c}) > \tilde{I}(\sigma\tilde{b} + (1-\sigma)\tilde{c})$, $\sigma \in (0,1)$,
3) if $\tilde{a} \geq \tilde{b}$ on $\mathcal{S}$, then $\tilde{I}(\tilde{a}) \geq \tilde{I}(\tilde{b})$.

Then, defining $\tilde{\eta}(\mathcal{H}) = \tilde{I}(\mathcal{H}^*)$ on $\tilde{\mathcal{F}}_{\mathcal{S}}$ we will have for all $\tilde{a} \in \mathcal{B}(\mathcal{K})$:

$$\tilde{I}(\tilde{a}) = \int_0^{\infty} \tilde{\eta}(\tilde{a} \geq \tilde{\delta}) d\tilde{\delta} + \int_{-\infty}^{0} \left(\tilde{\eta}(\tilde{a} \geq \tilde{\delta}) - 1\right) d\tilde{\delta}.$$

The proof consists in extending $\tilde{I}$ on $\mathcal{B}(\mathcal{K})$ to $\tilde{I}$ on $\mathcal{B}$ and showing that conditions of Theorem 4.3 are satisfied. As $\tilde{I}$ is homogeneous on $\mathcal{B}(\mathcal{K})$ it can be uniquely extended to a homogeneous function on $\mathcal{B}$. Next, by homogeneity, the extended functional $\tilde{I}$ satisfies monotonicity on $\mathcal{B}$. Comonotonic additivity of $\tilde{I}$ on $\mathcal{B}$ follows from the following Lemma and homogeneity property.

Lemma. *Given the conditions of the Corollary, let* $\tilde{a}$ *and* $\tilde{b}$ *in* $\mathcal{B}(\mathcal{K})$ *be comonotonic such that* $\tilde{d}_{fH}(\tilde{a}(\tilde{S}),0) \geq -1+\varepsilon$, $\tilde{d}_{fH}(\tilde{b}(\tilde{S}),0) \leq 1-\varepsilon$ *for some* $\varepsilon > 0$ *and let* $0 < \lambda < 1$. *Then* $\tilde{I}(\lambda\tilde{a} + (1-\lambda)\tilde{b}) = \lambda\tilde{I}(\tilde{a}) + (1-\lambda)\tilde{I}(\tilde{b})$.

Proof. Denote $\tilde{I}(\tilde{a}) = \tilde{\sigma}$ and $\tilde{I}(\tilde{b}) = \tilde{\beta}$. By the condition of the Lemma, and by (i) and (iii) of the Corollary it is true that $\tilde{\sigma}\mathcal{S}^*, \tilde{\beta}\mathcal{S}^* \in \mathcal{B}(\mathcal{K})$, $\tilde{I}(\tilde{\sigma}\mathcal{S}^*) = \tilde{\sigma}$ and $\tilde{I}(\tilde{\beta}\mathcal{S}^*) = \tilde{\beta}$.

We need to prove that $\tilde{I}(\lambda\tilde{a} + (1-\lambda)\tilde{b}) = \lambda\tilde{I}(\tilde{a}) + (1-\lambda)\tilde{I}(\tilde{b})$. Suppose that $\tilde{I}(\lambda\tilde{a} + (1-\lambda)\tilde{b}) > \lambda\tilde{I}(\tilde{a}) + (1-\lambda)\tilde{I}(\tilde{b})$ (the case of the other inequality is treated in a similar manner).

Let $0 < \xi < \varepsilon$. Then by (i) $\tilde{I}(\tilde{\sigma}) < \tilde{I}((\tilde{\sigma}+\xi)\mathcal{S}^*)$, $\tilde{I}(\tilde{b}) < \tilde{I}((\tilde{\beta}+\xi)\mathcal{S}^*)$. Now

$$\lambda\tilde{\sigma} + (1-\lambda)\tilde{\beta} + \xi = \tilde{I}(\lambda(\tilde{\sigma}+\xi)\mathcal{S}^* + (1-\lambda)(\tilde{\beta}+\xi)\mathcal{S}^*) >$$
$$> \tilde{I}(\lambda\tilde{a} + (1-\lambda)(\tilde{\beta}+\xi)\mathcal{S}^*) > \tilde{I}(\lambda\tilde{a} + (1-\lambda)\tilde{b}).$$

The equality follows from (i) and each of the two inequalities follows from (ii). The inequality above holds for any ξ $(0<\xi<\varepsilon)$, so we get the required contradiction. The proof is completed.

Remark 4.3. The Corollary holds if $\mathcal{B}(\mathcal{K})$ is replaced by $\mathcal{B}_0(\mathcal{K})$ the set of bounded, $\tilde{\mathcal{F}}_S$-measurable, fuzzy finite step functions on $\mathcal{S}$ with values in $\mathcal{K}$. The same is true for the Lemma.

Given the above mentioned auxiliary results on a fuzzy number-valued Choquet integral [79] of a fuzzy number-valued function with respect to a fuzzy number-valued fuzzy measure $\tilde{\eta}$ we can prove the theorems 4.1 and 4.2. Let us proceed to the proof of theorem 4.1

Proof of Theorem 4.1

Step 1. At this step we show the existence of an affine fuzzy-number-valued function defined over $\mathcal{Y}$.

Affinity of $\tilde{u}$ implies $\tilde{u}(\tilde{y})=\sum_{\tilde{X}\in\mathcal{X}}\tilde{y}(\tilde{X})\tilde{u}(\tilde{X})$ defined as follows:

$$\mu_{\tilde{u}(\tilde{y})}(u(\tilde{y}))=\sup\min_{\substack{u(\tilde{y})=\sum_{\tilde{X}\in\mathcal{X}}y(\tilde{X})u(\tilde{X})\\ \sum_{\tilde{X}\in\mathcal{X}}y(\tilde{X})=1}}(\mu_{\tilde{u}(\tilde{X})}(u(\tilde{X})),\mu_{\tilde{y}(\tilde{X})}(y(\tilde{X})))$$

Positive linear transformation $\tilde{u}'$ of $\tilde{u}$ implies $\tilde{u}'=\tilde{A}\tilde{u}(\tilde{y})+\tilde{B},\ \ \tilde{A}\in\mathcal{E}^1_+,\tilde{B}\in\mathcal{E}^1$, where addition and multiplication is defined on the base of Zadeh's extension principle.

Using the implications from von Neumann-Morgenstern theorem, we suppose that there exists a fuzzy-number-valued function $\tilde{u}$ representing LPR $\succsim_l$ induced on $\mathcal{Y}$. Now, from nondegeneracy axiom it follows that there exist such $\tilde{f}^*$ and $\tilde{f}_*$ in $\mathcal{A}_0$ that $\tilde{f}^*\succ_l\tilde{f}_*$. From monotonicity axiom it follows existence of a state $\tilde{S}$ in $\mathcal{S}$ such that $\tilde{f}^*(\tilde{S})\equiv\tilde{y}^*\succ_l\tilde{f}_*(\tilde{S})\equiv\tilde{y}_*$. Since $\tilde{u}$ is given up to a positive linear transformation, suppose that $\tilde{u}(\tilde{y}_*)=-\tilde{v}$ and $\tilde{u}(\tilde{y}^*)=\tilde{v}$, $\tilde{v}\in\mathcal{E}^1_+$. We denote $\mathcal{K}=\tilde{u}(\mathcal{Y})$ which is a convex subset [56] of $\mathcal{E}^1$ with $-\tilde{v},\tilde{v}\in\mathcal{K}$.

Step 2. At this step we show the existence of an affine fuzzy-number-valued function defined over $\mathcal{A}_0$.

Denote by $\mathcal{M}_{\tilde{f}}=\{\sigma\tilde{f}+(1-\sigma)\tilde{y}^S \mid \tilde{y}\in\mathcal{Y}\ and\ \sigma\in[0,1]\}$ for an arbitrary $\tilde{f}\in\mathcal{A}_0$. It is clear that $\mathcal{M}_{\tilde{f}}$ is convex and any two acts in $\mathcal{M}_{\tilde{f}}$ are comonotonic. So we can claim that there exists an affine fuzzy number-valued function over $\mathcal{M}_{\tilde{f}}$ representing the corresponding LPR $\succsim_l$. By using positive linear transformation we can define for this function denoted $\tilde{J}_{\tilde{f}}$: $\tilde{J}_{\tilde{f}}(\tilde{y}_*^S) = -\tilde{v}$ and $\tilde{J}_{\tilde{f}}(\tilde{y}^{*S}) = \tilde{v}$. For any $\tilde{h} \in \mathcal{M}_{\tilde{f}} \cap \mathcal{M}_{\tilde{g}}$, $\tilde{J}_{\tilde{f}}(\tilde{h}) = \tilde{J}_{\tilde{g}}(\tilde{h})$ holds. This allows to define fuzzy number-valued function $\tilde{J}(\tilde{f}) = \tilde{J}_{\tilde{f}}(\tilde{f})$ on $\mathcal{A}_0$, which represents the LPR $\succsim_l$ on $\mathcal{A}_0$ and satisfies for all $\tilde{y}$ in $\mathcal{Y}$: $\tilde{J}(\tilde{y}^S) = \tilde{u}(\tilde{y})$.

Step 3. At this step we show the existence of a fuzzy-number-valued functional [75] defined on the base of $\tilde{u}$ and $\tilde{J}_{\tilde{f}}$.

Denote by $\mathcal{B}_0(\mathcal{K})$ the $\tilde{\mathcal{F}}_S$-measurable, $\mathcal{K}$-valued fuzzy finite step functions on $\mathcal{S}$. By means of $\tilde{u}$ let us define onto function $\tilde{\Phi}: \mathcal{A}_0 \to \mathcal{B}_0(\mathcal{K})$ as $\tilde{\Phi}(\tilde{f})(\tilde{S}) = \tilde{u}(\tilde{f}(\tilde{S}))$, $\tilde{S} \in \mathcal{S}, \tilde{f} \in \mathcal{A}_0$. If $\tilde{\Phi}(\tilde{f}) = \tilde{\Phi}(\tilde{g})$ then $\tilde{f} \sim_l \tilde{g}$ (it follows from monotonicity). So, $\tilde{\Phi}(\tilde{f}) = \tilde{\Phi}(\tilde{g})$ implies $\tilde{J}(\tilde{f}) = \tilde{J}(\tilde{g})$.

Define a fuzzy number-valued function $\tilde{I}$ on $\mathcal{B}_0(\mathcal{K})$ as follows: $\tilde{I}(\tilde{a}) = \tilde{J}(\tilde{f})$ for $\tilde{a} \in \mathcal{B}_0(\mathcal{K})$, where $\tilde{f} \in \mathcal{A}_0$ is such that $\tilde{\Phi}(\tilde{f}) = \tilde{a}$. $\tilde{I}$ is well defined as $\tilde{J}$ is constant fuzzy number-valued function (that is, $\exists \tilde{v} \in \mathcal{E}^1$, $\tilde{J}(\tilde{f}) = \tilde{v}, \forall \tilde{f} \in \mathcal{A}$) on $\tilde{\Phi}^{-1}(\tilde{a})$.

Fuzzy number-valued function $\tilde{I}$ satisfies the following conditions:

(i) for all $\tilde{\sigma}$ in $\mathcal{K}$: $\tilde{I}(\tilde{\sigma}\mathcal{S}^*) = \tilde{\sigma}$. Indeed, let $\tilde{y} \in \mathcal{Y}$ be such that $\tilde{u}(\tilde{y}) = \tilde{\sigma}$, hence $\tilde{J}(\tilde{y}^S) = \tilde{\sigma}$ and $\tilde{\Phi}(\tilde{y}^S) = \tilde{\sigma}\mathcal{S}^*$ implying $\tilde{I}(\tilde{\sigma}\mathcal{S}^*) = \tilde{\sigma}$;

(ii) for all pairwise comonotonic functions $\tilde{a}$, $\tilde{b}$ and $\tilde{c}$ in $\mathcal{B}_0(\mathcal{K})$ and $\sigma \in [0,1]$: if $\tilde{I}(\tilde{a}) > \tilde{I}(\tilde{b})$, then $\tilde{I}(\sigma\tilde{a} + (1-\sigma)\tilde{c}) > \tilde{I}(\sigma\tilde{b} + (1-\sigma)\tilde{c})$. This is true because $\tilde{\Phi}$ preserves comonotonicity;

(iii) if $\tilde{a}(\tilde{S}) \geq \tilde{b}(\tilde{S})$ on $\mathcal{S}$ for $\tilde{a}$ and $\tilde{b}$ in $\mathcal{B}_0(\mathcal{K})$ then $\tilde{I}(\tilde{a}) \geq \tilde{I}(\tilde{b})$. This is true because $\tilde{\Phi}$ preserves monotonicity.

Step 4. This step completes the proof of the Theorem 4.1.

From the Corollary and Remark 4.4 for a fuzzy number-valued function on $\mathcal{B}_0(\mathcal{K})$, which satisfies conditions (i), (ii), and (iii) above, it follows that the fuzzy number-valued fuzzy measure $\tilde{\eta}$ on $\tilde{\mathcal{F}}_S$ defined by $\tilde{\eta}(\mathcal{H}) = \tilde{I}(\mathcal{H}^*)$ satisfies

$$\tilde{I}(\tilde{a}) \geq \tilde{I}(\tilde{b}) \text{ iff } \int_{\mathcal{S}} \tilde{a}\, d\tilde{\eta} \geq \int_{\mathcal{S}} \tilde{b}\, d\tilde{\eta},\ \forall \tilde{a}, \tilde{b} \in \mathcal{B}_0(\mathcal{K}) \tag{4.6}$$

Hence, for all $\tilde{f}$ and $\tilde{g}$ in $\mathcal{A}_0$:

$$\tilde{f} \succsim_l \tilde{g} \text{ iff } \int_S \tilde{\Phi}(\tilde{f})d\tilde{\eta} \geq \int_S \tilde{\Phi}(\tilde{g})d\tilde{\eta}.$$

The proof is completed.

Proof of Theorem 4.2

Step 1. At this step we show that LPR, which is induced by $\tilde{u}$ and $\tilde{\eta}$ on $\mathcal{A}_0$, satisfies axioms (i)-(v).

To prove this theorem we use Remarks 4.1-4.3, Theorem 4.3 and other results given above, which show that $\tilde{I}$ on $\mathcal{B}_0(\mathcal{K})$ defined by (4.6) satisfies conditions (i)-(iii). Secondly, we can see that $\tilde{J}$ is defined as a combination of $\tilde{\Phi}$ and $\tilde{I}$. Thus, the LPR on $\mathcal{A}_0$ induced by $\tilde{J}$ satisfies all the required conditions because $\tilde{\Phi}$ preserves monotonicity and comonotonicity and $\int_S \tilde{a}d\tilde{\eta}$ is a (sup) norm continuous function on $\tilde{a}$ in metrics $\tilde{d}_{fH}$ (this is based on the analogous property of endpoints of α-cuts of $\int_S \tilde{a}d\tilde{\eta}$ that are classical functionals of the type considered in [68]).

Step 2. At this step we show the uniqueness of the fuzzy utility representation.

In order to prove the uniqueness property of the utility representation suppose that there exists an affine fuzzy number-valued function $\tilde{u}'$ on $\mathcal{Y}$ and a fuzzy number-valued fuzzy measure $\tilde{\eta}'$ on $\tilde{\mathcal{F}}_S$ such that for all $\tilde{f}$ and $\tilde{g}$ in $\mathcal{A}_0$:

$$\tilde{f} \succsim_l \tilde{g} \text{ iff } \int_S \tilde{u}'(\tilde{f}(\tilde{S}))d\tilde{\eta}' \geq \int_S \tilde{u}'(\tilde{g}(\tilde{S}))d\tilde{\eta}' \tag{4.7}$$

Monotonicity of $\tilde{\eta}'$ can be derived. Considering (4.7) for all $\tilde{f}, \tilde{g} \in \mathcal{A}_c$ we obtain, based on implications of von Neumann and Morgenstern theorem and Zadeh's extension principle, that $\tilde{u}'$ is a positive linear transformation of $\tilde{u}$. But (4.7) is preserved for positive linear transformation of a fuzzy utility. Hence, to prove that $\tilde{\eta}' = \tilde{\eta}$ we may assume w.l.o.g. that $\tilde{u}' = \tilde{u}$. For an arbitrary $\mathcal{H}$ in $\tilde{\mathcal{F}}_S$ let $\tilde{f}$ in $\mathcal{A}_0$ be such that $\tilde{\Phi}(\tilde{f}) = \tilde{\lambda}\mathcal{H}^*, \tilde{\lambda} \in \mathcal{E}^1$. Then $\int_S \tilde{\Phi}(\tilde{f})d\tilde{\eta} = \tilde{\lambda}\tilde{\eta}(\mathcal{H})$ and $\int_S \tilde{\Phi}(\tilde{f})d\tilde{\eta}' = \tilde{\lambda}\tilde{\eta}'(\mathcal{H})$. Let $\tilde{y}$ in $\mathcal{Y}$ be such that $\tilde{u}(\tilde{y}) = \tilde{\lambda}\tilde{\eta}(\mathcal{H})$. Then $\tilde{f} \sim_l \tilde{y}^S$, which implies $\tilde{u}(\tilde{y}) = \tilde{u}'(\tilde{y}) = \int_S \tilde{u}'(\tilde{y})d\tilde{\eta}' = \tilde{\lambda}\tilde{\eta}'(\mathcal{H})$. So, $\tilde{\lambda}\tilde{\eta}(\mathcal{H}) = \tilde{\lambda}\tilde{\eta}'(\mathcal{H})$, and therefore, $\tilde{\eta}(\mathcal{H}) = \tilde{\eta}'(\mathcal{H})$. The proof is completed.

In brief, a value of a fuzzy utility function $\tilde{U}$ for action $\tilde{f}$ is determined as a fuzzy number-valued Choquet integral [6,7,15]:

$$\tilde{U}(\tilde{f}) = \int_{\mathcal{S}} \tilde{u}(\tilde{f}(\tilde{S}))d\tilde{\eta}_{\tilde{P}^l} = \sum_{i=1}^{n} \left(\tilde{u}(\tilde{f}(\tilde{S}_{(i)})) -_h \tilde{u}(\tilde{f}(\tilde{S}_{(i+1)}))\right) \cdot \tilde{\eta}_{\tilde{P}^l}(\mathcal{H}_{(i)}) \tag{4.8}$$

Here $\tilde{\eta}_{\tilde{P}^l}()$ is a fuzzy number-valued fuzzy measure obtained from linguistic probability distribution over $\mathcal{S}$ [6,7,13,15] and $\tilde{u}(\tilde{f}(\tilde{S}))$ is a fuzzy number-valued utility function used to describe NL-based evaluations of utilities, (i) means that utilities are ranked such that $\tilde{u}(\tilde{f}(\tilde{S}_{(1)})) \geq ... \geq \tilde{u}(\tilde{f}(\tilde{S}_{(n)}))$, $\mathcal{H}_{(i)} = \{\tilde{S}_{(1)},...,\tilde{S}_{(i)}\}$, $\tilde{u}(\tilde{f}_j(\tilde{S}_{(n+1)})) = 0$, and for each (i) there exists $\tilde{u}(\tilde{f}(\tilde{S}_{(i)})) -_h \tilde{u}(\tilde{f}(\tilde{S}_{(i+1)}))$. Mutliplication $\cdot$ is realized in the sense of the Zadeh's extension principle. An optimal $\tilde{f}^* \in \mathcal{A}$, that is $\tilde{f}^* \in \mathcal{A}$ for which $\tilde{U}(\tilde{f}^*) = \max_{\tilde{f} \in \mathcal{A}} \left\{ \int_{\mathcal{S}} \tilde{u}(\tilde{f}(\tilde{S}))d\tilde{\eta}_{\tilde{P}^l} \right\}$, can be determined by using a suitable fuzzy ranking method.

Note that for a special case the suggested decision making model and utility representation reduces to the model and representation suggested by Schmeidler in [68].

Fuzzy-Valued Measure Construction from Linguistic Probabilities

The crucial problem in the determination of an overall fuzzy utility of an alternative is a construction of a fuzzy number-valued fuzzy measure $\tilde{\eta}$. We will consider $\tilde{\eta}$ as a fuzzy number-valued lower probability constructed from linguistic probability distribution $\tilde{P}^l$. Linguistic probability distribution $\tilde{P}^l$ implies that a state $\tilde{S}_i \in \mathcal{S}$ is assigned a linguistic probability $\tilde{P}_i$ that can be described by a fuzzy number defined over [0,1]. However, fuzzy probabilities $\tilde{P}_i$ cannot initially be assigned for all $\tilde{S}_i \in \mathcal{S}$ [15,16]. Initial data are represented by fuzzy probabilities for $n-1$ fuzzy states of nature whereas for one of the given fuzzy states the probability is unknown. Subsequently, it becomes necessary to determine unknown fuzzy probability $\tilde{P}(\tilde{S}_j) = \tilde{P}_j$ [10,15,16]. In the framework of Computing with Words [9,32,33,51], the problem of obtaining the unknown fuzzy probability for state $\tilde{S}_j$ given fuzzy probabilities of all other states is a problem of propagation of generalized constraints [82,83,88]. Formally this problem is formulated as [83]:

$$\text{Given } \tilde{P}\left(\tilde{S}_i\right) = \tilde{P}_i;\ \tilde{S}_i \in \mathcal{E}^n,\ \tilde{P}_i \in \mathcal{E}^1_{[0,1]},\ i = \{1,...,j-1,j+1,...,n\} \tag{4.9}$$

$$\text{find unknown } \tilde{P}\left(\tilde{S}_j\right) = \tilde{P}_j,\ \tilde{P}_j \in \mathcal{E}^1_{[0,1]} \tag{4.10}$$

It reduces to a variational problem of constructing the membership function $\mu_{\tilde{P}_j}(\cdot)$ of an unknown fuzzy probability $\tilde{P}_j$:

$$\mu_{\tilde{P}_j}(p_j) = \sup_\rho \min_{i=\{1,...,j-1,j+1,...,n\}} (\mu_{\tilde{P}_i}(\int_S \mu_{\tilde{S}_i}(s)\rho(s)ds)) \tag{4.11}$$

$$\text{subject to } \int_S \mu_{\tilde{S}_j}(s)\rho(s)ds = p_j,\ \int_S \rho(s)ds = 1 \tag{4.12}$$

here $\mu_{\tilde{S}_j}(s)$ is the membership function of a fuzzy state $\tilde{S}_j$.

When $\tilde{P}_j$ has been determined, linguistic probability distribution $\tilde{P}^l$ for all states $\tilde{S}_i$ is determined:

$$\tilde{P}^l = \tilde{P}_1/\tilde{S}_1 + \tilde{P}_2/\tilde{S}_2 + ... + \tilde{P}_n/\tilde{S}_n$$

If we have linguistic probability distribution over fuzzy values of some fuzzy set-valued random variable $\tilde{S}$, the important problem that arises is the verification of its consistency, completeness, and redundancy [1,21].

Let the set of linguistic probabilities $\tilde{P}^l = \left\{\tilde{P}_1,...,\tilde{P}_n\right\}$ correspond to the set of linguistic values $\left\{\tilde{S}_1,...,\tilde{S}_n\right\}$ of the fuzzy set-valued random variable $\tilde{S}$. For special case, a fuzzy probability distribution $\tilde{P}^l$ is inconsistent when the condition

$$p_i = \int_S \mu_{\tilde{S}_i}(s)\rho(s)ds \tag{4.13}$$

or

$$\mu_{\tilde{P}_i}\left(\int_S \mu_{\tilde{S}}(s)\rho(s)ds\right) = 1 \tag{4.14}$$

is not satisfied for any density ρ from the set of evaluations of densities.

The degree of inconsistency (denoted **contr**) of a linguistic probability distribution $\tilde{P}^l$ could be determined as

$$\operatorname{contr}\tilde{P}^l = \min_{\rho}\left[1 - \int_S \rho(s)ds\right] \tag{4.15}$$

where ρ satisfies conditions (4.13) and (4.14). Obviously, $\operatorname{contr}\tilde{P}^l = 0$ if the required density ρ exists.

Let a fuzzy probability distribution $\tilde{P}^l$ be consistent, that is $\operatorname{contr}\tilde{P}^l = 0$. If this distribution is given as a set of crisp probabilities p_i, then its incompleteness (denoted **in**) and redundancy (denoted **red**) can be expressed as

$$\operatorname{in}\tilde{P}^l = \max\left\{0, 1 - \sum_i p_i\right\} \tag{4.16}$$

$$\operatorname{red}\tilde{P}^l = \max\left\{0, \sum_i p_i - 1\right\} \tag{4.17}$$

If $\tilde{P}^l$ is given using linguistic probabilitics $\tilde{P}_i$ then its incompleteness and redundancy can be expressed as

$$\operatorname{in}\tilde{P}_S = \max\left\{0, 1 - \sup_{\gamma\in\Gamma}\gamma\right\} \tag{4.18}$$

$$\operatorname{red}\tilde{P}_S = \max\left\{0, \inf_{\gamma\in\Gamma}\gamma - 1\right\} \tag{4.19}$$

where $\Gamma = \{\gamma | \mu_\Lambda(\gamma) = 1\}$. Here Λ is a sum of linguistic probabilities $\tilde{P}_i \in \tilde{P}^l$

Given consistent, complete and not redundant linguistic probability distribution $\tilde{P}^l$ we can obtain from it a fuzzy set $\tilde{P}^\rho$ of possible probability distributions $\rho(s)$. We can construct a fuzzy measure from $\tilde{P}^\rho$ as its lower probability function (lower prevision) [58] by taking into account a degree of correspondence of $\rho(s)$ to $\tilde{P}^l$. Lower prevision is a unifying measure as opposed to the other existing additive and non-additive measures [72,73]. We denote the fuzzy-number-valued fuzzy measure by $\tilde{\eta}_{\tilde{P}^l}$ [6,7,10,15] because it is derived from the given linguistic probability distribution $\tilde{P}^l$. A degree of membership of an arbitrary probability distribution $\rho(s)$ to $\tilde{P}^\rho$ (a degree of correspondence of $\rho(s)$ to $\tilde{P}^l$) can be obtained by the formula

$$\pi_{\tilde{P}}(\rho(s)) = \min_{i=1,n}(\pi_{\tilde{P}_i}(p_i)) ,$$

where $p_i = \int_S \rho(s)\mu_{\tilde{S}_i}(s)ds$ is numeric probability of fuzzy state $\tilde{S}_i$ defined by $\rho(s)$. Furthermore, $\pi_{\tilde{P}_i}(p_i) = \mu_{\tilde{P}_i}\left(\int_S \rho(s)\mu_{\tilde{S}_i}(s)ds\right)$ is the membership degree of p_i to $\tilde{P}_i$.

To derive a fuzzy-number-valued fuzzy measure $\tilde{\eta}_{\tilde{P}^l}$ we use the following formulas [15]:

$$\eta_{\tilde{P}^l}(\mathcal{H}) = \bigcup_{\alpha\in(0,1]} \alpha\left[\eta_{\tilde{P}^l}{}_1^{\alpha}(\mathcal{H}), \eta_{\tilde{P}^l}{}_2^{\alpha}(\mathcal{H})\right], \tag{4.20}$$

where

$$\begin{aligned} \eta_1^{\alpha}(\mathcal{H}) &= \inf\left\{ \int_S \rho(s)\max_{S\in\mathcal{H}}\mu_S(s)ds \,\middle|\, \rho(s)\in P^{\rho^{\alpha}} \right\}, \\ \eta_2^{\alpha}(\mathcal{H}) &= \inf\left\{ \int_S \rho(s)\max_{S\in\mathcal{H}}\mu_S(s)ds \,\middle|\, \rho(s)\in core(\tilde{P}^{\rho}) \right\}, \\ P^{\rho^{\alpha}} &= \left\{ \rho(s) \,\middle|\, \min_{i=1,n}(\pi_{\tilde{P}_i}(p_i)) \geq \alpha \right\},\ core(\tilde{P}^{\rho}) = P^{\rho^{\alpha=1}},\ \mathcal{H}\subset\mathcal{S} \end{aligned} \tag{4.21}$$

The support of $\tilde{\eta}_{\tilde{P}^l}$ is defined as $\operatorname{supp}\tilde{\eta}_{\tilde{P}^l} = cl\left(\bigcup_{\alpha\in(0,1]} \eta_{\tilde{P}^l}^{\alpha}\right)$.

For special case, when states of nature are just some elements, fuzzy number-valued fuzzy measure $\tilde{\eta}_{\tilde{P}^l}$ is defined as

$$\tilde{\eta}_{\tilde{P}^l}(\mathcal{H}) = \bigcup_{\alpha\in(0,1]} \alpha\left[\eta_{\tilde{P}^l}{}_1^{\alpha}(\mathcal{H}), \eta_{\tilde{P}^l}{}_2^{\alpha}(\mathcal{H})\right],\ \mathcal{H}\subset S = \{s_1,\ldots,s_n\} \tag{4.22}$$

where

$$\begin{aligned} \eta_{\tilde{P}^l}^{\alpha}(\mathcal{H}) &= \inf\left\{ \sum_{s_i\in H} p(s_i) \,\middle|\, (p(s_1),\ldots,p(s_n))\in P^{\rho^{\alpha}} \right\}, \\ P^{\rho^{\alpha}} &= \left\{ (p(s_1),\ldots,p(s_n))\in P_1^{\alpha}\times\ldots\times P_n^{\alpha} \,\middle|\, \sum_{i=1}^{n} p(s_i) = 1 \right\}. \end{aligned} \tag{4.23}$$

Here $P_1^{\alpha},\ldots,P_n^{\alpha}$ are α-cuts of fuzzy probabilities $\tilde{P}_1,\ldots,\tilde{P}_n$ repectively, $p(s_1),\ldots,p(s_n)$ are basic probabilities for $\tilde{P}_1,\ldots,\tilde{P}_n$ respectively, $\times$ denotes the Cartesian product.

A General Methodology for Decision Making with Imperfect Information

The problem of decision making with imperfect information consists in determination of an optimal action $\tilde{f}^* \in \mathcal{A}$, that is $\tilde{f}^* \in \mathcal{A}$ for which $\tilde{U}(\tilde{f}^*) = \max_{\tilde{f}\in\mathcal{A}}\left\{\int_S \tilde{u}(\tilde{f}(\tilde{S}))d\tilde{\eta}_{\tilde{P}^l}\right\}$. In this section we present the methodology for solving this problem. The methodology consists of the several stages described below.

At the first stage it becomes necessary to assign linguistic utility values $\tilde{u}(\tilde{f}_j(\tilde{S}_i))$ to every action $\tilde{f}_j \in \mathcal{A}$ taken at a state $\tilde{S}_i \in \mathcal{S}$.

The second stage consists in construction of a fuzzy number-valued fuzzy measure over $\mathcal{F}_S$ based on partial knowledge available in form of linguistic probabilities. First, given known probabilities $P(\tilde{S}_i) = \tilde{P}_i;\ \tilde{S}_i \in \mathcal{E}^n,\ \ \tilde{P}_i \in \mathcal{E}^1_{[0,1]}$, $i = \{1,\ldots,j-1,j+1,\ldots,n\}$ one has to find an unknown probability $P(\tilde{S}_j) = \tilde{P}_j, \tilde{P}_j \in \mathcal{E}^1_{[0,1]}$ by solving the problem described by (4.11) - (4.12). As a result one would obtain a linguistic probability distribution $\tilde{P}^l$ expressed over all the states of nature. If some additional information about the probability over $\mathcal{S}$ is received (e.g. from indicator events), it is required to update $\tilde{P}^l$ on the base of this information by using fuzzy Bayes' formula [22]. Then based on the latest $\tilde{P}^l$ it is necessary to construct fuzzy number-valued fuzzy measure $\tilde{\eta}_{\tilde{P}^l}$ by solving the problem expressed by (4.20) – (4.21).

At the next stage the problem of calculation of a fuzzy-valued Choquet integral for every action $\tilde{f}_j$ is solved. At this stage, first it is required for an action $\tilde{f}_j$ to rearrange indices of states $\tilde{S}_i$ on the base of Definition 3.8 and find such new indices (i) that $\tilde{u}(\tilde{f}_j(\tilde{S}_{(1)})) \geq \ldots \geq \tilde{u}(\tilde{f}_j(\tilde{S}_{(n)}))$. Next it is needed to calculate fuzzy values $\tilde{U}(\tilde{f}_j)$ of a fuzzy-valued Choquet integral for every action $\tilde{f}_j$ by using the (4.8).

Finally, by using a suitable fuzzy ranking method, an optimal $\tilde{f}^* \in \mathcal{A}$, that is $\tilde{f}^* \in \mathcal{A}$ for which $\tilde{U}(\tilde{f}^*) = \max_{\tilde{f}\in\mathcal{A}}\left\{\int_S \tilde{u}(\tilde{f}(\tilde{S}))d\tilde{\eta}_{\tilde{P}^l}\right\}$, is determined.

Multicriteria Decision Analysis with Imperfect Information

As it was shown in [53], there exists a formal parallelism between decision making under uncertainty and multicriteria decision making [23,27,52,55,57,66,69,70]. Let us consider a problem of decision making with imperfect information

$D_{DMII} = (\mathcal{S},\mathcal{Y},\mathcal{A},\succsim_l)$ and multicriteria decision making $D_{MCDM} = (\mathcal{X},\succsim_l)$, with $\mathcal{X} = \mathcal{X}_1 \times ... \times \mathcal{X}_n$. As we consider $\mathcal{S}$ as a finite set, we can identify acts $\tilde{f} \in \mathcal{A}$ with the elements of $\mathcal{X}$ by considering them as $(\tilde{\varphi}_1,...,\tilde{\varphi}_n)$, where $\tilde{\varphi}_i \in \mathcal{X}_i$ denotes $\tilde{f}(\tilde{S}_i)$. Then the preference relation $\succsim_l$ over the acts becomes a preference relation over $\mathcal{X}$. For decision making with imperfect information, the preference is formally expressed as

$$\forall \tilde{f}, \tilde{g} \in \mathcal{A}, \tilde{f} \succsim_l \tilde{g} \Leftrightarrow (\tilde{\varphi}_1,...,\tilde{\varphi}_n) \succsim_l (\tilde{\psi}_1,...,\tilde{\psi}_n)$$

and for multicriteria decision making, the fact that an alternative $\tilde{X}$ is preferred to $\tilde{X}'$ is expressed as

$$\tilde{X} \succsim_l \tilde{X}' \Leftrightarrow (\tilde{X}_1,...,\tilde{X}_n) \succsim_l (\tilde{X}_1',...,\tilde{X}_n'), \tilde{X}_i, \tilde{X}_i' \in \mathcal{X}_i, \forall i \in I \ ,$$

The problem of decision analysis with imperfect information can be written as multiattribute decision making problem using the following identifications [53]:

1) States of the nature and criteria: $\mathcal{S} \leftrightarrow \mathcal{I}$; 2) acts and alternatives as values of criteria $\mathcal{A} \leftrightarrow \mathcal{X}$, and $\tilde{f} \succsim_l \tilde{g} \Leftrightarrow (\tilde{\varphi}_1,...,\tilde{\varphi}_n) \succsim_l (\tilde{\psi}_1,...,\tilde{\psi}_n)$.

Let us now recall the statement of the multiattribute decision making problem. Let us assume that unidimensional utility functions $\tilde{u}_i : \mathcal{X}_i \to \mathcal{E}^1$, for $i = 1,...,n$ are defined. In order to represent the preference relation $\succsim_l$ we need to find an appropriate aggregate operator $\mathrm{H}:\mathcal{E}^n \to \mathcal{E}^1$ such that $\tilde{X} \succsim_l \tilde{X}'$ holds iff

$$\tilde{U}(\tilde{X}) = \mathrm{H}(\tilde{u}_1(\tilde{X}_1),...,\tilde{u}_n(\tilde{X}_n)) \geq \tilde{U}(\tilde{X}') = \mathrm{H}(\tilde{u}_1(\tilde{X}_1'),...,\tilde{u}_n(\tilde{X}_n'))$$

In the framework of the methodology proposed here, we use the operator $\mathrm{H}(\cdot)$ as a fuzzy-valued Choquet integral and hence

$$\tilde{U}(\tilde{X}) = \sum_{i=1}^{n} \left[\tilde{u}_{(i)}(\tilde{X}_{(i)}) - \tilde{u}_{(i+1)}(\tilde{X}_{(i+1)})\right] \tilde{\eta}_{\tilde{P}^l}(\mathcal{H}_{(i)}) \geq \tilde{U}(\tilde{X}') =$$
$$= \sum^{n} \left[\tilde{u}_{(i)}(\tilde{X}_{(i)}') - \tilde{u}_{(i+1)}(\tilde{X}_{(i+1)}')\right] \tilde{\eta}_{\tilde{P}^l}(\mathcal{H}_{(i)}), \mathcal{H}_{(i)} \subset 2^{\mathcal{I}} \tag{4.24}$$

Here $\tilde{\eta}_{\tilde{P}^l}$ on $2^{\mathcal{I}}$ is obtained from linguistic probability $\tilde{P}^l$ by solving the problem (4.11) – (4.12).

4.2 Multi-agent Fuzzy Hierarchical Models for Decision Making

Economy as a complex system is composed of a number of agents interacting in distributed mode. Advances in distributed artificial intelligence, intelligent agent theory and soft computing technology make it possible for these agents as components of a complex system to interact, cooperate, contend and coordinate in order to form global behavior of economic system. Recently, there has been great interest in development of Intelligent Agents (IA) and Multi-agent systems in economics, in particular in decision analysis and control of economic systems [9,31,40,43,48,49,71,76,85].

It should be noted that economic agents often deal with incomplete, contradictory, missing and inaccurate data and knowledge [4,13,26,41]. Furthermore, the agents have to make decisions in uncertain situations, i.e. multi-agent economical systems in the real world function within an environment of uncertainty and imprecision.

In this section we consider two approaches to multi-agent economic system: conventional concept and alternative concept.

In conventional concept of multi-agent distributed intelligent systems the main idea is granulation of functions and powers from a central authority to local authorities. In these terms, economy is composed of several agents, which can perform their own functions independently, and, therefore, have information, authority and power necessary to perform only their own functions. These intelligent agents can communicate together to work, cooperate and be coordinated in order to reach a common goal of economy in a system [13].

An alternative concept of a multi-agent distributed intelligent decision making system with cooperation and competition among agents, distinguished from the conventional approach by the following [9,31]: each intelligent agent acts fully autonomously; each intelligent agent proposes full solution of the problem (not only for own partial problem); each agent has access to full available input information; total solution of the problem is determined as a proposal of one of the parallel functioning agents on the basis of a competition procedure (not by coordinating and integrating partial solutions of agents, often performed iteratively); agents' cooperation produces desired behavior of the system; cooperation and competition acts in the systems are performed simultaneously (not sequentially).

A similar idea of decomposition of the overall system into agents with cooperation and competition among them is implemented in [40,71,85].

Zhang proposed a way to synthesize final solutions in systems where different agents use different inexact reasoning models to solve a problem. In this approach a number of expert systems propose solutions to a given problem. These solutions are then synthesized using ego-altruistic approach [43].

Below we present the formalism of the conventional approach.

For simplicity, we will mainly consider systems with so-called “fan” structure in which the economic system consists of N agents in the lower level and one element in the higher level (which we call a center). A hierarchy has two levels: the focus or overall goal of the decision making problem at the top, and competing alternatives at the bottom.

The state of *i-th* agent ($i = 1,...,N$) is characterized by vector $\tilde{X}_i$. The vector $\tilde{X}_i$ should meet the local constraints

$$\tilde{X}_i \in \mathcal{X}_i \subset \mathcal{E}^{n_i} \tag{4.25}$$

where $\mathcal{X}_i$- is a set in n_i dimensional space $\mathcal{E}^{n_i}$. A specificity of these hierarchical systems is information aggregation at the higher level. This means that the only agent of the higher level, the center, is concerned not on individual values of variables $\tilde{X}_i$, but some indexes evaluating elements' activities produced from those values. Let's denote the vector of such indexes as:

$$\tilde{F}_i(\tilde{X}_i) = (\tilde{f}_{i1}(\tilde{X}_i),..., \tilde{f}_{im_i}(\tilde{X}_i)), i = 1,..., N. \tag{4.26}$$

The local concerns of i-th agent are represented by vector criteria $\tilde{\Phi}_i(\tilde{X}_i) = (\tilde{\varphi}_{i_1}(\tilde{X}_i),...,\tilde{\varphi}_{i_{K_i}}(\tilde{X}_i))$. Let's assume that the agents concern in increased values fuzzy of criteria: $\tilde{\varphi}_{i_k}(\tilde{X}_i), k = 1,..., K_i$. Sometimes the indexes may directly mimic the criteria, but in general, the indexes can be related to these criteria in a specific manner. It is worth noting that the number of indexes m_i and criteria K_i is much less than the dimension of vector $\tilde{X}_i$.

Consider a situation in which all parameters of the objective functions and the constraints are fuzzy numbers represented in any form of membership functions. The problem can be formulated as follows, in general.

$$\begin{cases} \max \ \left\langle \tilde{c}, \ \tilde{X} \right\rangle_F = \left(\sum_{i=1}^{n} \tilde{c}_{1i}\tilde{X}_i, \ \sum_{i=1}^{n} \tilde{c}_{2i}\tilde{X}_i, ..., \ \sum_{i=1}^{n} \tilde{c}_{Ki}\tilde{X}_i \right)^T \\ s.t. \quad \tilde{A}\tilde{X} \underset{=}{<} \tilde{B}, \tilde{X} \underset{=}{>} 0, \end{cases} \tag{4.27}$$

The state of the center is characterized by vector $\tilde{F}_0$, the components of which are the indexes of the agents of the lower level:

$$\tilde{F}_0 = (\tilde{F}_1,..., \tilde{F}_N), \text{ where } \tilde{F}_i = \tilde{F}_i(\tilde{X}_i) \tag{4.28}$$

The vector $\tilde{F}_0$ should satisfy the global constraints:

$$\tilde{F}_0 \in \mathcal{X}_0 \subset \mathcal{E}^{m_0}, \text{ where } m_0 = \sum_{i=1}^{N} m_i \tag{4.29}$$

Let's assume that the set $\mathcal{X}_0$ is defined on the constraint set:

$$\mathcal{X}_0 = \left\{ \tilde{F}_0 / \tilde{H}(\tilde{F}_0) \geq \tilde{B} \right\}, \tag{4.30}$$

where $\tilde{H}(\tilde{F}_0)$ is some fuzzy vector-function, $\tilde{B}=(\tilde{B}_1,\ldots,\tilde{B}_M)$ is a vector. The objective of the center is to maximize the vector criteria:

$$\tilde{\varphi}_0(\tilde{F}_0)=(\tilde{\varphi}_{01}(\tilde{F}_0),\ldots,\tilde{\varphi}_{0K_0}(\tilde{F}_0))\rightarrow \max$$

We have to use the following notions for solving the optimization problem.

Definition 4.7. Fuzzy Complete Optimal Solution [13,49]. $\tilde{X}^*$ is said to be a *fuzzy complete optimal solution*, if and only if there exists an $\tilde{X}^*\in\mathcal{X}$ such that $\tilde{f}_i(\tilde{X}^*)\geq\tilde{f}_i(\tilde{X})$, $i=1,\ldots,K$, for all $\tilde{X}\in\mathcal{X}$.

In general, such a complete optimal solution that simultaneously maximizes (or minimizes) all objective functions does not always exist when the objective functions conflict with each other. Thus, a concept of a *fuzzy Pareto optimal solution* is taken into consideration.

Definition 4.8. Fuzzy Pareto Optimal Solution [13,49]. $\tilde{X}^*$ is said to be a *fuzzyPareto optimal solution*, if and only if there does not exist another $\tilde{X}\in\mathcal{X}$ such that $\tilde{f}_i(\tilde{X})\geq\tilde{f}_i(\tilde{X}^*)$ for all i and $\tilde{f}_j(\tilde{X})\neq\tilde{f}_j(\tilde{X}^*)$ for at least one j.

Definition 4.9. Weak Fuzzy Optimal Solution [13,49]. $\tilde{X}^*$ is said to be a *weak fuzzy Pareto optimal solution*, if and only if there does not exist another $\tilde{X}\in\mathcal{X}$ such that $\tilde{f}_i(\tilde{X})>\tilde{f}_i(\tilde{X}^*)$, $i=1,\ldots,K$.

Let $\tilde{X}^{CO}$, $\tilde{X}^{P}$ or $\tilde{X}^{WP}$ denote fuzzy complete optimal, fuzzy Pareto optimal, or weak fuzzy Pareto optimal solution sets respectively. Then from above definitions, we can easily get the following relations:

$$\tilde{X}^{CO}\subseteq\tilde{X}^{P}\subseteq\tilde{X}^{WP} \tag{4.31}$$

A fuzzy satisfactory solution is a reduced subset of the feasible set that exceeds all of the aspiration levels of each attribute. A set of satisfactory solutions is composed of acceptable alternatives. Satisfactory solutions do not need to be non-dominated. And a preferred solution is a non-dominated solution selected as the final choice through decision makers' involvement in the information processing.

Commonly, the decision making process implies the existence of a person making the final decision (Decision Maker-agent) at the higher level.

As our primary goal is to coordinate the center and the agents of the lower level to align their objectives, we consider the objective function of the center as known:

$$\tilde{\tilde{H}}(\tilde{\Phi}_0(\tilde{F}_0))=\tilde{H}_0(\tilde{F}_0)\rightarrow \max \tag{4.32}$$

or

$$\tilde{\tilde{H}}(\tilde{\Phi}_0(\tilde{F}_0)) = \int_0^T \tilde{H}_0(\tilde{F}_0)dt \rightarrow \max.$$

The coordination (overall optimization) problem can be formulated as follows:

$$\tilde{H}_0(\tilde{F}_1,...,\tilde{F}_N) \rightarrow \max; \tag{4.33}$$

$$\tilde{H}(\tilde{F}_1,...,\tilde{F}_N) \geq \tilde{B}; \tag{4.34}$$

$$\tilde{F}_i \in \mathcal{I}_i^{\tilde{F}} = \left\{\tilde{F}_i / \tilde{F}_i = \tilde{F}_i(\tilde{X}_i), \tilde{X}_i \in P_i^{\tilde{X}} \left(\text{or } \tilde{X}_i \in R_i^{\tilde{X}}\right)\right\}, \tag{4.35}$$

where $P_i^{\tilde{X}}$ ($R_i^{\tilde{X}}$) is the set of effective (semi-effective) solutions of the problem

$$\tilde{\Phi}_i(\tilde{X}_i) = (\tilde{\varphi}_{i1}(\tilde{X}_i),...,\tilde{\varphi}_{iK_i}(\tilde{X}_i)) \rightarrow \max; \tag{4.36}$$

$$\tilde{X}_i \in \mathcal{X}_i \tag{4.37}$$

Because the sets $P_i^{\tilde{X}}$ ($R_i^{\tilde{X}}$) and accordingly the set $\mathcal{I}_i^{\tilde{F}}$ can have rather complex structures, the condition (4.35) can be replaced by

$$\tilde{F}_i \in Q_i^{\tilde{F}} \tag{4.38}$$

where $Q_i^{\tilde{F}}$ is the set satisfying the condition:

$$\mathcal{I}_i^{\tilde{F}} \subset Q_i^{\tilde{F}} \subset \mathcal{J}_i^{\tilde{F}} = \left\{\tilde{F}_i / \tilde{F}_i = \tilde{F}_i(\tilde{X}_i), \tilde{X}_i \in \mathcal{X}_i\right\} \tag{4.39}$$

We suggest two methods to solution to (4.33)-(4.35)

a) Non-iterative method

Non-iterative optimization method includes three main phases. At the first stage, the local problems of vector optimization are dealt with. The solutions of these problems are sets $P_i^{\tilde{X}}$ and sets $\mathcal{I}_i^{\tilde{F}}$ ($Q_i^{\tilde{F}}$) or any approximations of these sets. The second stage implies the implementation of the center's task (4.33)-(4.35) as result of which we get the optimal values of the agents′ criteria $\tilde{F}^* = (\tilde{F}_1^* ..., \tilde{F}_N^*)$. Vector $\tilde{F}_i^*$ is then passed to i-th agent which implements the third stage by solving the problem:

$$\tilde{F}_i(\tilde{X}_i) = \tilde{F}_i^*; \tilde{X}_i \in \mathcal{X}_i \tag{4.40}$$

A solution of (4.40) represents local variables $\tilde{X}_i^*$. In case when several solutions exist, one is selected based on preferences of the element.

Note that the center receives the information only about the indexes $\tilde{F}_i$, not about the vector $\tilde{X}_i$. Because the dimension of $\tilde{F}_i$ is usually significantly less than the dimension of vector $\tilde{X}_i$, it considerably reduce amount of data circulating between the levels.

b) Iterative method

Let $\Omega_i \subset \mathcal{E}^{K_i}$ be a subset in the space of criteria. Let's call the elements $\tilde{\omega}_i \in \Omega_i$ from this subset as coordinating signals.

Definition 4.10. Coordinating Function [13]. The function $\tilde{\bar{F}}_i(\tilde{\omega}_i) = (\tilde{\bar{f}}_{i1}(\tilde{\omega}_i), ..., \tilde{\bar{f}}_{im_i}(\tilde{\omega}_i))$ is named a coordinating function if the following conditions are satisfied:

a) For $\forall \tilde{\omega}_i \in \Omega_i$ there exists such element $\tilde{\bar{X}}_i(\tilde{\omega}_i) \in R_i^{\tilde{X}}$, for which $\tilde{\bar{F}}_i(\tilde{\omega}_i) = \tilde{F}_i(\tilde{\bar{X}}_i(\tilde{\omega}_i))$;

b) Inversely, for any element $\tilde{X}_i^0$ – of subset $P_i^{\tilde{X}}$, there exists such coordinating signal $\tilde{\omega}_i^0 \in \Omega_i$, for which $\tilde{\bar{F}}_i(\tilde{\omega}_i^0) = \tilde{F}_i(\tilde{X}_i^0)$.

From the definition above, it follows that the problem (4.33)-(4.35) is equivalent to the problem given below:

$$\left.\begin{aligned} &\tilde{H}_0(\bar{F}_1(\tilde{\omega}_1), ..., \bar{F}_N(\tilde{\omega}_N)) \to \max; \\ &\tilde{H}(\tilde{\bar{F}}_i(\tilde{\omega}_i), ..., \tilde{\bar{F}}_N(\tilde{\omega}_N)) \geq \tilde{B}; \\ &\tilde{\omega}_i \in \Omega_i, i = 1, ..., N. \end{aligned}\right\} \tag{4.41}$$

Thus, the variables of problem (4.41) are coordinating signals $\tilde{\omega}_i$, defined on the set of acceptable coordinating signals Ω_i. The rationale for such transformation is simpler structure of set Ω_i compared to the set of effective elements (points).

Choosing among different coordinating functions $\tilde{\bar{F}}_i(\tilde{\omega}_i)$ and different solution approach of problem (4.41) it is possible to construct a large number of iterative decomposition coordinating methods [12,13]. It can be shown that for crisp case the known decomposition algorithms such as Dantzig-Wolf algorithm or the algorithm based on the interaction prediction principle are special cases of the more general suggested algorithm [13].

Let's consider the procedure of optimization in two-level multi-agent decision making systems [13]. The agents send their solutions to the center for further global decision making. Each solution represents a vector of criteria of the agent, acceptable within the local constraints. On the basis of the received alternatives the center creates a solution optimal to the system as a whole. This solution is forwarded to the agents who then implements it. In this case the task of the center is in the determination of the optimal values for weight coefficients of the agents' solutions and can be written as follows:

$$\tilde{H}_0 = \sum_{i=1}^{N}\sum_{j=1}^{R_i} \lambda_{ij}\tilde{Z}_{0ij} \to \max; \tag{4.42}$$

$$\tilde{H}_m = \sum_{i=1}^{N}\sum_{j=1}^{R_i} \lambda_{ij}\tilde{Z}_{mij} \geq \tilde{B}_m, \ m = 1,...,M; \tag{4.43}$$

$$\sum_{j=1}^{R_i} \lambda_{ij} = 1, \ i = [1,...,N], \tag{4.44}$$

here λ_{ij} is the weight coefficient sought for *j*-th alternative of an i-th agent; $\tilde{Z}_{mij} =< \tilde{a}_{mij}, \tilde{F}_{ij} >, m = 0,...,M;$ $\tilde{a}_{mij}$ are constant coefficients in the model of the center; $\tilde{F}_{ij}$ is the vector criteria of the solution of *i*-th agent in j-th alternative.

Definition 4.11 Fuzzy Optimality [13]. Vector λ^* giving the maximum of the membership function for the set $\tilde{D}$, "desirability of λ the viewpoint of satisfaction of all indices" is fuzzy optimal solution to (4.42) – (4.44).

Thus, the decision making process in the center is reduced to solving the fuzzy linear programming problem (4.42)-(4.44). Agents' alternatives are created by Pareto optimality (PO) procedures. The complexity of the solution of (4.42)-(4.44) rapidly increases as the number of agents' alternatives increases. In the interactive procedure given below we use the fuzzy optimality principle in the decision making process on the base of (4.42)-(4.44).

Many decision makers prefer an interactive approach to find an optimal solution for a decision problem as such approach enables decision makers to directly engage in the problem solving process. In this section, we propose an interactive method, which not only allows decision makers to give their fuzzy goals, but also allows them to continuously revise and adjust their fuzzy goals. In this way, decision makers can explore various optimal solutions under their goals, and then choose the most satisfactory one.

First of all Decision Maker analyzes agents' individual solutions received by the center. At this preliminary stage the opinion of Decision Maker on the agents' activity, which can be evaluated qualitatively, is taken into consideration.

Based on the extension principle, for a fixed $\lambda = \{\lambda_{ij}, i = 1,...,N, \ j = 1,...,R_i\}$ it is possible to obtain the "value of *m*-th criteria for this λ":

$$\tilde{H}_{m\lambda} = \lambda_{11}\tilde{Z}_{m11} \oplus ... \oplus \lambda_{NR_N}\tilde{Z}_{mNR_N}$$

Then the fuzzy values $\tilde{B}_m$ are constructed having sense of "the value of m-th criteria desirable for Decision Maker", $m = 1,...,M$. Note that in accordance with the principle of Bellmann-Zadeh [19], the $\tilde{H}_0$, which is the objective of the problem (4.42) - (4.44) and $\tilde{H}_m, m=1,...,M$ (constraints), are represented by fuzzy sets. The sets $\tilde{b}_m$ are constructed by processing preferences of Decision Maker such as "it is necessary that m-th criteria be not less than c_m^1 and it is desired that it is not less than c_m^2, while the allowable range is $c_m \pm \Delta c_m$". The membership function of the fuzzy set $\tilde{B}_m$ corresponding to the last clause can be set as follows:

$$\mu(B_m) = \begin{cases} 0, \; if \;\; B_m \le c_m - \Delta c_m \;\; or \;\; B_m \ge c_m + \Delta c_m; \\ \dfrac{B_m - c_m + \Delta c_m}{\Delta c_m}, \; if \;\; c_m - \Delta c_m \le B_m \le c_m; \\ \dfrac{c_m + \Delta c_m - B_m}{\Delta c_m}, \;\; if \;\; c_m \le B_m \le c_m + \Delta c_m. \end{cases}$$

Then for each a fixed λ it can be related the fuzzy set $\tilde{\Phi}_{m\lambda} \triangleq \tilde{H}_{m\lambda} \cap \tilde{B}_m$. The membership function of this set $\mu_{\tilde{\Phi}_{m\lambda}}()$ for argument $\Phi_{m\lambda}$ is equal to the smaller of the two values: the first value defines the possibility of the fact that m-th criteria would take the value of $\tilde{\Phi}_{m\lambda}$ considering the fuzziness of the parameters that determine it and the second value defines the degree of desire ability of this value for the Decision Maker.

Let's consider the fuzzy sets $\tilde{D}_m$ such as "desirability of λ from the viewpoint of $m-$th criteria", $m=[0,...,M]$, defined on the domain set λ given by (4.44).

The membership function of a fuzzy set $\tilde{D}_m$ denoted as $\tilde{v}_m(\lambda)$ is determined as the height of fuzzy set $\tilde{\Phi}_{m\lambda}$

$$v_m(\lambda) \triangleq \max \mu_{\tilde{\Phi}_{m\lambda}}(\Phi_{m\lambda}) = \mu_{\tilde{\Phi}_{m\lambda}}(\Phi^*_{m\lambda}).$$

Let fuzzy sets $\tilde{H}_{m\lambda}$ and $\tilde{B}_m$ are fuzzy numbers $\tilde{H}_{m\lambda} = \left(H_{m\lambda}, \underline{H}_{m\lambda}, \overline{H}_{m\lambda}\right)$ and $\tilde{B}_m = \left(B_m, \underline{B}_m, \overline{B}_m\right)$. Then from formula (4.44), for determining the height of intersection of two fuzzy numbers, we have:

$$\mu\left(\hat{\mathrm{O}}^*_{m\lambda}\right) = \begin{cases} R\left[\dfrac{B_m - H_{m\lambda}}{\overline{H}_{m\lambda} - \underline{B}_m}\right], \; if \; B_m - H_{m\lambda} \ge 0; \\ L\left[\dfrac{H_{m\lambda} - B_m}{\underline{H}_{m\lambda} - \overline{B}_m}\right], \; if \; B_m - H_{m\lambda} \le 0. \end{cases}$$

Thus, we obtained the formulas for the membership functions of $\tilde{D}_m$ "desirability of λ from the viewpoint of m-th criterion", $m=0,...,M$. Then it is necessary to construct the membership function of the fuzzy set $\tilde{D}_m$ "desirability of λ from the viewpoint of satisfaction of all indices". Let's assume that the membership function of this fuzzy set which we will denote $v_m(\lambda)$ is defined via membership functions $v_m(\lambda)$, $m=0,...,M$ and vector $\pi=(\pi_0,...,\pi_M)$, where π_M is the degree of importance of m-th index or the function $v_m(\lambda)$ related to this index:

$$v(\lambda,\pi)=\bar{v}(\mu_0(\lambda),...,\mu_M(\lambda),\pi_0,...,\pi_M).$$

Functions $\bar{v}$ can take different forms, such as, for example,

$$\bar{v}^1=\sum_k \pi_k\mu_k(\lambda);\ \bar{v}^2=\min_k\{\pi_k\mu_k(\lambda)\};$$

$$\bar{v}^3=\prod_k \mu_k(\lambda)^{\pi_k};\ \bar{v}^4=\min_k\{\mu_k(\lambda)^{\pi_k}\}.$$

The weight coefficients π_k is determined by way of processing of Decision Maker responses to stated questions. The success of such human-centered procedure depends mainly on understandability of the questions to the Decision Maker. In our view, the procedure in which Decision Maker is presented two alternatives q_1 and q_2 for comparison by using a predefined set of linguistic terms is appropriate. We can use the following set: "Can not say which of the alternatives is better (worse)", "Alternative q_1 is somewhat better (worse) than alternative q_2", "Alternative q_1 is noticeably better (worse) that alternative q_2", "Alternative q_1 is considerably better (worse) that alternative q_2".

The linguistic expressions presented to Decision Maker can be considered as linguistic labels for corresponding fuzzy sets, defined on the universe of discourse $R=\{r/r=v(\lambda_1,\pi)-v(\lambda_2,\pi)/\lambda_1,\lambda_2 \text{ are allowable}\}$.

Because $0\le v(\lambda,\pi)\le 1$, then $R\subset[-1,1]$. Hence, the membership function for the expression "Alternatives q_1 and q_2 are equivalent" can be given by formula:

$$\rho_4(r)=\begin{cases}1-25r^2,\ if\ |r|\le 0,2;\\ 0,\ if\ |r|\ge 0,2;\end{cases}$$

And the membership function for the expression "Alternative q_1 is somewhat better than alternative q_2" can be given by formula:

$$\rho_5(r)=\begin{cases}1-25(r-0.3)^2,\ if\ 0.1\le r\le 0,5;\\ 0,\ if\ |r|\ge 0,5;\end{cases}$$

Membership functions $\rho_s(r)$ of linguistic terms labeled by *s=1,2,...,7* (*4* – "alternatives q_1 and q_2 are equivalent"; *5* – "alternative q_1 is somewhat better than alternative q_2") for comparison of alternatives are shown in Fig. 4.1.

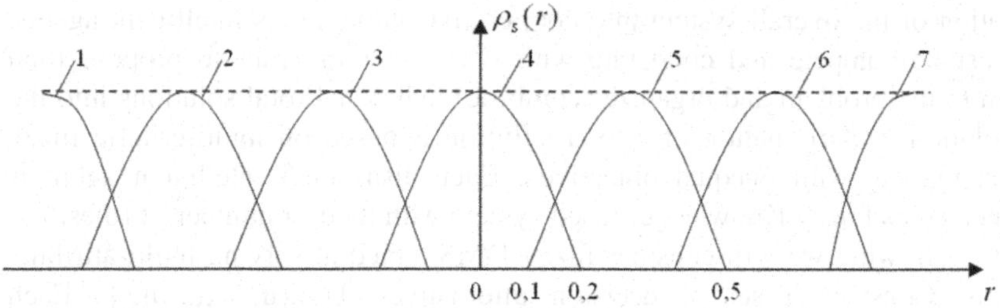

Fig. 4.1 Linguistic terms for comparison of alternatives

Let a Decision Maker compare Q pairs of alternatives. The alternative q is characterized by the vector of decision variables λ corresponding to this vector by values of membership functions $\nu_m(\lambda)$ of fuzzy sets $\tilde{D}_m$ "desirability of λ from the viewpoint of *m*-th criterion" as well as fuzzy sets $\tilde{H}_{m\lambda}$ "value of *m*-th criterion for current λ".

If on comparing alternatives q_1 and q_2 Decision Maker chooses linguistic label s with the membership function $\rho_s(r)$, then a fuzzy set $\rho_q(\pi)$ is defined in space of weight coefficients π, which can be called as "consistent with the q-th response of Decision Maker". The corresponding membership function is defined as:

$$\overline{\rho}_q(\pi) = \rho_s(r),\ r = \nu(\lambda_{q_1}, \pi) - \nu(\lambda_{q_2}, \pi)$$

Then, consecutively we determine:

1) Intersection of fuzzy sets $\overline{p}_q(\pi)$:

$$\overline{p}(\pi) = \bigcap_{q \in [1:Q]} \overline{p}_q(\pi)$$

2) Vector π^* giving the maximum of the function $\overline{\rho}(\pi)$

$$\pi^* = Arg \max_{\pi} \overline{p}(\pi^*)$$

3) Membership function $\nu(\lambda, \pi^*)$ for fuzzy set $\tilde{D}$ "desirability of λ from the viewpoint of satisfaction of all objectives"

4) Vector λ^* giving the maximum of this membership function

$$\lambda^* = Arg \max_{\lambda} \nu(\lambda, \pi^*).$$

By revising fuzzy goals, this method will provide decision makers with a series of optimal solutions. Hence, decision makers can select the most suitable one on the basis of their preference, judgment, and experience.

Now we consider an alternative concept.

The main idea of the proposed decision making system (DMS) is based on granulation of the overall system into cooperative autonomous intelligent agents. These agents compete and cooperate with each other in order to propose total solution to the problem and organize (combine) individual total solutions into the final solution. Determination of a total solution is based on multi-criteria fuzzy decision making with unequal objectives. Each distributed intelligent agent is implemented as Fuzzy Knowledge Based system with modest number of rules.

In the following we will consider fuzzy DMS aimed at solving multi-attribute problems. Consider a set of decision alternatives $U=\{u_1,u_2,...,u_N\}$. Each alternative is characterized by M criteria (attributes) $C=\{c_1,c_2,...,c_M\}$ according to which the desirability of the solution is determined.

Let $\mu_{c_i}(u_j)\in[0,1]$ be rating estimation value (membership value) of alternative u_j on criteria c_i. The objective is to determine the optimal alternative u_j which is better than the others in terms of criteria c_i.

Figure 4.2 shows the structure of the considered multi-agent distributed intelligent

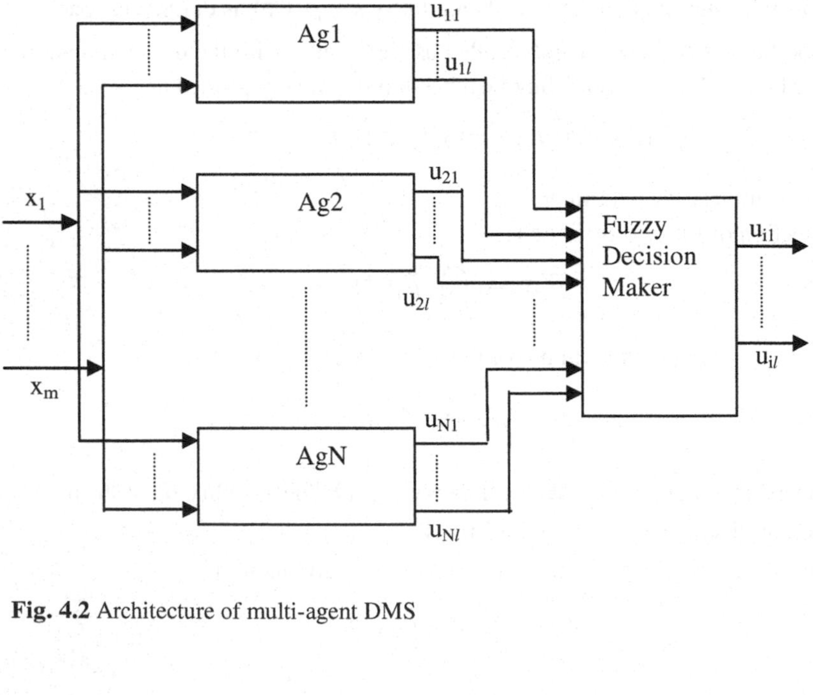

Fig. 4.2 Architecture of multi-agent DMS

DMS. The system is composed of N parallel agents. All agents receive the same fuzzy (or partly non-fuzzy) input information $x_1, x_1, ..., x_m$. Each agent, which is a knowledge-based system, performs inference and produces its own solution of the full problem (not partial problem as in classical distributed systems) $u_j = [u_{j1}, u_{j2}, ..., u_{jl}]^T, j = \overline{1,N}$.

The architecture of agent is given in Fig. 4.3. Note that granulation fine, i.e., number of agents in DMS is determined by the system designer.

Each agent Ag_j, $j = \overline{1,N}$ is an autonomous intelligent agent, capable of performing inference to produce solution. The agents are knowledge-based systems with modest number of fuzzy rules related through "ALSO":

$$R^k: \text{IF } x_1 \text{ is } A_{k1} \text{ AND } x_2 \text{ is } A_{k2} \text{ AND } ... \text{AND } x_m \text{ is } A_{km} \text{ THEN} \qquad (4.45)$$
$$u_{k1} \text{ is } B_{k1} \text{ AND } u_{k2} \text{ is } B_{k2} \text{ AND} ... \text{ AND } u_{kl} \text{ is } B_{kl}, \; k = \overline{1,K}$$

where x_i, $i = \overline{1,m}$ and u_j, $j = \overline{1,l}$ are common input and individual agent output variables, A_{kj}, B_{kj} are fuzzy sets, and K is the number of rules. Note, that inputs $x_1, x_2, ..., x_m$ may be crisp or fuzzy variables. If input data are crisp, then fuzzifier (Fig. 4.3) will map these data into fuzzy sets. The decision of each agent is made by the composition rule, which is the basis of inference mechanism, as follows:

$$\tilde{U}_j = R_j \circ \tilde{X}, \qquad j = \overline{1,N}, \qquad (4.46)$$

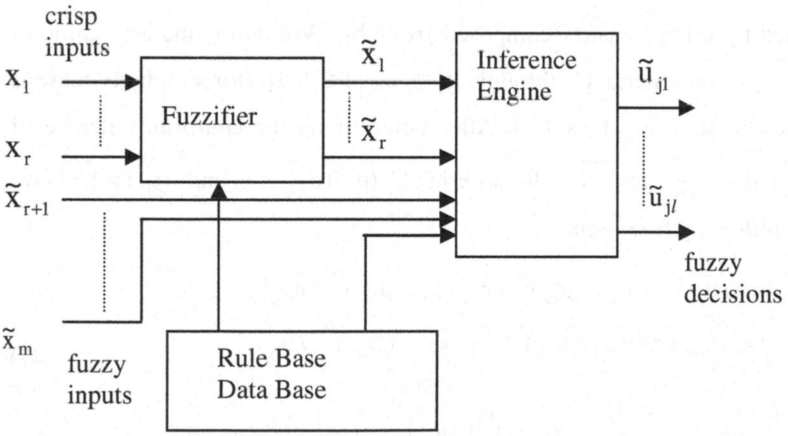

Fig. 4.3 Structure of a competitive agent

where $\tilde{U}_j$ is fuzzy value of decision of j-th agent, R_j is the fuzzy relation corresponding to the fuzzy model (4.45), and $\tilde{X}$ is total input information after fuzzification.

Every agent's solution is evaluated by the fuzzy decision maker on the basis of their achievements on M criteria C_i, $i=\overline{1,M}$. Among N agents the best agent ("winner" agent) is the one that satisfies criteria C_i, $i=\overline{1,M}$ best. The solution of this agent will be taken as total solution of the full system.

Decision Generation in Multi-agent Distributed Intelligent DSS

A set of solutions of DMS $\{u_{Ag\,1}, u_{Ag\,2}, \ldots, u_{AgN}\}$ is formed through fuzzy reasoning by parallel working agents. Each decision alternative is characterized by M criteria, $\{C_1, C_2, \ldots, C_M\}$, according to which the desirability of a solution is determined. The problem here is to determine optimal alternative (winner), which is better than other alternatives in terms of criteria C_i. This problem is alternative selection under condition of uncertainty, which was considered in [14,22,44,77,78,89].

We will use the method suggested by Yager [78] for selection of the winner alternative. The proposals of agents are evaluated in accordance with the criteria C_i, $i=\overline{1,\ M}$. In real decision situation criteria C_i have different importance $\alpha_i, i=\overline{1,M}$. The relative importance of each criterion is determined using the criteria comparison procedure. For example, as it is shown in [78], equal importance $b_{ij}=1$, weak importance $b_{ij}=3$, strong importance $b_{ij}=5$, ($b_{ij}=\frac{1}{b_{ji}}$) etc. are determined on the scale of importance evaluation. $\alpha_i, i=\overline{1,M}$ are obtained by using a matrix composed from b_{ij}. We define the estimation of alternative u_{Ag_i} on criteria C_i through $\mu_{C_i}(u_{Ag_j}) \in [0,1]$ (for simplicity reasons we will denote $\mu_{C_i}(u_{Ag_j})$ as μ_{ij}). After determining the conformity degree of each alternative $u_{Ag_j}, j=\overline{1,N}$, to criteria C_i $(\mu_{C_i}(u_{Ag_j}))$ and $\alpha_i, i=\overline{1,M}$ we derive the following fuzzy sets:

$$
\begin{aligned}
C_1^{\alpha_1} &= \{(\mu_{11})^{\alpha_1}/u_{Ag1}, (\mu_{12})^{\alpha_1}/u_{Ag2}, \ldots, (\mu_{1N})^{\alpha_1}/u_{AgN}\} \\
C_1^{\alpha_2} &= \{(\mu_{11})^{\alpha_2}/u_{Ag1}, (\mu_{12})^{\alpha_2}/u_{Ag2}, \ldots, (\mu_{1N})^{\alpha_2}/u_{AgN}\} \\
&\vdots \\
C_1^{\alpha_M} &= \{(\mu_{M1})^{\alpha_M}/u_{Ag1}, (\mu_{M2})^{\alpha_M}/u_{Ag2}, \ldots, (\mu_{MN})^{\alpha_M}/u_{AgN}\}
\end{aligned}
\tag{4.47}
$$

It is known that optimal alternative selection may be performed by:

$$\mu_D(u^*_{Ag}) = \max_j \min_i \mu^{\alpha_i}_{C_i}(u_{Ag_j}), \tag{4.48}$$

where u^*_{Ag} is the optimal alternative.

As an example, here we consider a decision marketing system in oligopolistic industry. In such industry the number of firms is limited to a few, where the actions of one firm have an impact on the industry demand. A number of firms compete for three products (*x*, *y*, and *z*). Each firm is managed by a team, which pursues profitability and market share goals. The teams make marketing, production, and financial decisions each quarter using DMS incorporating data and models. Marketing decisions include the decisions on mix and amounts of goods to produce, pricing, advertising expenses, and other variables. The firm's price and advertising strategies, as well as prices and advertising expenditures of the competitors are the major factors influencing profits, market share, and other marketing variables.

The existing DMS are oriented to econometrics model of the industry history. In particular, as is shown in [67], existing econometrics models deal inadequately with the information about competitors' future behavior. The models assume the competitor actions are known when the firm makes its decisions. However, the information about competitor future actions is at best vague. [67] presents procedure to model the uncertainties of competitors' behavior under fuzzy information.

The proposed multi-agent distributed intelligent decision making marketing system consists of 5 knowledge based agents, each of which has 9 fuzzy rules. Input information are fuzzy variables of average price $\tilde{x}_1$ (AvgPrice) and average advertising $\tilde{x}_2$ (Avg-Adv) of competitors and are the same for all 5 agents. Using fuzzy inference rule each agent produces its output solutions: firms own price ($\tilde{u}_{i1}$), $i=\overline{1,5}$, and firm's own advertising $\tilde{u}_{i2}, i=\overline{1,5}$. Fuzzy rules in knowledge base of each agent are of the following type (for example, for the first agent):

IF Avg Price is HIGH and AvgAdv is LOW, THEN
Price is HIGH and Advertising is MEDIUM
IF Avg Price is LOW and AvgAdv is HIGH , THEN
Price is MEDIUM and Advertising is HIGH
IF AvgPrice is MEDIUM and Avg Adv is MEDIUM, THEN
Price is HIGH and Advertising is MEDIUM

⋮

For the second agent:
IF AvgPrice is HIGH and Avg Adv is LOW, THEN
Price is HIGH and advertising is LOW
IF Avg Price is LOW and AvgAdv is HIGH, THEN
Price is MEDIUM and Advertising is HIGH

IF AvgPrice is MEDIUM and Avg Adv is MEDIUM, THEN
Price is HIGH and Advertising is LOW

⋮

For the third agent:
IF Avg Price is HIGH and AvgAdv is LOW, THEN
Price is LOW and Advertising is LOW
IF Avg Price is LOW and AvgAdv is HIGH, THEN
Price is LOW and Advertising is HIGH
IF AvgPrice is MEDIUM and Avg Adv is MEDIUM, THEN
Price is LOW and Advertising is MEDIUM

⋮

For the fourth agent:
IF Avg Price is HIGH and AvgAdv is LOW, THEN
Price is HIGH and Advertising is LOW
IF Avg Price is LOW and AvgAdv is HIGH, THEN
Price is LOW and Advertising is HIGH
IF AvgPrice is MEDIUM and Avg Adv is MEDIUM, THEN
Price is MEDIUM and Advertising is MEDIUM

⋮

Finally, for the fifth agent:
IF Avg Price is HIGH and AvgAdv is LOW, THEN
Price is HIGH and Advertising is MEDIUM
IF Avg Price is LOW and AvgAdv is HIGH, THEN
Price is LOW and Advertising is HIGH
IF AvgPrice is MEDIUM and Avg Adv is MEDIUM, THEN
Price is MEDIUM and Advertising is HIGH

⋮

In this example, the firm's management must decide price and advertising levels for each quarter of operations. For the first quarter solutions of each agent for the situation where AvgPrice is about \$325 and AvgAdv is about \$60,000 are shown in Table 4.1.

Table 4.1 Solutions proposed by five agents

Agent#	Price	Advertising
Agent 1	\$331.59	\$80,000
Agent 2	\$331.67	\$53,000
Agent 3	\$313.61	\$60,000
Agent 4	\$331.59	\$53,000
Agent 5	\$331.59	\$70,000

The agents in the considered multi-agent distributed intelligent marketing DSS are characterized by the following criteria:

- C_1 - conformity to situations;
- C_2 - confidence factor (CF);
- C_3 - track record.

Conformity to situations specifies the degree to which an agent's expertise is adequate for the situation at hand. For example, certain agent's expertise may be most adequate to the situations when there is excess inventories and low demand in the industry. Confidence factor measures an agent's confidence in the submitted proposals. Proposals with high degree of confidence tend to be superior to proposals with low degree of confidence. Track record indicates an agent's past performance. Agents whose proposals led to profitable decisions in the past acquire strong track record.

Fuzzy sets characterizing alternatives on criteria C_i have the following form:

$$C_1 = \{0.7/u_{Ag1}, 0.8/u_{Ag2}, 0.6/u_{Ag3}, 0.7/u_{Ag4}, 0.4/u_{Ag5}\}$$

$$C_2 = \{0.8/u_{Ag1}, 0.7/u_{Ag2}, 0.5/u_{Ag3}, 0.6/u_{Ag4}, 0.3/u_{Ag5}\}$$

$$C_3 = \{0.8/u_{Ag1}, 0.6/u_{Ag2}, 0.4/u_{Ag3}, 0.4/u_{Ag4}, 0.25/u_{Ag5}\}$$

The pairwise comparisons of the above criteria resulted in the following:

- C_1 and C_2; C_1 is more important than C_2 with the intensity 7.
- C_1 and C_3; C_1 is more important than C_3 with the intensity 5.
- C_3 and C_2; C_3 is more important than C_2 with the intensity 3.

This results in the following matrix:

$$\begin{bmatrix} 1 & 7 & 5 \\ 1/7 & 1 & 1/3 \\ 1/5 & 3 & 1 \end{bmatrix}$$

The corresponding eigenvector V with $\lambda = 3.065$ is:

$$\begin{bmatrix} 2.192 \\ 0.243 \\ 0.565 \end{bmatrix}$$

Therefore, we derive:

$$\begin{aligned} C_1^{2.192} &= \{0.458/u_{Ag1}, 0.613/u_{Ag2}, 0.326/u_{Ag3}, 0.458/u_{Ag4}, 0.134/u_{Ag5}\} \\ C_2^{0.243} &= \{0.947/u_{Ag1}, 0.917/u_{Ag2}, 0.845/u_{Ag3}, 0.883/u_{Ag4}, 0.746/u_{Ag5}\} \\ C_3^{0.565} &= \{0.882/u_{Ag1}, 0.749/u_{Ag2}, 0.596/u_{Ag3}, 0.596/u_{Ag4}, 0.457/u_{Ag5}\} \end{aligned} \quad (4.49)$$

On the basis of the formula (4.48), and using (4.46) and (4.49) we derive:

$$\begin{aligned} \mu_D(u^*_{Ag}) = \{\max[&\min(0.458, 0.947, 0.882)/u_{Ag_1}, \min(0.613, 0.917, 0.749)/u_{Ag_2}, \\ &\min(0.326, 0.845, 0.596)/u_{Ag_3}, \min(0.458, 0.883, 0.596)/u_{Ag_4}, \\ &\min(0.134, 0.746, 0.457)/u_{Ag_5}]\} \end{aligned}$$

$$\begin{aligned} &\mu_D(u^*_{Ag}) = \max \ [0.458/u_{Ag_1}, 0.613/u_{Ag_2}, 0.326/u_{Ag_3}, \\ &0.458/u_{Ag_4}, 0.134/u_{Ag_5}] = \{0.613/u_{Ag_2}\} \end{aligned}$$

The optimal alternative is u_{Ag_2}, So Ag_2 is the "winner" agent with the proposal

Price =$331.67
Advertising=$53.000

as the solution of the system.

The final solution of the problem (price and advertising for quarters 1,2,3, and 4) consists of the sequence of solutions of agent 2 for quarter 1, agent 5 for quarter 2, agent 5 for quarter 3, and agent 3 for quarter 4. All five agents competed for the total solution of the problem. In each quarter the most suitable agent for the situation was the "winner" while the other four agents were losers. For example, for the situation in quarter 2 agent 5 was more suitable than the rest of the agents.

The suggest approach leads to a more effective decision support since it facilitates generation of distinct alternatives by different agents. Furthermore, the choice of the final decision is supported through consideration of several criteria by the evaluator. The suggested multi-agent DMS also promotes user's learning and understanding through different agent's viewpoints regarding the situation at hand.

4.3 Decision Making on the Basis of Fuzzy Optimality

In the realm of decision making under uncertainty, the general approach is the use of the utility theories. The main disadvantage of this approach is that it is based on an evaluation of a vector-valued alternative by means of a scalar-valued quantity. This transformation is counterintuitive and leads to loss of information. The latter is related to restrictive assumptions on preferences underlying utility models like independence, completeness, transitivity etc. Relaxation of these assumptions results into more adequate but less tractable models.

In contrast, humans conduct direct comparison of alternatives as vectors of attributes' values and don't use artificial scalar values. Although vector-valued utility function-based methods exist, a fundamental axiomatic theory is absent and the problem of direct comparison of vectors remains a challenge with wide scope of research and applications.

There also exist situations for which utility function cannot be applied (for example, lexicographic ordering).

In realm of multicriteria decision making there exist approaches like TOPSIS and AHP to various extent utilizing components-wise comparison of vectors. Basic principle of such comparison is the Pareto optimality which is based on counterintuitive assumption that all alternatives within Pareto optimal set are considered equally optimal. The above mentioned circumstances mandate necessity to develop new decision approaches which, from one side, may be based on direct pairwise comparison of vector-valued alternatives. From the other side, new approaches should be based on linguistic comparison of alternatives to be able to deal with vague vector-valued alternatives because real-life alternatives are almost always a matter of a degree. Linguistic modeling of preferences will help to reduce Pareto optimal set of alternatives arriving at one optimal alternative or a narrowed subset of optimal alternatives when all relevant information is described in NL. For this purpose, a fuzzy optimality concept [30] can be used as obtained from the ideas of Computing with Words (CW)-based redefinitions of the existing scientific concepts [30,83].

The existing classical decision theories, especially in economics, use utility function to transform of a vector to an appropriate scalar value. For many real decision problems, information is imperfect that complicates encoding preferences by a utility function. From the other side, in the existing theories there are so much restrictive assumptions that facilitate decision making where, really, utility function does not exists. For such cases, scoring of alternatives may be conducted by direct ranking of alternatives. Then, as a rule, instead of a utility function it is used binary relations that provide finding an optimal or near to optimal alternative(s). There exists a spectrum of works in this area.

For performing outranking of alternatives, some methods were developed for the traditional multiattribute decision making (MADM) problem [49,84]. One of the first among these methods was ELECTRE method [37,64,65]. The general scheme of this method can be described as follows. A decision maker (DM) assigns weights for each criterion and concordance and discordance indices are constructed. After this, a decision rule is constructed including construction of a binary relation on the base of information received form a DM. Once a binary relation is constructed, a DM is provided with a set of non-dominated alternatives and chooses an alternative from among them as a final decision.

Similar methods like TOPSIS [60,78], VIKOR [60,78] are based on the idea that optimal vector-valued alternative(s) should have the shortest distance from the positive ideal solution and farthest distance from the negative ideal solution. Optimal alternative in the VIKOR is based on the measure of "closeness" to the positive ideal solution. TOPSIS and VIKOR methods use different aggregation functions and different normalization methods. There exists a series of works based on applying interval and fuzzy techniques to extend AHP [28,49], TOPSIS [45,49,74] and VIKOR [46,61] for these methdos to be able to deal with fuzziness and imprecision of information.

In [74] it is suggested fuzzy balancing and ranking method for MADM. They appraise the performance of alternatives against criteria via linguistic variables as TFNs. Selection of an optimal alternative is implemented on the base of a direct comparison of alternatives by confronting their criteria values.

Main drawbacks of the existing approaches mentioned above are the following: 1) detailed and complete information on alternatives provided by a DM and intensive involving the latter into choosing an optimal or suboptimal alternative(s) are required; 2) TOPSIS, ELECTRE, VIKOR and similar approaches require to pose the weights of effective decision criteria that may not be always implementable; 3) Almost all these methods are based on the use of numerical values to evaluate alternatives with respect to criteria; 4) Classical Pareto optimality principle which underlies these methods makes it necessary to deal with a large space of non-dominated alternatives that sufficiently complicates the choice; 5) All these methods are developed only for MADM problems and not for problems of decision making under risk or uncertainty. Although there exists parallelism between these problems, these methods cannot be directly applied for decision making under uncertainty; 6) Absence of a significant impact of fuzzy sets theory for decision making. The existing decision approaches are mainly fuzzy extensions of the well-known MADM approaches like AHP [42], TOPSIS [39,47,87], PROMETHEE, ELECTRE and others. These are approaches where fuzzy sets are applied to the original numerical techniques to account for impreciseness and vagueness. Some of these fuzzy extensions suffer from disadvantages of direct, artificial replacement of numerical quantities by fuzzy numbers. This leads to loss of the properties of the original numerical approaches (for example loss of consistency and transitivity of preference matrices in AHP when replacing their numerical elements by fuzzy numbers). Also, when directly fuzzifying existing approaches one deals with hard computational problems or even problems which don't have solutions (some fuzzy equations), or arrives at counterintuitive formal constructs. From the other side, in some approaches they use defuzzification at an initial or a final stage of computations which lead to loss of information. The decision which is obtained by disregarding real-world imperfect information may not be reliable and may not be consistent with human choices which we try to model. At the same time, it is necessary to mention that there exist mathematically correct and tractable extensions of the existing approaches to the fuzzy environment. For example, these are constraint-based method [59] or an approach suggested in [62] as fuzzy versions of AHP. However, these methods don't resolve disadvantages of the original crisp approaches but only model imprecision within these approaches. These all means that there is a need in applying fuzzy theory to model how humans actually think and reason with perceptions in making decisions, i.e. to use original capabilities of fuzzy theory to describe human-like intelligent decision making.

In order to address the above mentioned problems, it is adequate to use CW-based approaches which are able to cope with imperfect information and to provide an intuitive, human friendly way to decision making.

In [30] they suggested a fuzzy optimality definition for multicriteria decision problems. In this approach each Pareto optimal alternative is assigned a degree of optimality reflecting to what extent the alternative is optimal. To calculate this degree, they directly compare alternatives and arrive at total degrees to which one alternative is better than, is equivalent to and is worse than another one. These degrees are determined as graded sums of differences between criteria values for considered alternatives. Such comparison is closer to the way humans compare alternatives by confronting their criteria values. On the base of these degrees an overall degree of optimality of an alternative is determined. Thus, the Pareto optimal set becomes a fuzzy set reflecting fuzzy constraint on a set of optimal solutions and such constraint is closer to human intuition than a crisp constraint used in classical Pareto optimality.

Let $\mathcal{S}=\{\tilde{S}_1,...,\tilde{S}_M\}\subset\mathcal{E}^n$ be a set of fuzzy states of the nature and $\mathcal{X}\subset\mathcal{E}^n$ be a set of fuzzy outcomes. Fuzziness of states of nature is used for a fuzzy granulation of objective conditions when pure partitioning of the latter is impossible due to vagueness of the relevant information described in NL. A set of alternatives is considered as a set $\mathcal{A}$ of fuzzy functions $\tilde{f}$ from $\mathcal{S}$ to $\mathcal{X}$ [15,16]. Linguistic information on likelihood $\tilde{P}^l$ of the states of nature is represented by fuzzy probabilities $\tilde{P}_j$ of the states $\tilde{S}_j$:

$$\tilde{P}^l=\tilde{P}_1/\tilde{S}_1+\tilde{P}_2/\tilde{S}_2+...+\tilde{P}_M/\tilde{S}_M,$$

where $\tilde{P}_j\in\mathcal{E}^1_{[0,1]}$.

Vague preferences over a set $\mathcal{A}$ of imprecise alternatives is modeled by linguistic preference relation over $\mathcal{A}$. For this purpose it is adequate to introduce a linguistic variable "degree of preference" [21,81] with term-set $\mathcal{T}=(T_1,...,T_K)$. Terms can be labeled as, for example "equivalence", "little preference", "high preference", and each can be described by a fuzzy number defined over some scale, for example [0,1] or [0,10] etc. The fact that $\tilde{f}_i$ is linguistically preferred to $\tilde{f}_k$ is written as $\tilde{f}_i\succsim_l\tilde{f}_k$. The latter means that there exist some $T_i\in\mathcal{T}$ as a linguistic degree $Deg(\tilde{f}_i\succsim_l\tilde{f}_k)$ to which $\tilde{f}_i$ is preferred to $\tilde{f}_k$: $Deg(\tilde{f}_i\succsim_l\tilde{f}_k)\approx T_i$.

So, a framework of decision making with fuzzy imperfect information can be formalized as a 4-tuple $(\mathcal{S},\mathcal{X},\mathcal{A},\succsim_l)$. The problem of decision making consist in determination of an optimal alternative as an alternative $\tilde{f}^* \in \mathcal{A}$ for which $Deg(\tilde{f}^* \succsim_l \tilde{f}_i) \geq Deg(\tilde{f}_i \succsim_l \tilde{f}^*)\ \forall \tilde{f}_i \in \mathcal{A}$. $Deg(\tilde{f}_i \succsim_l \tilde{f}_k)$ is to be determined on the base of degrees of optimality of an alternatives $\tilde{f}_i$ and $\tilde{f}_k$. Degree of optimality denoted $do(\tilde{f}_i)$ is an overall degree to which $\tilde{f}_i$ dominates all the other alternatives [30].

The fuzzy optimality (FO) formalism suggested in [30] is developed for perfect information structure, i.e. when all the decision relevant information is represented by precise numerical evaluations. From the other side, this approach is developed for multiattribute decision making. We will extend the FO formalism for the considered framework of decision making with imperfect information. The method of solution is described below.

At the first stage it is needed for fuzzy probabilities $\tilde{P}_j$ to be known for each fuzzy state of nature $\tilde{S}_j$. However, it can be given only partial information represented by fuzzy probabilities for all fuzzy states except one. The unknown fuzzy probability cannot be assigned but must be computed based on the known fuzzy probabilities. Computation of unknown fuzzy probability, as was shown in the Section 4.1 is an variational problem as it requires construction of a membership function.

The important problem that arises for the obtained $\tilde{P}^l$, is the verification of its consistency, completeness and redundancy (see Section 4.1, formulas (4.13)-(4.19)).

At the second stage, given consistent, complete and not-redundant distribution of fuzzy probabilities over all states of nature it is needed to determine the total degrees of statewise superiority, equivalence and inferiority of $\tilde{f}_i$ with respect to $\tilde{f}_k$ taking into account fuzzy probability $\tilde{P}_j$ of each fuzzy state of nature $\tilde{S}_j$. The total degrees of superiority nbF (number of better fuzzy alternatives), equivalence neF (number of equivalent fuzzy alternatives), and statewise inferiority nwF (number of worse fuzzy alternatives) of $\tilde{f}_i$ with respect to $\tilde{f}_k$ are determined on the base of differences between fuzzy outcomes of $\tilde{f}_i$ and $\tilde{f}_k$ at each fuzzy state of nature as follows:

$$nbF(\tilde{f}_i, \tilde{f}_k) = \sum_{j=1}^{M} \mu_b^j(gmv((\tilde{f}_i(\tilde{S}_j) - \tilde{f}_k(\tilde{S}_j)) \cdot \tilde{P}_j)), \tag{4.50}$$

$$neF(\tilde{f}_i, \tilde{f}_k) = \sum_{j=1}^{M} \mu_e^j(gmv((\tilde{f}_i(\tilde{S}_j) - \tilde{f}_k(\tilde{S}_j)) \cdot \tilde{P}_j)), \tag{4.51}$$

$$nwF(\tilde{f}_i, \tilde{f}_k) = \sum_{j=1}^{M} \mu_w^j(gmv((\tilde{f}_i(\tilde{S}_j) - \tilde{f}_k(\tilde{S}_j)) \cdot \tilde{P}_j)). \tag{4.52}$$

where $\mu_b^j, \mu_e^j, \mu_w^j$ are membership functions for linguistic evaluations "better", "equivalent" and "worse" respectively, determined as in [30]. For j-th state $\mu_b^j, \mu_e^j, \mu_w^j$ are constructed such that Ruspini condition holds, which, in turn, results in the following condition [30]:

$$nbF(\tilde{f}_i,\tilde{f}_k)+neF(\tilde{f}_i,\tilde{f}_k)+nwF(\tilde{f}_i,\tilde{f}_k)=\sum_{j=1}^{M}(\mu_b^j+\mu_e^j+\mu_w^j)=M \tag{4.53}$$

On the base of $nbF(\tilde{f}_i,\tilde{f}_k)$, $neF(\tilde{f}_i,\tilde{f}_k)$, and $nwF(\tilde{f}_i,\tilde{f}_k)$ we calculate $(1-kF)$-dominance as a dominance in the terms of its degree. This concepts suggests that $\tilde{f}_i$ $(1-kF)$-dominates $\tilde{f}_k$ iff

$$neF(\tilde{f}_i,\tilde{f}_k)<M\ ,\ nbF(\tilde{f}_i,\tilde{f}_k)\geq\frac{M-neF(\tilde{f}_i,\tilde{f}_k)}{kF+1}, \tag{4.54}$$

with $kF\in[0,1]$.

In order to determine the greatest kF such that $\tilde{f}_i$ $(1-kF)$-dominates $\tilde{f}_k$, a function d is introduced:

$$d(\tilde{f}_i,\tilde{f}_k)=\begin{cases}0, & \text{if } nbF(\tilde{f}_i,\tilde{f}_k)\leq\dfrac{M-neF(\tilde{f}_i,\tilde{f}_k)}{2}\\ \dfrac{2\cdot nbF(\tilde{f}_i,\tilde{f}_k)+neF(\tilde{f}_i,\tilde{f}_k)-M}{nbF(\tilde{f}_i,\tilde{f}_k)}, & \text{otherwise}\end{cases} \tag{4.55}$$

Given d, the desired greatest kF is found as $1-d(\tilde{f}_i,\tilde{f}_k)$.

$d(\tilde{f}_i,\tilde{f}_k)=1$ implies Pareto dominance of $\tilde{f}_i$ over $\tilde{f}_k$ whereas $d(\tilde{f}_i,\tilde{f}_k)=0$ means no Pareto dominance of $\tilde{f}_i$ over $\tilde{f}_k$.

In contrast to determine whether $\tilde{f}^*$ is Pareto optimal in FO formalism we determine whether $\tilde{f}^*$ is a Pareto optimal with the considered degree kF. $\tilde{f}^*$ is kF optimal if and only if there is no $\tilde{f}_i\in\mathcal{A}$ such that $\tilde{f}_i$ $(1-kF)$-dominates $\tilde{f}^*$.

The main idea of fuzzy optimality concept suggests to consider $\tilde{f}^*$ in terms of its degree of optimality $do(\tilde{f}^*)$ determined as follows:

$$do(\tilde{f}^*) = 1 - \max_{\tilde{f}_i \in \mathcal{A}} d(\tilde{f}_i, \tilde{f}^*). \tag{4.56}$$

So, $do(\tilde{f}^*)$ is a degree resulted from degrees of preferences of $\tilde{f}^*$ to all the other alternatives.

Function do can be considered as the membership function of a fuzzy set describing the notion of kF -optimality.

We call kF -Optimal Set $\mathbb{S}_{kF}$ and kF -Optimal Front $\mathbb{F}_{kF}$ the set of kF -optimal solutions in the design domain and the objective domain respectively.

Let $\mu_{\tilde{D}}(\tilde{f}_i, \tilde{f}_k)$ be a membership function defined as follows:

$$\mu_{\tilde{D}}(\tilde{f}_i, \tilde{f}_k) = \varphi_{\mu_{\tilde{D}}}(nbF(\tilde{f}_i, \tilde{f}_k), neF(\tilde{f}_i, \tilde{f}_k), nwF(\tilde{f}_i, \tilde{f}_k)). \tag{4.57}$$

Then $\mu_{\tilde{D}}(\tilde{f}_i, \tilde{f}_k)$ is a fuzzy dominance relation if for any $\alpha \in [0,1]$ $\mu_{\tilde{D}}(\tilde{f}_i, \tilde{f}_k) > \alpha$ implies that $\tilde{f}_i$ $(1-kF)$ -dominates $\tilde{f}_k$.

Particularly, $\varphi_{\mu_{\tilde{D}}}$ is defined as follows:

$$\varphi_{\mu_{\tilde{D}}} = \frac{2 \cdot nbF(\tilde{f}_i, \tilde{f}_k) + neF(\tilde{f}_i, \tilde{f}_k)}{2M} \tag{4.58}$$

A membership function $\mu_{\tilde{D}}(\tilde{f}_i, \tilde{f}_k)$ represents the fuzzy optimality relation if for any $0 \le kF \le 1$ $\tilde{f}^*$ belongs to the kF -cut of $\mu_{\tilde{D}}$ if and only if there is no $\tilde{f}_i \in \mathcal{A}$ such that

$$\mu_{\tilde{D}}(\tilde{f}_i, \tilde{f}^*) > kF \tag{4.59}$$

At the third stage, on the base of values of $nbF(\tilde{f}_i, \tilde{f}_k)$, $neF(\tilde{f}_i, \tilde{f}_k)$, and $nwF(\tilde{f}_i, \tilde{f}_k)$, the value of degree of optimality $do(\tilde{f}_i)$, as a degree of membership to fuzzy Pareto optimal set, is determined by using formulas (4.54)-(4.58) for each $\tilde{f}_i \in \mathcal{A}$. The obtained $do()$ allows for justified determination of linguistic preference relation $\succsim_l$ over $\mathcal{A}$.

At the fourth stage, the degree $Deg(\tilde{f}_i \succsim_l \tilde{f}_k)$ of preference of $\tilde{f}_i$ to $\tilde{f}_k$ for any $\tilde{f}_i, \tilde{f}_k \in \mathcal{A}$ should be determined based on $do()$. For simplicity, one can calculate $Deg(\tilde{f}_i \succsim_l \tilde{f}_k)$ as follows:

$$Deg(\tilde{f}_i \succsim_l \tilde{f}_k) = do(\tilde{f}_i) - do(\tilde{f}_k).$$

4.4 An Operational Approach to Decision Making under Interval and Fuzzy Uncertainly

As it is mentioned in prefers chapter of the book traditional decision theory is based on a simplifying assumption that for each two alternatives, a user can always meaningfully decide which of them is preferable. In reality, often, when the alternatives are close or vauge, the user is either completely unable to select one of these alternatives, or selects one of the alternatives only ``to some extent".

In the session 4.1 we proposed a natural generalization of the usual decision theory axioms to interval and fuzzy cases, and described decision coming from this generalization.

In this section, we make the resulting decisions more intuitive by providing commonsense operational explanation [11]. First, we recall the main assumption behind the traditional decision theory. We then consider the case when in addition to deciding which of the two alternatives is better, the user can also reply that he/she is unable to decide between the two close alternatives; this leads to interval uncertainty. Finally, we consider the general case when the user makes fuzzy statements about preferences.

Let us assume that for every two alternatives A' and A'', a user can tell:

- whether the first alternative is better for him/her; we will denote this by $A'' < A'$;
- or the second alternative is better; we will denote this by $A' < A''$;
- or the two given alternatives are of equal value to the user; we will denote this by $A' = A''$.

Under the above assumption, we can form a natural numerical scale for describing attractiveness of different alternatives. Namely, let us select a very bad alternative A_0 and a very good alternative A_1, so that most other alternatives are better than A_0 but worse than A_1. Then, for every probability $p \in [0,1]$, we can form a lottery $L(p)$ in which we get A_1 with probability p and A_0 with the remaining probability $1-p$.

When $p=0$, this lottery simply coincides with the alternative A_0: $L(0)=A_0$. The larger the probability p of the positive outcome increases, the better the result, i.e., $p' < p''$ implies $L(p') < L(p'')$. Finally, for $p=1$, the lottery coincides with the alternative A_1: $L(1)=A_1$. Thus, we have a continuous scale of alternatives $L(p)$ that monotonically goes from A_0 to A_1.

We have assumed that most alternatives A are better than A_0 but worse than A_1: $A_0 < A < A_1$. Since $A_0 = L(0)$ and $A_1 = L(1)$, for such alternatives, we thus get $L(0) < A < L(1)$. We assumed that every two alternatives can be compared. Thus, for each such alternative A, there can be at most one value p for which $L(p) = A$; for others, we have $L(p) < A$ or $L(p) > A$. Due to monotonicity of $L(p)$ and transitivity of preference, if $L(p) < A$, then $L(p') < A$ for all $p' \leq p$; similarly, if $A < L(p)$, then $A < L(p')$ for all $p' > p$. Thus, the supremum (= least upper bound) $u(A)$ of the set of all p for which $L(p) < A$ coincides with the infimum (= greatest lower bound) of the set of all p for which $A < L(p)$. For $p < u(A)$, we have $L(p) < A$, and for for $p > u(A)$, we have $A < L(p)$. This value $u(A)$ is called the *utility* of the alternative A.

It may be possible that A is equivalent to $L(u(A))$; however, it is also possible that $A \neq L(u(A))$. However, the difference between A and $L(u(A))$ is extremely small: indeed, no matter how small the value $\varepsilon > 0$, we have $L(u(A) - \varepsilon) < A < L(u(A) + \varepsilon)$. We will describe such (almost) equivalence by ≡, i.e., we write that $A \equiv L(u(A))$.

How can we actually find utility values. The above definition of utility is somewhat theoretical, but in reality, utility can be found reasonably fast by the following iterative bisection procedure.

We want to find the probability $u(A)$ for which $L(u(A)) \equiv A$. On each stage of this procedure, we have the values $\underline{u} < \overline{u}$ for which $L(\underline{u}) < A < L(\overline{u})$. In the beginning, we have $\underline{u} = 0$ and $\overline{u} = 1$, with $|\overline{u} - \underline{u}| = 1$.

To find the desired probability $u(A)$, we compute the midpoint $\tilde{u} = \dfrac{\underline{u} + \overline{u}}{2}$ and compare the alternative A with the corresponding lottery $L(\tilde{u})$. Based on our assumption, there are three possible results of this comparison:

- if the user concludes that $L(\tilde{u}) < A$, then we can replace the previous lower bound $\underline{u}$ with the new one $\tilde{p}$;
- if the user concludes that $A < L(\tilde{u})$, then we can replace the original upper bound $\overline{u}$ with the new one $\tilde{u}$;
- finally, if $A = L(\tilde{u})$, this means that we have found the desired probability $u(A)$.

In this third case, we have found $u(A)$, so the procedure stops. In the first two cases, the new distance between the bounds $\underline{u}$ and $\overline{u}$ is the half of the original distance. By applying this procedure k times, we get values $\underline{u}$ and $\overline{u}$ for which $L(\underline{u}) < A < L(\overline{u})$ and $|\overline{u} - \underline{u}| \leq 2^{-k}$. One can easily check that the desired value $u(A)$ is within the interval $[\underline{u}, \overline{u}]$, so the midpoint $\tilde{u}$ of this interval is an $2^{-(k+1)}$-approximation to the desired utility value $u(A)$.

In other words, for any given accuracy, we can efficiently find the corresponding approximation to the utility $u(A)$ of the alternative A.

How to Make a Decision Based on Utility Values. If we know the utilities $u(A')$ and $u(A'')$ of the alternatives A' and A'', then which of these alternatives should we choose? By definition of utility, we have $A' \equiv L(u(A'))$ and $A'' \equiv L(u(A''))$. Since $L(p') < L(p'')$ if and only if $p' < p''$,we can thus conclude that A' is preferable to A'' if and only if $u(A') > u(A'')$. In other words, we should always select an alternative with the largest possible value of utility.

How to estimate utility of an action: why expected utility. To apply the above idea to decision making, we need to be able to compute utility of different actions. For each action, we usually know possible outcomes $S_1, \ldots, S_n$, and we can often estimate the probabilities $p_1, \ldots, p_n$, $\sum_{i=1}^{n} p_i$, of these outcomes. Let $u(S_1), \ldots, u(S_n)$ be utilities of the situations $S_1, \ldots, S_n$. What is then the utility of the action?

By definition of utility, each situation S_i is equivalent (in the sense of the relation $\equiv$) to alottery $L(u(S_i))$ in which we get A_1 with probability $u(S_i)$ and A_0 with the remaining probability $1 - u(S_i)$. Thus, the action in which we get S_i with probability p_i is equivalent to complex lotteryin which:

- first, we select one of the situations S_i with probability $p_i : P(S_i) = p_i$;
- then, depending on the selected situation S_i, we get A_1 with probability $u(S_i)$ and A_0 with probability $1 - u(S_i)$: $P(A_1 | S_i) = u(S_i)$ and $P(A_0 | S_i) = 1 - u(S_i)$.

In this complex lottery, we end up either with the alternative A_1 or with the alternative A_0. The probability of getting A_1 can be computed by using the complete probability formula:

$$P(A_1) = \sum_{i=1}^{n} P(A_1|S_i) \cdot P(S_i) = \sum_{i=1}^{n} u(S_i) \cdot p_i .$$

Thus, the original action is equivalent to a lottery in which we get A_1 with probability $\sum_{i=1}^{n} p_i \cdot u(S_i)$ and A_0 with the remaining probability. By definition of utility, this means that the utility of our action is equal to $\sum_{i=1}^{n} p_i \cdot u(S_i)$.

In probability theory, this sum is known as the expected value of utility $u(S_i)$. Thus, we can conclude that the utility of each action is equal to its expected utility; in other words, among several possible actions, we should select the one with the largest value of expected utility.

Non-uniqueness of utility. The above definition of utility depends on a selection of two alternatives A_0 and A_1. What if we select different alternatives A_0' and A'? How will utility change? In other words, if A is an alternative with utility $u(A)$ in the scale determined by A_0 and A_1, what is its utility $u'(A)$ in the scale determined by A_0' and A_1'? Let us first consider the case when $A_0' < A_0 < A_1 < A_1'$. In this case, since A_0 is in between A_0' and A_1', for each of them, there exists a probability $u'(A_0')$ for which A_0 is equivalent to a lottery $L'(u'(A_0))$ in which we get A_1 with probability $u'(A_0)$ and A_0' with the remaining probability $1 - u'(A_0)$. Similarly, there exists a probability $u'(A_1)$ for which A_1 is equivalent to a lottery $L'(u'(A_1))$ in which we get A_1' with probability $u'(A_1)$ and A_0' with the remaining probability $1 - u'(A_1)$.

By definition of the utility $u(A)$, the original alternative A is equivalent to a lottery in which we get A_1 with probability $u(A)$ and A_0 with the remaining probability $1-u(A)$. Here, A_1 is equivalent to the lottery $L'(u'(A_1))$, and A_0 is equivalent to the lottery $L'(u'(A_0))$. Thus, the alternative A is equivalent to a complex lottery, in which:

- first, we select A_1 with probability $u(A)$ and A_0 with probability $1 - u(A)$;
- then, depending on the selection A_i, we get A_1' with probability $u'(A_i)$ and A_0' with the remaining probability $1 - u'(A_i)$.

In this complex lottery, we end up either with the alternative A_1' or with the alternative A'. The probability $u'(A) = P(A_1')$ of getting A_1' can be computed by using the complete probability formula:

$u'(A)=P(A_1')=P(A_1' \mid A_1)\cdot P(A_1)+P(A_1' \mid A_0)\cdot P(A_0)=u'(A_1)\cdot u(A)+u(A_0)\cdot(1-u(A)) = u(A)\cdot(u(A_1)-u(A_0))+u(A_0)$.

Thus, the original alternative A is equivalent to a lottery in which we get A_1' with probability $u'(A) = u(A)\cdot(u'(A_1) - u'(A_0)) + u'(A_0)$. By definition of utility, this means that the utility $u'(A)$ of the alternative A in the scale determined by the alternatives A_0' and A_1' is equal to $u'(A) = u(A)\cdot(u'(A_1) - u'(A_0)) + u'(A_0)$.

Thus, in the case when $A_0' < A_0 < A_1 < A_1'$, when we change the alternatives A_0 and A_1, the new utility values are obtained from the old ones by a linear transformation. In other cases, we can use auxiliary events A_0'' and A_1'' for which $A_0'' < A_0, A_0'$ and $A_1, A_1' < A''$.

1. In this case, as we have proven, transformation from $u(A)$ to $u''(A)$ is linear and transformation from $u'(A)$ to $u''(A)$ is also linear. Thus, by combining linear transformations $u(A) \to u''(A)$ and $u''(A) \to u'(A)$, we can conclude that the transformation $u(A) \to u'(A)$ is also linear. So, in general, utility is defined modulo an (increasing) linear transformation $u' = a\cdot u+b$, with $a > 0$.

Comment. So far, once we have selected alternatives A_0 and A_1, we have defined the corresponding utility values $u(A)$ only for alternatives A for which $A_0 < A < A_1$. For such alternatives, the utility value is always a number from the interval $[0,1]$.

For other alternatives, we can define their utility $u'(A)$ with respect to different pairs A_0' and A_1', and then apply the corresponding linear transformation to re-scale to the original units. The resulting utility value $u(A)$ can now be an arbitrary real number. Subjective probabilities. In our derivation of expected utility, we assumed that we know the probabilities p_i of different outcomes. In practice, we often do not know these probabilities, we have to rely on a subjective evaluation of these probabilities. For each event E, a natural way to estimate its subjective probability is to compare the lottery $\ell(E)$ in which we get a fixed prize (e.g.,$1) if the event E occurs and 0 is it does not occur, with a lottery $\ell(p)$ in which we get the same amount with probability p. Here, similarly to the utility case, we get a value $ps(E)$ for which $\ell(E)$ is (almost) equivalent to $\ell(ps(E))$ in the sense that $\ell(ps(E) - \varepsilon) < \ell(E) < \ell(ps(E) + \varepsilon)$ for every $\varepsilon > 0$. This value $ps(E)$ is called the *subjective probability* of the event E.

From the viewpoint of decision making, each event E is equivalent to an event occurring with the probability $ps(E)$. Thus, if an action has n possible $s_1,\ldots,s_n$, in which S_i happens if the event E_i occurs, then the utility of this action is equal to $\sum_{i=1}^{n} pS(E_i)\cdot u(S_i)\cdot$

Beyond Traditional Decision Making: Towards a More Realistic Description. Instead of assuming that a user can always decide which of the two alternatives A' and A'' is better, let us now consider a more realistic situation in which a user is allowed to say that he or she is unable to meaningfully decide between the two alternatives; we will denote this option by $A' \,||\, A''$.

In mathematical terms, this means that the preference relation is no longer a *total* (linear) order, it can be a *partial* order.

From Utility to Interval-Valued Utility. Similarly to the traditional decision making approach, we can select two alternatives $A_0 < A_1$ and compare each alternative A which is better than A_0 and worse than $A1$ with lotteries $L(p)$. The main difference is that here, the supremum $u(A)$ of all the values p for which $L(p) < A$ is, in general, smaller than the infimum $u(A)$ of all the values p for which $A < L(p)$. Thus, for each alternative A, instead of a single value $u(A)$ of the utility, we now have an *interval* $[u(A), u(A)]$ such that:

- if $p < u(A)$, then $L(p) < A$;
- if $p > u(A)$, then $A < L(p)$; and
- if $u(A) < p < u(A)$, then $A||L(p)$.

We will call this interval the *utility* of the alternative A.

How to Efficiently Find the Interval-Valued Utility. To elicit the corresponding utility interval from the user, we can use a slightly modified version of the above bisection procedure. At first, the procedure is the same as before: namely, we produce a narrowing interval $[u, u]$ for which $L(u) < A < L(u)$.

We start with the interval $[\underline{u},\overline{u}]=[0,1]$, and we repeatedly compute the midpoint $\tilde{u} = \dfrac{\underline{u}+\overline{u}}{2}$ and compare A with $L(\tilde{u})$. If $L(\tilde{u}) < A$, we replace $\underline{u}$ with $\tilde{u}$; if $A < L(\tilde{u})$, we replace $\overline{u}$ with $\tilde{u}$. If we get $A||L(\tilde{p})$, then we switch to the new second stage of the iterative algorithm. Namely, now, we have *two* intervals:

- an interval $[\underline{u}_1,\overline{u}_1]$ (which is currently equal to $[\underline{u},\tilde{u}]$) for which $L(\underline{u}_1)<A$ and $L(\tilde{u}_1)||A$, and
- an interval $[\underline{u}_2,\overline{u}_2]$ (which is currently equal to $[\tilde{u},\overline{u}]$) for which $L(\underline{u}_2)||A$ and $A < L(\underline{u}_2)$.

Then, we perform bisection of each of these two intervals. For the first interval, we compute the midpoint $\tilde{u}_1 = \dfrac{\underline{u}_1+\overline{u}_1}{2}$, and compare the alternative A with the lottery $L(\tilde{u}_1)$:

- if $L(\tilde{u}_1) < A$, then we replace $\underline{u}_1$ with $\tilde{u}_1$;
- if $L(\tilde{u}_1) || A$, then we replace $\overline{u}_1$ with $\tilde{u}_1$.

As a result, after k iterations, we get the value $\underline{u}(A)$ with accuracy $2-k$.

Similarly, for the second interval, we compute the midpoint $\tilde{u}_2 = \frac{\underline{u}_2 + \overline{u}_2}{2}$, and compare the alternative A with the lottery $L(\tilde{u}_2)$:

- if $L(\tilde{u}_2) || A$, then we replace $\underline{u}_2$ with $\tilde{u}_2$;
- if $A < L(\tilde{u}_2)$, then we replace $\overline{u}_2$ with $\tilde{u}_2$.

As a result, after k iterations, we get the value $\overline{u}(A)$ with accuracy $2-k$.

Interval-Valued Subjective Probability. Similarly, when we are trying to estimate the probability of an event E, we no longer get a single value $ps(E)$, we get an interval $[\underline{ps}(E), \overline{ps}(E)]$ of possible values of probability.

Need for Decision Making under Interval Uncertainty. In the traditional approach, for each alternative A, we produce a number $u(A)$ – the utility of this alternative. Then, an alternative A' is preferable to the alternative A'' if and only if $u(A') > u(A'')$.

How can we make a similar decision in situations when we only know interval-valued probabilities? At first glance, the situation may sound straightforward: if $A' || A''$, it does not matter whether we select A' or A''. However, this is *not* a good way to make a decision. For example, let us assume that there is an alternative A about which we know nothing. In this case, we have no reason to prefer A or $L(p)$, so we have $A || L(p)$ for all p. By definition of $\underline{u}(A)$ and $\overline{u}(A)$, this means that we have $\underline{u}(A) = 0$ and $\overline{u}(A) = 1$, i.e., the alternative A is characterized by the utility interval [0, 1].

In this case, the alternative A is indistinguishable both from a good lottery $L(0.999)$ (in which the good alternative $A1$ appears with probability 99.9%) and from a bad lottery $L(0.001)$ (in which the bad alternative $A0$ appears with probability 99.9%). If we recommend, to the user, that A is equivalent both to to $L(0.999)$ and $L(0.001)$, then this user will feel comfortable exchanging his chance to play in the good lottery with A, and then – following the same logic – exchanging A with a chance to play in a bad lottery. As a result, following our recommendations, the user switches from a very good alternative to a very bad one.

This argument does not depend on the fact that we assumed complete ignorance about A. Every time we recommend that the alternative A is equivalent to $L(p)$ and $L(p')$ with two different values $p < p'$, we make the user vulnerable to a similar switch from a better alternative $L(p')$ to a worse one $L(p)$. Thus, there should be only a single value p for which A can be reasonably exchanged with $L(p)$.

In precise terms: we start with the utility interval $[\underline{u}(A), \overline{u}(A)]$, and we need to select a single utility value u for which it is reasonable to exchange the alternative A with a lottery $L(u)$. How can we find this value u?

How to make decisions under interval uncertainty? We will use Hurwicz optimism-pessimism criterion. The problem of decision making under such interval uncertainty was first handled by the future Nobelist L. Hurwicz [38].

We need to assign, to each interval $[\underline{u}, \overline{u}]$, a utility value $u(\underline{u}, \overline{u})$.

No matter what value u we get from this interval, this value will be larger than or equal to $\underline{u}$ and smaller than or equal to $\overline{u}$. Thus, the equivalent utility value $u(\underline{u}, \overline{u})$ must satisfy the same inequalities: $\underline{u} \le u(\underline{u}, \overline{u}) \le \overline{u}$. In particular, for $\underline{u} = 0$ and $\overline{u} = 1$, we get $0 \le \alpha_H \le 1$, where we denoted $\alpha_H \stackrel{def}{=} u(0,1)$.

We have mentioned that the utility is determined modulo a linear transformation $u' = a \cdot u + b$. It is therefore reasonable to require that the equivalent utility does not depend on what scale we use, i.e., that for every $a > 0$ and b, we have

$$u(a \cdot \underline{a} + b, a \cdot \overline{u} + b) = a \cdot u(\underline{u}, \overline{u}) + b.$$

In particular, for $\underline{u} = 0$ and $\overline{u} = 1$, we get

$$u(b, a+b) = a \cdot u(0,1) + b = a \cdot \alpha_H + b.$$

So, for every $\underline{u}$ and $\overline{u}$, we can take $b = \underline{u}$, $a = \overline{u} - \underline{u}$, and get

$$u(\underline{u}, \overline{u}) = \underline{u} + \alpha_H \cdot (\overline{u} - \underline{u}) = \alpha_H \cdot \overline{u} + (1 - \alpha_H) \cdot \underline{u}.$$

This expression is called *Hurwicz optimism-pessimism criterion*, because:

- when $\alpha_H = 1$, we make a decision based on the most optimistic possible values $u = \overline{u}$;
- when $\alpha_H = 0$, we make a decision based on the most pessimistic possible values $u = \underline{u}$;
- for intermediate values $\alpha_H \in (0,1)$, we take a weighted average of the optimistic and pessimistic values.

So, if we have two alternatives A' and A'' with interval-valued utilities $[\underline{u}(A'), \overline{u}(A')]$ and $[\underline{u}(A''), \overline{u}(A'')]$, we recommend an alternative for which the equivalent utility value is the largest. In other words, we recommend to select A' if $\alpha_H \cdot \overline{u}(A') + (1-\alpha_H) \cdot \underline{u}(A') > \alpha_H \cdot \overline{u}(A'') + (1-\alpha_H) \cdot \underline{u}(A'')$ and A'' otherwise.

Which value α_H should we choose? An argument in favor of $\alpha_H = 0.5$.

To answer this question, let us take an event E about which we know nothing. For a lottery L^+ in which we get A_1 if E and A_0 otherwise, the utility interval is $[0,1]$, thus, from a decision making viewpoint, this lottery should be equivalent to an event with utility $\alpha_H \cdot 1 + (1-\alpha_H) \cdot 0 = \alpha_H$.

Similarly, for a lottery L^- in which we get A_0 if E and A_1 otherwise, the utility interval is $[0,1]$, thus, this lottery should also be equivalent to an event with utility $\alpha_H \cdot 1 + (1-\alpha_H) \cdot 0 = \alpha_H$.

We can now combine these two lotteries into a single complex lottery, in which we select either L^+ or L^- with equal probability 0.5. Since L^+ is equivalent to a lottery $L(\alpha_H)$ with utility α_H and L^- is also equivalent to a lottery $L(\alpha_H)$ with utility α_H, the complex lottery is equivalent to a lottery in which we select either $L(\alpha_H)$ or $L(\alpha_H)$ with equal probability 0.5, i.e., to $L(\alpha_H)$. Thus, the complex lottery has an equivalent utility α_H.

On the other hand, no matter what is the event E, in the above complex lottery, we get A_1 with probability 0.5 and A_0 with probability 0.5. Thus, this complex lottery coincides with the lottery $L(0.5)$ and thus, has utility 0.5. Thus, we conclude that $\alpha_H = 0.5$.

Which action should we choose? Suppose that an action has n possible outcomes $S_1, \ldots, S_n$, with utilities $[\underline{u}(S_i), \overline{u}(S_i)]$, and probabilities $[\underline{p}_i, \overline{p}_i]$. How do we then estimate the equivalent utility of this action?

We know that each alternative is equivalent to a simple lottery with utility $u_i = \alpha_H \cdot \overline{u}(S_i) + (1-\alpha_H) \cdot \underline{u}(S_i)$, and that for each i, the i-th event is -- from the viewpoint of decision making -- equivalent to $p_i = \alpha_H \cdot \overline{p}_i + (1-\alpha_H) \cdot \underline{p}_i$. Thus, from the viewpoint of decision making, this action is equivalent to a situation in which we get utility u_i with probability p_i. We know that the utility of such a situation is equal to $\sum_{i=1}^{n} p_i \cdot u_i$. Thus, the equivalent utility of the original action is equivalent to

$$\sum_{i=1}^{n} p_i \cdot u_i = \sum_{i=1}^{n} (\alpha_H \cdot \overline{p}_i + (1-\alpha_H) \cdot \underline{p}_i) \cdot (\alpha_H \cdot \overline{u}(S_i) + (1-\alpha_H) \cdot \underline{u}(S_i)).$$

The resulting decision depends on the level of detail. We make a decision in a situation when we do not know the exact values of the utilities and when we do not know the exact values of the corresponding probabilities. Clearly, if gain new information, the equivalent utility may change. For example, if we know nothing about an alternative A, then its utility is $[0,1]$ and thus, its equivalent utility is α_H. Once we narrow down the utility of A, e.g., to the interval $[0.5,0.9]$, we get a different equivalent utility $\alpha_H \cdot 0.9 + (1-\alpha_H) \cdot 0.5 = 0.5 + 0.4 \cdot \alpha_H$. On this example, the fact that we have different utilities makes perfect sense.

However, there are other examples where the corresponding difference is not as intuitively clear. Let us consider a situation in which, with some probability p, we gain a utility u, and with the remaining probability $1-p$, we gain utility 0. If we know the exact values of u and p, we can then compute the equivalent utility of this situation as the expected utility value $p \cdot u + (1-p) \cdot 0 = p \cdot u$.

Suppose now that we only know the interval $[\underline{u}, \overline{u}]$ of possible values of utility and the interval $[\underline{p}, \overline{p}]$ of possible values of probability. Since the expression $p \cdot u$ for the expected utility of this situation is an increasing function of both variables:

- the largest possible utility of this situation is attained when both p and u are the largest possible: $u = \overline{u}$ and $p = \overline{p}$, and
- the smallest possible utility is attained when both p and u are the smallest possible: $u = \underline{u}$ and $p = \underline{p}$.

In other words, the resulting amount of utility ranges from $\underline{p} \cdot \underline{u}$ to $\overline{p} \cdot \overline{u}$.

If we know the structure of the situation, then, according to our derivation, this situation has an equivalent utility $u_k = (\alpha_H \cdot \overline{p} + (1-\alpha_H) \cdot \underline{p}) \cdot (\alpha_H \cdot \overline{u} + (1-\alpha_H) \cdot \underline{u})$ (k for *k*now). On the other hand, if we do not know the structure, if we only know that the resulting utility is from the interval $[\underline{p} \cdot \underline{u}, \overline{p} \cdot \overline{u}]$, then, according to the Hurwicz criterion, the equivalent utility is equal to $u_d = \alpha_H \cdot \overline{p} \cdot \overline{u} + (1-\alpha_H) \cdot \underline{p} \cdot \underline{u}$ (d for *d*on't know). One can check that

$$u_d - u_k = \alpha_H \cdot \overline{p} \cdot \overline{u} + (1-\alpha_H) \cdot \underline{p} \cdot \underline{u} - \alpha_H^2 \cdot \overline{p} \cdot \overline{u} - \alpha_H \cdot (1-\alpha_H) \cdot (\underline{p} \cdot \overline{u} + \overline{p} \cdot \underline{u}) -$$
$$(1-\alpha_H)^2 \cdot \underline{p} \cdot \underline{u} = \alpha_H \cdot (1-\alpha_H) \cdot \overline{p} \cdot \overline{u} + \alpha_H \cdot (1-\alpha_H) \cdot \underline{p} \cdot \underline{u} - \alpha_H \cdot (1-\alpha_H) \times$$
$$(\underline{p} \cdot \overline{u} + \overline{p} \cdot \underline{u}) = \alpha_H \cdot (1-\alpha_H) \cdot (\overline{p} - \underline{p}) \cdot (\overline{u} - \underline{u}).$$

This difference is always positive, meaning that additional knowledge decreases the utility of the situation. (This is maybe what the Book of Ecclesiastes means by ``For with much wisdom comes much sorrow"?)

From Intervals to General Sets. In the ideal case, we know the exact situation s in all the detail, and we can thus determine its utility $u(s)$. Realistically, we have an imprecise knowledge, so instead of a single situation s, we only know a *set* S of possible situations s. Thus, instead of a single value of the utility, we only know that the actual utility belongs to the set $U = \{u(s) : s \in S\}$. If this set S is an interval $[\underline{u}, \overline{u}]$, then we can use the above arguments to come up with its equivalent utility value $\alpha_H \cdot \overline{u} + (1 - \alpha_H) \cdot \underline{u}$.

What is U is a generic set? For example, we can have a 2-point set $U = \{\underline{u}, \overline{u}\}$. What is then the equivalent utility?

Let us first consider the case when the set U contains both its infimum $\underline{u}$ and its supremum $\overline{u}$. The fact that we only know the set of possible values and have no other information means that *any* probability distribution on this set is possible (to be more precise, it is possible to have any probability distribution on the set of possible situations S, and this leads to the probability distribution on utilities). In particular, for each probability p, it is possible to have a distribution in which we have $\overline{u}$ with probability p and $\underline{u}$ with probability $1 - p$. For this distribution, the expected utility is equal to $p \cdot \overline{u} + (1 - p) \cdot \underline{u}$. When p goes from 0 to 1, these values fill the whole interval $[\underline{u}, \overline{u}]$. Thus, every value from this interval is the possible value of the expected utility. On the other hand, when $u \in [\underline{u}, \overline{u}]$, the expected value of the utility also belongs to this interval -- no matter what the probability distribution. Thus, the set of all possible utility values is the whole interval $[\underline{u}, \overline{u}]$ and so, the equivalent utility is equal to $\alpha_H \cdot \overline{u} + (1 - \alpha_H) \cdot \underline{u}$.

When the infimum and/or supremum are not in the set S, then the set S contains points as close to them as possible. Thus, the resulting set of possible values of utility is as close as possible to the interval $[\underline{u}, \overline{u}]$ -- and so, it is reasonable to assume that the equivalent utility is as close to $u_0 = \alpha_H \cdot \overline{u} + (1 - \alpha_H) \cdot \underline{u}$ as possible – i.e., coincides with this value u_0.

From Sets to Fuzzy Sets. What if utility is a fuzzy number, described by a membership function $\mu(u)$? One of the natural interpretations of a fuzzy set is via its nested intervals α-cuts $u(\alpha)=[\underline{u}(\alpha_H),\overline{u}(\alpha_H)]=\{u:\mu(u)\geq\alpha\}$. For example, when we are talking about a measurement error of a given measuring instrument, then we know the guaranteed upper bound, i.e., the guaranteed interval that contains all possible values of the measurement error. In addition to this guaranteed interval, experts can usually pinpoint a narrower interval that contains the measurement error with some certainty; the narrower the interval, the smaller our certainty. Thus, we are absolutely sure (with certainty 1) that the actual value u belongs to the α-cut $u(0)$; also, with a degree of certainty $1-\alpha$, we claim that $x\in u(\alpha)$. Thus, if we select some small value $\Delta\alpha$ and take $\alpha=0,\Delta\alpha,2\Delta,...,n\Delta,...$, we conclude that:

- with probability $\alpha=0$, the set of possible values of u is the interval $[\underline{u}(0),\overline{u}(0)]$;
- with probability $\alpha=\Delta\alpha$, the set of possible values of u is the interval $[\underline{u}(\Delta\alpha),\overline{u}(\Delta\alpha)]$;
- ...
- with probability $\alpha=n\Delta\alpha$, the set of possible values of u is the interval $[\underline{u}(n\Delta\alpha),\overline{u}(n\Delta\alpha)]$;
- ...

For each interval, the equivalent utility value is $\alpha\cdot\overline{u}(\alpha)+(1-\alpha)\underline{u}(\alpha)$. The entire situation is a probabilistic combination of such intervals, so the resulting equivalent utility is equal to the expected value of the above utility, i.e., to

$$\alpha\cdot\int_0^1\overline{u}(\alpha)d\alpha+(1-\alpha)\cdot\int_0^1\underline{u}(\alpha)d\alpha$$

References

1. Alexeyev, A.V., Borisov, A.N., Glushkov, V.I., Krumberg, O.A., Merkuryeva, G.V., Popov, V.A., Slyadz, N.N.: A linguistic approach to decision-making problems. Fuzzy Sets and Systems 22, 25–41 (1987)
2. Aliev, R.A.: Decision and Stability Analysis in Fuzzy Economics. In: Annual Meeting of the North American Fuzzy Information Processing Society (NAFIPS 2009), Cincinnati, USA, pp. 1–2 (2009)
3. Aliev, R.A.: Decision Making Theory with Imprecise Probabilities. In: Proceedings of the Fifth International Conference on Soft Computing and Computing with Words in System Analysis, Decision and Control (ICSCCW 2009), pp. 1 (2009)
4. Aliev, R.A., Aliev, R.R.: Fuzzy Distributed Intelligent Systems for Continuous Production. Application of Fuzzy Logic. In: Jamshidi, M., Titli, M., Zadeh, L., Boverie, S. (eds.) Towards High Machine Intelligence Quotient Systems, pp. 301–320 (1997)

5. Aliev, R.A., Aliev, R.R.: Soft Computing and its Application. World Scientific, New Jersey (2001)
6. Aliev, R.A., Aliyev, B.F., Gardashova, L.A., Huseynov, O.H.: Selection of an Optimal Treatment Method for Acute Periodontitis Disease. Journal of Medical Systems 36(2), 639–646 (2012)
7. Aliev, R.A., Alizadeh, A.V., Guirimov, B.G., Huseynov, O.H.: Precisiated information-based approach to decision making with imperfect information. In: Proc. of the Ninth International Conference on Application of Fuzzy Systems and Soft Computing (ICAFS 2010), Prague, Czech Republic, pp. 91–103 (2010)
8. Aliev, R.A., Bonfig, K.W., Aliev, F.T.: Messen, Steuern und Regeln mit Fuzzy- Logik. Franzis-Verlag, München (1993) (in German)
9. Aliev, R.A., Fazlollahi, B., Vahidov, R.M.: Soft Computing Based Multi-Agent Marketing Decision Support Systems. Journal of Intelligent and Fuzzy Systems 9, 1–9 (2000)
10. Aliev, R.A., Huseynov, O.H., Aliev, R.R.: Decision making with imprecise probabilities and its application. In: Proc. 5th International Conference on Soft Computing and Computing with Words in System Analysis, Decision and Control (ICSCCW 2009), pp. 1–5 (2009)
11. Aliev, R.A., Huseynov, O.H., Kreinovich, V.: Decision making with imprecise probabilities and its application. In: Proc. of the Ninth International Conference on Application of Fuzzy Systems and Soft Computing (ICAFS 2012), pp. 145–152 (2012)
12. Aliev, R.A., Krivosheev, V.P., Liberzon, M.I.: Optimal decision coordination in hierarchical systems. News of Academy of Sciences of USSR, Tech. Cybernetics 2, 72–79 (1982) (in English and Russian)
13. Aliev, R.A., Liberzon, M.I.: Coordination methods and algorithms for integrated manufacturing systems, vol. 208. Radio I svyaz, Moscow (1987) (in Russian)
14. Aliev, R.A., Mamedova, G.M., Aliev, R.R.: Fuzzy Sets and its Application. Tabriz University, Tabriz (1993)
15. Aliev, R.A., Pedrycz, W., Fazlollahi, B., Huseynov, O.H., Alizadeh, A.V., Guirimov, B.G.: Fuzzy logic-based generalized decision theory with imperfect information. Information Sciences 189, 18–42 (2012)
16. Aliev, R.A., Pedrycz, W., Huseynov, O.H.: Decision theory with imprecise probabilities. International Journal of Information Technology & Decision Making 11(2), 271–306 (2012), doi:10.1142/S0219622012400032
17. Alo, R., de Korvin, A., Modave, F.: Using Fuzzy functions to select an optimal action in decision theory. In: Proc. of the North American Fuzzy Information Processing Society (NAFIPS), pp. 348–353 (2002)
18. Azadeh, I., Fam, I.M., Khoshnoud, M., Nikafrouz, M.: Design and implementation of a fuzzy expert system for performance assessment of an integrated health, safety, environment (HSE) and ergonomics system: The case of a gas refinery. Information Sciences 178(22, 15), 4280–4300 (2008)
19. Belman, R.E., Zadeh, L.A.: Decision Making in a Fuzzy environment. Management Sci. 17, 141–164 (1970)
20. Billot, A.: An existence theorem for fuzzy utility functions: a new elementary proof. Fuzzy Sets and Systems 74, 271–276 (1995)
21. Borisov, A.N., Alekseyev, A.V., Merkuryeva, G.V., Slyadz, N.N., Gluschkov, V.I.: Fuzzy information processing in decision making systems. Radio i Svyaz, Moscow (1989)

22. Borisov, A.N., Krumberg, O.A., Fedorov, I.P.: Decision making on the basis of fuzzy models. Examples for utilization, Riga, Zinatne (1984)
23. Chen, T.-Y.: Signed distanced-based TOPSIS method for multiple criteria decision analysis based on generalized interval-valued fuzzy numbers, making approach. International Journal of Information Technology & Decision Making 10(6), 1131–1159 (2011), doi:10.1142/S0219622011004749
24. de Wilde, P.: Fuzzy utility and equilibria. IEEE Transactions on Systems, Man, and Cybernetics, Part B: Cybernetic 34(4), 1774–1785 (2004)
25. Diamond, P., Kloeden, P.: Metric spaces of fuzzy sets. Theory and applications. World Scientific, Singapoure (1994)
26. Dowling, J.M., Chin Fang, Y.: Modern Developments in Behavioral Economics. Social Science Perspectives on Choice and Decision making, vol. 446. World Scientific Publishing Co. Pte. Ltd., Singapore (2007)
27. Dubois, D., Prade, H.: A review of fuzzy sets aggregation connectives. Information Sciences 36, 85–121 (1985)
28. Enea, M., Piazza, T.: Project Selection by Constrained Fuzzy AHP. Fuzzy Optimization and Decision Making 3(1), 39–62 (2004)
29. Epstein, L.G., Schneider, M.: Ambiguity, information quality and asset pricing. Journal of Finance 63(1), 197–228 (2008)
30. Farina, M., Amato, P.: A fuzzy definition of "optimality" for many-criteria optimization problems. IEEE Transactions on Systems, Man and Cybernetics, Part A: Systems and Humans 34(3), 315–326 (2004)
31. Fazlollahi, B., Vahidov, R.M., Aliev, R.A.: Multi-Agent Distributed Intelligent Systems Based on Fuzzy Decision-Making. International Journal of Intelligent Systems 15, 849–858 (2000)
32. Ferson, S., Ginsburg, L., Kreinovich, V., et al.: Uncertainty in risk analysis. Towards a general second-order approach combining interval, probabilistic, and fuzzy techniques. Proceedings of FUZZ-IEEE 2, 1342–1347 (2002)
33. Fodor, J., Roubens, M.: Fuzzy Preference Modelling and Multicriteria Decision Support. Kluwer, Dordrecht (1994)
34. Gil, M.A., Jain, P.: Comparison of Experiments in Statistical Decision Problems with Fuzzy Utilities. IEEE Transactions on Systems, Man, And Cyberneticts 22(4), 662–670 (1992)
35. Guo, P.: One-Shot Decision Theory. IEEE Transactions on Systems, Man and Cybernetics – Part A: Systems and Humans 41(5), 917–926 (2011)
36. Guo, P., Tanaka, H.: Decision Analysis based on Fused Double Exponential Possibility Distributions. European Journal of Operational Research 148, 467–479 (2003)
37. Huangm, W.C., Chen, C.H.: Using the ELECTRE II method to apply and analyze the differentiation theory. In: Proc. of the Eastern Asia Society for Transportation Studies, vol. 5, pp. 2237–2249 (2005)
38. Hurwicz, L.: Optimality Criteria for Decision Making Under Ignorance, Cowles Commission Discussion Paper, Statistics, No. 370 (1951)
39. Jahanshahloo, G.R., Lotfi, F., Hosseinzadeh, I.M.: An algorithmic method to extend TOPSIS for decision-making problems with interval data. Appl. Math. Comput. 175(2), 1375–1384 (2006)
40. Jong, Y., Liang, W., Reza, L.: Multiple Fuzzy Systems for Function Approximation. In: Proceedings of NAFIPS 1997, Syracuse, New York, pp. 154–159 (1997)
41. Kahneman, D., Slovic, P., Tversky, A.: Judgment Under Uncertainty: Heuristics and Biases, p. 544. Cambridge University Press (1982)

42. Kapoor, V., Tak, S.S.: Fuzzy application to the analytic hierarchy process for robot selection. Fuzzy Optimization and Decision Making 4(3), 209–234 (2005)
43. Khan, N.A., Jain, R.: Uncertainty Management in a Distributed Knowledge Base system. In: Proceedings of International Joint Conference on Artificial Intelligence, pp. 318–320 (1985)
44. Klir, C.J., Yuan, B.: Fuzzy Sets and Fuzzy Logic: Theory and Applications. Prentice Hall, PTR (1995)
45. Lakshmikantham, V., Mohapatra, R.: Theory of fuzzy differential equations and inclusions. Taylor & Francis, London (2003)
46. Liu, P., Wang, M.: An extended VIKOR method for multiple attribute group decision making based on generalized interval-valued trapezoidal fuzzy numbers. Scientific Research and Essays 6(4), 766–776 (2011)
47. Liu, W.J., Zeng, L.: A new TOPSIS method for fuzzy multiple attribute group decision making problem. J Guilin Univ. Electron. Technol. 28(1), 59–62 (2008)
48. Loia, V. (ed.): Soft Computing Agents. A New Perspective for Dynamic Information Systems, p. 254. IOS Press, Amsterdam (2002)
49. Lu, J., Zhang, G., Ruan, D., Wu, F.: Multi-objective group decision making. In: Methods, Software and Applications with Fuzzy Set Techniques. Series in Electrical and Computer Engineering, vol. 6. Imperial College Press, London (2007)
50. Mathieu Nicot, B.: Fuzzy Expected Utility. Fuzzy Sets and Systems 20(2), 163–173 (1986)
51. Mendel, J.M.: Computing with words and its relationships with fuzzistics. Information Sciences 179(8), 988–1006 (2007)
52. Mikhailov, L., Didehkhani, H., Sadi Nezhad, S.: Weighted prioritization models in the fuzzy analytic hierarchy process. International Journal of Information Technology & Decision Making 10(4), 681–694 (2011), doi:10.1142/S0219622011004518
53. Modave, F., Grabisch, M., Dubois, D., Prade, H.: A Choquet Integral Representation in Multicriteria Decision Making. Technical Report of the Fall Symposium of Association for the Advancement of Artificial Intelligence (AAAI), pp. 22–29. AAAI Press, Boston (1997)
54. Mordeson, J.N., Nair, P.S.: Fuzzy mathematics: an introduction for engineers and scientists. Physica-Verlag, Heidelberg (2001)
55. Murofushi, T.: Semiatoms in Choquet Integral Models of Multiattribute Decision Making. Journal of Advanced Computational Intelligence and Intelligent Informatics 9(5), 477–483 (2005)
56. Nanda, S.: Fuzzy linear spaces over valued fields. Fuzzy Sets and Systems 42(3), 351–354 (1991)
57. Nazari Shirkouhi, S., Ansarinejad, A., Miri Nargesi, S.S., Majazi Dalfard, V., Rezaie, K.: Information systems outsourcing decisions under fuzzy group decision making approach. International Journal of Information Technology & Decision Making 10(6), 989–1022 (2011), doi:10.1142/S0219622011004683
58. Nguyen, H.T., Walker, E.A.: A first Course in Fuzzy logic. CRC Press, Boca Raton (1996)
59. Ohnishi, S., Dubois, D., Prade, H., Yamanoi, T.: A fuzzy constraint-based approach to the analytic hierarchy process. In: Bouchon-Meunier, B., et al. (eds.) Uncertainty and Intelligent Information Systems, pp. 217–228. World Scientific, Singapore (2008)
60. Opricovic, S., Tzeng, G.H.: Compromise solution by MCDM methods: a comparative analysis of VIKOR and TOPSIS. Eur. J. Operat. Res. 156(2), 445–455 (2004)

61. Park, J.H., Cho, H.J., Kwun, Y.C.: Extension of the VIKOR method for group decision making with interval-valued intuitionistic fuzzy information. Fuzzy Optimization and Decision Making 10(3), 233–253 (2011)
62. Ramík, J., Korviny, P.: Inconsistency of pair-wise comparison matrix with fuzzy elements based on geometric mean. Fuzzy Sets and Systems 161, 1604–1613 (2010)
63. Roe, J.: Index theory, coarse geometry, and topology of manifolds. In: CBMS: Regional Conf. Ser. in Mathematics. The American Mathematical Society, Rhode Island (1996)
64. Roy, B.: Multicriteria Methodology for Decision Aiding. Kluwer Academic Publishers, Dordrecht (1996)
65. Roy, B., Berlier, B.: La Metode ELECTRE II. In: Sixieme Conf. Internationale de Rechearche Operationelle, Dublin (1972)
66. Ruan, D.: Computational intelligence in complex decision systems. World Scientific (2010)
67. Schott, B., Whalen, T.: Fuzzy uncertainty in imperfect competition. J. Information Sciences 76(3-4), 339–354 (1994)
68. Schmeidler, D.: Subjective probability and expected utility without additivity. Econometrita 57(3), 571–587 (1989)
69. Su, Z.X.: A hybrid fuzzy approach to fuzzy multi-attribute group decision-making. International Journal of Information Technology & Decision Making 10(4), 695–711 (2011), doi:10.1142/S021962201100452X
70. Thomaidis, N.S., Nikitakos, N., Dounias, G.D.: The evaluation of information technology projects: a fuzzy multicriteria decision-making approach. Journal of Advanced Computational Intelligence and Intelligent Informatics 5(1), 89–122 (2006), doi:10.1142/S0219622006001897
71. Tsuji, T., Jazidie, A., Kaneko, M.: Distributed trajectory generation for cooperative multi-arm robots via virtual force interactions. IEEE Transactions on Systems, Man, and Cybernetics- Part B: Cybernetics 27(5), 862–867 (1997)
72. Walley, P.: Measures of uncertainty in expert systems. Artificial Intelligence 83(1), 1–58 (1996)
73. Walley, P., de Cooman, G.: A behavioral model for linguistic uncertainty. Information Sciences 134(1-4), 1–37 (2001)
74. Wang, Y.M., Elhag, T.M.S.: Fuzzy TOPSIS method based on alpha level sets with an application to bridge risk assessment. Expert Syst. Appl. 31(2), 309–319 (2006)
75. Wang, G., Li, X.: On the convergence of the fuzzy valued functional defined by μ-integrable fuzzy valued functions. Fuzzy Sets and Systems 107(2), 219–226 (1999)
76. Whinston, A.: Intelligent Agents as a Basis for Decision Support Systems. Decision Support Systems 20(1), 1 (1997)
77. Yager, R.R.: Multiple Objective Decision Making Using Fuzzy Sets. International Journal of Man-Machine Studies 9(4), 375–382 (1977)
78. Yager, R.R.: Fuzzy Decision Making Including Unequal Objectives. J. Fuzzy Sets & Systems 1, 87–95 (1978)
79. Yang, R., Wang, Z., Heng, P.A., Leung, K.S.: Fuzzy numbers and fuzzification of the Choquet integral. Fuzzy Sets and Systems 153(1), 95–113 (2005)
80. Yoon, K.: A reconciliation among discrete compromise solutions. J. of Operat. Res. Soc. 38(3), 272–286 (1987)
81. Zadeh, L.A., Klir, G., Yuan, B.: Fuzzy sets, fuzzy logic, and fuzzy systems: selected papers By Lotfi Asker Zadeh. World Scientific Publishing Co. (1996)

82. Zadeh, L.A.: Fuzzy logic = Computing with Words. IEEE Transactions on Fuzzy Systems 4(2), 103–111 (1996)
83. Zadeh, L.A.: Generalized theory of uncertainty (GTU) – principal concepts and ideas. Computational statistics & Data Analysis 51, 15–46 (2006)
84. Zeleny, M.: Multiple Criteria Decision Making. McGraw-Hill, New York (1982)
85. Zhang, C.: Cooperation Under Uncertainty in Distributed Expert Systems. Artificial Intelligence 56, 21–69 (1992)
86. Zhang, G.Q.: Fuzzy number-valued fuzzy measure and fuzzy number-valued fuzzy integral on the fuzzy set. Fuzzy Sets and Systems 49, 357–376 (1992)
87. Zhu, H.P., Zhang, G.J., Shao, X.Y.: Study on the application of fuzzy topsis to multiple criteria group decision making problem. Ind. Eng. Manage. 1, 99–102 (2007)
88. Zadeh, L.A.: Is there a need for fuzzy logic? Information Sciences 178, 2751–2779 (2008)
89. Zimmermann, H.J.: Fuzzy Sets Theory and its Application. Kluwer Academic Publishers (1990)

Chapter 5
Extention to Behavioral Decision Making

5.1 Decision Making with Combined States

Decision making is a behavioral process. During the development of decision theories scientists try to take into account features of human choices in formal models to make the latter closer to human decision activity. Risk issues were the first basic behavioral issues which became necessary to consider in construction of decision methods. Three main categories of risk-related behaviors: risk aversion, risk seeking and risk neutrality were introduced. Gain-loss attitudes [28] and ambiguity attitudes [26] were revealed as other important behavioral features. Prospect theory, developed for decision under risk [28] was the first decision theory incorporating both risk and gain-loss attitudes into a single utility model. Cumulative Prospect theory [45] (CPT), as its development, can be applied both for decision under risk and uncertainty and is one of the most successful decision theories. CPT is based on the use of Choquet integrals and, as a result, is able to represent not only risk and gain-loss attitudes but also ambiguity attitudes. Choquet Expected Utility (CEU) is a well-known typical model which can be used both for ambiguity and risk situations.

The first model developed for ambiguity aversion was Maximin expected utility [26] (MMEU). Its generalization, α-MMEU, is able to represent both ambiguity aversion and ambiguity seeking [25]. Smooth ambiguity model [29] is a more advanced decision model for describing ambiguity attitudes.

A large stream of investigations led to development of parametric and non-parametric decision models taking into account such important psychological, moral and social aspects of decisions as reciprocity [20,22], altruism [11,21], trust [14,20] and others.

A large area of research in modeling decision makers (DMs) (agents) in line with nature (environment) is mental-level models [16-18,40], idea of which was suggested in [35,37]. In these models a DM is modeled by a set of states. Each state describes his/her possible decision-relevant condition and is referred to as "mental state", "state of mind" etc. In these models, they consider relations between mental state and state of environment [19,47] (nature). In [16,42] a mental state and a state of nature compose a state of the whole system called a "global state". Within the scope of mental-level models there are two main research areas of a mental state modeling: internal modeling [16,50] and

R.A. Aliev: *Fuzzy Logic-Based Generalized Theory of Decisions*, STUDFUZZ 293, pp. 191–216.
DOI: 10.1007/ 978-3-642-34895-2_5

implementation-independent (external) modeling [16]. The first is based on modeling a mental state by a set of characteristics (variables) and the second is based on modeling a mental state on the base of beliefs, preferences and decision criterion [16,39,41].

Now we observe a significant progress in development of a series of successful decision theories based on behavioral issues. Real-life human choices are based on simultaneous influence of main aspects of decision situations like risk, ambiguity and others. The question arises of how to adequately model joint influence of these factors on human choices and whether we should confine ourselves to assuming that these determinants influence choices independently. Due to highly constrained computational ability of human brain, independent influence of these factors can hardly be met. Humans conduct an intelligent, substantive comparison of real-life alternatives in whole, i.e. as some mixes of factors without pure partitioning of them. This implies interaction of the mentioned aspects in their influence on human choices. However, one of the disadvantages of the existing theories of decisions under uncertainty is an absence of a due attention to interaction of the factors. A vector of variables describing the factors is introduced into a decision model without fundamental consideration of how these factors really interact, they are considered separately. Also, information on intensity of the factors and their interaction is rather uncertain and vague and can mainly be described qualitatively and not quantitatively. The mentioned issues are the main reasons of why humans are not completely rational but partially, or bounded rational DMs and why the existing decision models based on pure mathematical formalism become inconsistent with human choices.

Motivation

The necessity to take into account that humans are not fully rational DMs was first conceptually addressed by Herbert Simon [44]. He proposed the concept of bounded rationality which reflects notable limitations of humans' knowledge and computational abilities. Despite their significant importance, the ideas of bounded rationality did not found its mathematical fundamentals to form a new consistent formal basis adequate to real decisions. The theory which can help to form an adequate mathematical formalism for bounded rationality-based decision analysis is the fuzzy set theory suggested by L.A. Zadeh [34,52]. The reason for this is that fuzzy set theory and its developments deal with the formalization of linguistically (qualitatively) described imprecise or vague information and partial truth. Indeed, limitation of human knowledge, taken as one of the main aspects in bounded rationality, in real world results in the fact that humans use linguistic evaluations because the latter as, opposed to precise numbers, are tolerant for impreciseness and vagueness of real decision-relevant information. In fuzzy sets theory this is formalized by using fuzzy sets and fuzzy numbers. The other aspect – limitation of computational ability of humans – leads to the fact that humans think and reason in terms of propositions in natural language (NL), but not in terms of pure mathematical expressions. Such activity results in arriving at approximate solutions and satisfactory results but not at precise optimal solutions. This coincides with what is stated in bounded rationality ideas. In fuzzy logic, this is termed as approximate reasoning [12,53]. Fuzzy sets theory was initially suggested for an analysis of humanistic systems where perceptions play a pivotal role. Perceptions are imprecise, they have fuzzy boundaries [38], and, as a result, are often described linguistically. Fuzzy sets theory and its successive technologies [53,54] as tools for correct formal processing of perception-based information may help to arrive at perceptions-friendly and mathematically consistent decisions. So, there is an evident connection between ideas of bounded rationality theory and fuzzy set theory [38].

So, in addition to missing interaction of behavioral factors in the existing theories they don't extensively take into account that information on a DM's behavior is imperfect. To be more concrete, in CPT they imperatively consider that a DM is risk averse when dealing with gains and risk seeking when dealing with losses. However it is too simplified view and in reality we don't have such complete information concerning risk attitudes of a considered DM in a considered situation. In α -MMEU they consider a balance of ambiguity aversion and ambiguity seeking that drives a DM's choices, but this is modeled by precise value α , whereas real information about the ambiguity attitudes is imprecise. This all means that it is needed to model possibilistic uncertainty reflecting incomplete and imprecise relevant information on decision variables and not only probabilistic uncertainty.

Necessity of considering interaction of behavioral factors under imperfect information is the main insight for development of new decision approaches. Following this, we suggest considering a space of vectors of variables describing behavioral factors (for example, risk and ambiguity attitudes) as composed of main subspaces each describing one principal DM's behavior. Each subspace we suggest to consider as a DM's state in which he/she may be when making choices [3]. Such formalization is in the direction of internal modeling of DMs (or agent) within the scope of mental-level models. However, we suggest to consider these subspaces as not exclusive, but as some overlapping sets to reflect the facts that principal behaviors have indeed some similarity and proximity, which should not be disregarded because the state of a DM is uncertain itself and cannot be sharply bounded. For this case we suggest to use fuzzy granulation of the considered space, i.e. granulation into fuzzy sets each describing one state of a DM. This helps to closer model a DM's condition as the relevant information is mainly described in linguistic (qualitative) form and could not be reliably described by precise dependencies.

In our approach, uncertainty related to what state of a DM is likely to occur is described by a linguistic (fuzzy) probability. Fuzzy probability describes impreciseness of beliefs coming from uncertainty and complexity of interaction of the factors, from absence of ideal information.

Concerning states of nature, in many real problems there also is no sufficient information to consider them as "mutually exclusive": for example, if one considers states of economy, the evaluations like "moderate growth" and "strong growth" don't have sharp boundaries and, as a result, may not be "exclusive" – they may overlap. Observing some actual situation an expert may conclude that to a larger extent it concerns the moderate growth and to a smaller extent to the strong growth. An appropriate way to model this is the use of fuzzy sets. In real-life it is often impossible to construct exclusive and exhaustive states of nature, due to uncertainty of relevant information [24]. In general, a DM cannot exhaustively determine each objective condition that may be faced and precisely differentiate them. Each state of nature is, essentially, some area which collects similar objective conditions, that is some set of "elementary" states or quantities. Unfortunately, in the existing decision theories a small attention paid to the essence and structure of states of nature.

We suggest to consider the space of states of nature and space of DM's states as constituting a single space of combined states [3,6] i.e. to considering Cartesian product of these two important spaces as basis for comparison of alternatives. Likelihood of occurrence of each combined space as a pair consisting of one state of nature and one DM's state is to be described by fuzzy probability of their joint occurrence. This fuzzy joint probability (FJP) is to be found on the base of fuzzy marginal probabilities of state of nature and state of a DM and, if possible, on the base of some information about dependence of these states. Utilities of outcomes are also to be distributed over the combined states reflecting naturally various evaluation of the outcomes by a DM in his/her various states.

Consideration of DM's behavior by space of states and its Cartesian product with space of states of nature [3,6] will allow for transparent analysis of decisions. In contrast, in the existing utility models human attitudes to risk, ambiguity and others are included using complex mathematical expressions – nonlinear transformations, second-order probabilities etc. Indeed, most of the existing decision theories are based on parametric modeling of behavioral features. As a result, they cannot adequately describe human decision activity; they are mathematically complex and not transparent. The existing non-parametric approaches are more fundamental, but, they also are based on perfect and precise description of human decision activity. This is non-realistic because human thinking and motivation are perception-based [15] and cannot be captured by precise techniques. Real-life information related to a DM behavior and objective conditions is intrinsically imperfect. We mostly make decisions under vagueness, impreciseness, partial truth etc. of decision-relevant information and ourselves think in categories of 'smooth' concepts even under perfect information.

So, our approach is based on three justifications: necessity of considering dependence between various behavioral determinants, necessity to take into account uncertainty of what behavior will be present in making choices (e.g. what risk attitude), a need for construction transparent behavioral model of decision analysis.

In the present study we develop investigations started in [2,3,6-8]. We suggest a new approach to behavioral decision making under imperfect information, namely under mix of probabilistic and possibilistic uncertainties. We show that the expected utility [36](EU), CEU [40] and CPT are special cases of the combined states-based approach. For a representation in the suggested model we adopt the generalized fuzzy Choquet-like aggregation with respect to a fuzzy-valued bi-capacity.

Let $\mathcal{S}=\{\tilde{S}_1,\tilde{S}_2,...,\tilde{S}_M\}\subset\mathcal{E}^n$ be a space of fuzzy states of nature and $\mathcal{X}$ be a space of fuzzy outcomes as a bounded subset of $\mathcal{E}^n$. Denote by $\mathrm{H}=\{\tilde{h}_1,\tilde{h}_2,...,\tilde{h}_N\}\subset\mathcal{E}^n$ a set of fuzzy states of a DM [3,7]. Then we call $\Omega=\mathcal{S}\times\mathrm{H}$ a space "nature-DM", elements of which are combined states $\tilde{w}=(\tilde{S},\tilde{h})$ where $\tilde{S}\in\mathcal{S},\tilde{h}\in\mathrm{H}$.

Denote $\mathcal{F}_\Omega$ a σ-algebra of subsets of Ω. Then consider $\mathcal{A}=\{\tilde{f}\subset\mathcal{A}\,|\,\tilde{f}:\Omega\to\mathcal{X}\}$ the set of fuzzy actions as the set of all $\tilde{\mathcal{F}}_\Omega$-measurable fuzzy functions from Ω to $\mathcal{X}$ [3,7].

A problem of behavioral decision making with combined states under imperfect information (BDMCSII) can be denoted as $D_{BDMCSII}=(\Omega,\mathcal{X},\mathcal{A},\succsim_l)$ where $\succsim_l$ are linguistic preferences of a DM.

In general, it is not known which state of nature will take place and what state of a DM will present at the moment of decision making. Only some partial knowledge on probability distributions on $\mathcal{S}$ and $\mathcal{H}$ is available. An information relevant to a DM can be formalized as a linguistic probability distribution over his/her states: $\tilde{P}_1/\tilde{h}_1+\tilde{P}_2/\tilde{h}_2+...+\tilde{P}_N/\tilde{h}_N$, where $\tilde{P}_i$ is a linguistic belief degree or a linguistic probability. So, $\tilde{P}_i/\tilde{h}_i$ can be formulated as, for example, "a probability that a DM's state is $\tilde{h}_i$ is $\tilde{P}_i$".

For closer description of human behavior and imperfect information on Ω we use a fuzzy number-valued bi-capacity $\tilde{\eta}=\tilde{\eta}(\tilde{V},\tilde{W})$, $\tilde{V},\tilde{W}\subset\Omega$. A fuzzy-valued bi-capacity is defined as follows.

Definition 5.1. Fuzzy Number-Valued Bi-capacity. A fuzzy number-valued bi-capacity on $\mathcal{F}^2=\mathcal{F}\times\mathcal{F}$ is a fuzzy number-valued set function $\tilde{\eta}:\mathcal{F}^2\to\mathcal{E}^1_{[-1,1]}$ with the following properties:

(1) $\tilde{\eta}(\varnothing,\varnothing)=0$;
(2) if $\mathcal{V}\subset\mathcal{V}'$ then $\tilde{\eta}(\mathcal{V},\mathcal{W})\le\tilde{\eta}(\mathcal{V}',\mathcal{W})$;
(3) if $\mathcal{W}\subset\mathcal{W}'$ then $\tilde{\eta}(\mathcal{V},\mathcal{W})\ge\tilde{\eta}(\mathcal{V},\mathcal{W}')$;
(4) $\tilde{\eta}(\Omega,\varnothing)=1$ and $\tilde{\eta}(\varnothing,\Omega)=-1$.

In special case, values of a fuzzy-valued bi-capacity $\tilde{\eta}(\mathcal{V},\mathcal{W})$ can be determined as the difference of values of two fuzzy-valued measures $\tilde{\eta}_1(\mathcal{V})-\tilde{\eta}_2(\mathcal{W})$, where "$-$" is defined on the base of Zadeh's extension principle.

Value or utility of an outcome $\tilde{X}=\tilde{f}(\tilde{S},\tilde{h})$ in various DM's states will also be various, and then can be formalized as a function $\tilde{u}(\tilde{X})=\tilde{u}(\tilde{f}(\tilde{S},\tilde{h}))$. We can claim that the value function of Kahneman and Tversky $v=v(f(S))$ [28] appears then as a special case. So, an overall utility $\tilde{U}(\tilde{f})$ of an action $\tilde{f}$ is to be determined as a fuzzy number-valued bi-capacity-based aggregation of $\tilde{u}(\tilde{f}(\tilde{S},\tilde{h}))$ over space Ω. Then the BDMCSII problem consists in determination of an optimal action as an action $\tilde{f}^*\in\mathcal{A}$ with $\tilde{U}(\tilde{f}^*)=\max_{\tilde{f}\in\mathcal{A}}\int_\Omega \tilde{U}(\tilde{f}(\tilde{w}))d\tilde{\eta}$.

Axiomatization. As the basis for our model we use the framework of bi-capacity [30] formulated by Labreuche and Grabisch. The bi-capacity is a natural generalization of capacities and is able to describe interaction between attractive and repulsive values (outcomes, criteria values), particularly, gains and losses. We extend this framework to the case of imperfect information by using linguistic preference relation [1,9]. The linguistic preference means that the preference among actions $\tilde{f}$ and $\tilde{g}$ is modeled by a degree $Deg(\tilde{f}\succsim_l\tilde{g})$ to which $\tilde{f}$ is at least as good as $\tilde{g}$ and a degree $Deg(\tilde{g}\succsim_l\tilde{f})$ to which $\tilde{g}$ is at least as good as $\tilde{f}$. The degrees $Deg()$ are from [0,1]. The closer $Deg(\tilde{f}\succsim_l\tilde{g})$ to 1 the more $\tilde{f}$ is preferred to $\tilde{g}$. These degrees are used to represent vagueness of preferences, that is, situations when decision relevant information is too vague to definitely determine preference of one alternative against another. For special case, when $Deg(\tilde{g}\succsim_l\tilde{f})=0$ and $Deg(\tilde{f}\succsim_l\tilde{g})\neq 0$ we have the classical preference, i.e. we say that $\tilde{f}$ is preferred to $\tilde{g}$.

We use bi-capacity-adopted integration [30] at the space "nature-DM" for determination of an overall utility of an alternative. The base for our model is composed by intra-combined state information and inter-combined states information. Intra-combined state information is used to form utilities representing preference over outcomes $\tilde{f}(\tilde{w}_i)=\tilde{X}_i$, where $\tilde{w}_i=(\tilde{S}_{i_1},\tilde{h}_{i_2})$ of an act $\tilde{f}\in\mathcal{A}$ with understanding that these are preferences at state of nature $\tilde{S}_{i_1}$ conditioned by a state $\tilde{h}_{i_2}$ of a DM.

Inter-combined states information will be used to form fuzzy-valued bi-capacity representing dependence between combined states as human behaviors under incomplete information.

Proceeding from these assumptions, for an overall utility $\tilde{U}$ of action $\tilde{f}$ we use an aggregation operator based on the use of a bi-capacity. Bi-capacity is a more powerful tool to be used in a space "nature-DM". More concretely, we use a fuzzy-valued generalized Choquet-like aggregation with respect to fuzzy-valued bi-capacity over Ω :

$$\tilde{U}(\tilde{f})=\sum_{l=1}^{N}(\tilde{u}(\tilde{f}(\tilde{w}_{(l)}))-_h \tilde{u}(\tilde{f}(\tilde{w}_{(l+1)})))\tilde{\eta}(\{\tilde{w}_{(1)},...,\tilde{w}_{(l)}\}\cap N^+,\{\tilde{w}_{(1)},...,\tilde{w}_{(l)}\}\cap N^-), \quad (5.1)$$

provided $\tilde{u}(\tilde{f}(\tilde{w}_{(l)}))\geq \tilde{u}(\tilde{f}(\tilde{w}_{(l+1)}))$; $N^+=\{\tilde{w}\in\Omega:\tilde{u}(\tilde{f}(\tilde{w}))\geq 0\}$, $N^-=\Omega\setminus N^+$, $\tilde{\eta}(\cdot,\cdot)$ is a fuzzy number-valued bi-capacity.

In (5.1) under level α we have an interval $U^\alpha(\tilde{f})=[U_1^\alpha(\tilde{f}),U_2^\alpha(\tilde{f})]$ of possible precise overall utilities, where $U_1^\alpha(\tilde{f}),\ U_2^\alpha(\tilde{f})$ are described as follows:

$$U_1^\alpha(\tilde{f})=(u_1^\alpha(\tilde{f}(\tilde{w}_{(1)}))-u_1^\alpha(\tilde{f}(\tilde{w}_{(2)})))\eta_1^\alpha(\{\tilde{w}_{(1)}\}\cap N^+,\{\tilde{w}_{(1)}\}\cap N^-)+$$
$$+(u_1^\alpha(\tilde{f}(\tilde{w}_{(2)}))-u_1^\alpha(\tilde{f}(\tilde{w}_{(3)})))\eta_1^\alpha(\{\tilde{w}_{(1)},\tilde{w}_{(2)}\}\cap N^+,\{\tilde{w}_{(1)},\tilde{w}_{(2)}\}\cap N^-)+$$
$$+...+u_1^\alpha(\tilde{f}(\tilde{w}_{(n)}))\eta_1^\alpha(\{\tilde{w}_{(1)},\tilde{w}_{(2)},...,\tilde{w}_{(n)}\}\cap N^+,\{\tilde{w}_{(1)},\tilde{w}_{(2)},...,\tilde{w}_{(n)}\}\cap N^-),$$

$$U_2^\alpha(\tilde{f})=(u_2^\alpha(\tilde{f}(\tilde{w}_{(1)}))-u_2^\alpha(\tilde{f}(\tilde{w}_{(2)})))\eta_2^\alpha(\{\tilde{w}_{(1)}\}\cap N^+,\{\tilde{w}_{(1)}\}\cap N^-)+$$
$$+(u_2^\alpha(\tilde{f}(\tilde{w}_{(2)}))-u_2^\alpha(\tilde{f}(\tilde{w}_{(3)})))\eta_2^\alpha(\{\tilde{w}_{(1)},\tilde{w}_{(2)}\}\cap N^+,\{\tilde{w}_{(1)},\tilde{w}_{(2)}\}\cap N^-)+$$
$$+...+u_2^\alpha(\tilde{f}(\tilde{w}_{(n)}))\eta_2^\alpha(\{\tilde{w}_{(1)},\tilde{w}_{(2)},...,\tilde{w}_{(n)}\}\cap N^+,\{\tilde{w}_{(1)},\tilde{w}_{(2)},...,\tilde{w}_{(n)}\}\cap N^-),$$

provided that $u_1^\alpha(\tilde{f}(\tilde{w}_{(1)}))\geq...\geq u_1^\alpha(\tilde{f}(\tilde{w}_{(n)}))$ and $u_2^\alpha(\tilde{f}(\tilde{w}_{(1)}))\geq...\geq u_2^\alpha(\tilde{f}(\tilde{w}_{(n)}))$. So, $U_i^\alpha(\tilde{f}),\ i=1,2$ is a common Choquet-like precise bi-capacity based functional [30], with u_i^α and η_i^α $i=1,2$ being a precise utility function and a precise bi-capacity respectively. This representation captures impreciseness of both a utility and a bi-capacity arising from impreciseness of outcomes and probabilities in real-world decision problems.

An optimal action $\tilde{f}^*\in\mathcal{A}$, that is $\tilde{f}^*\in\mathcal{A}$ for which $\tilde{U}(\tilde{f}^*,\tilde{c})=$ $=\max_{\tilde{f}\in\mathcal{A}}\left\{\int_\Omega \tilde{u}(\tilde{f}(\tilde{S},\tilde{h})d\tilde{\eta}\right\}$ is found by a determination of $Deg(\tilde{f}\succsim_l \tilde{g}),\ \tilde{f},\tilde{g}\in\mathcal{A}$: optimal action $\tilde{f}^*\in\mathcal{A}$ is an action for which $Deg(\tilde{f}^*\succsim_l \tilde{f})\geq Deg(\tilde{f}\succsim_l \tilde{f}^*)$ is satisfied for all $\tilde{f}\in\mathcal{A},\tilde{f}\neq\tilde{f}^*$. The determination of $Deg(\tilde{f}\succsim_l \tilde{g})$ is based on comparison of $U(\tilde{f})$ and $U(\tilde{g})$ as the basic values of $\tilde{U}(\tilde{f})$ and $\tilde{U}(\tilde{g})$ respectively as follows. Membership functions of $\tilde{U}(\tilde{f})$ and $\tilde{U}(\tilde{g})$ describe possibilities of their various basic values $U(\tilde{f})$ and $U(\tilde{g})$ respectively, that is, possibilities for various precise values of overall utilities of $\tilde{f}$ and $\tilde{g}$. In accordance with these membership functions, there is possibility $\alpha\in(0,1]$ that precise overall utilities of $\tilde{f}$ and $\tilde{g}$ are equal to $U_1^\alpha(\tilde{f}), U_2^\alpha(\tilde{f})$ and $U_1^\alpha(\tilde{g}), U_2^\alpha(\tilde{g})$ respectively. Therefore, we can state that there is possibility $\alpha \in (0,1]$ that the difference between precise overall utilities of $\tilde{f}$ and $\tilde{g}$ is $U_i^\alpha(\tilde{f}) - U_j^\alpha(\tilde{g}), i, j = 1,2$. As $\tilde{f}$ is preferred to $\tilde{g}$ when overall utility of $\tilde{f}$ is larger than that of $\tilde{g}$, we will consider only positive $U_i^\alpha(\tilde{f}) - U_i^\alpha(\tilde{g})$. Consider now the following functions:

$$\sigma(\alpha) = \sum_{i=1}^{2}\sum_{j=1}^{2} \max(U_i^\alpha(\tilde{f}) - U_j^\alpha(\tilde{g}), 0);$$

$$\delta_{ij}(\alpha) = \begin{cases} \dfrac{\max(U_i^\alpha(\tilde{f}) - U_j^\alpha(\tilde{g}), 0)}{\left|U_i^\alpha(\tilde{f}) - U_j^\alpha(\tilde{g})\right|}, & if\ U_i^\alpha(\tilde{f}) - U_j^\alpha(\tilde{g}) \neq 0 \\ 0, & else \end{cases} \quad i, j = 1,2.$$

$$\delta(\alpha) = \sum_{i}^{2}\sum_{j}^{2} \delta_{ij}(\alpha)$$

$\sigma(\alpha)$ shows the sum of all positive differences between $U_1^\alpha(\tilde{f}), U_2^\alpha(\tilde{f})$ and $U_1^\alpha(\tilde{g}), U_2^\alpha(\tilde{g})$ and $\delta(\alpha)$ shows the number of these differences. Consider now the quantity $\dfrac{\int_0^1 \alpha\sigma(\alpha)d\alpha}{\int_0^1 \alpha\delta(\alpha)d\alpha}$ as a weighted average of differences $U_i^\alpha(\tilde{f}) - U_j^\alpha(\tilde{g}), i, j = 1,2$ where weights are their possibilities $\alpha \in (0,1]$. The degree $Deg(\tilde{f} \succ_l \tilde{g})$ is determined then as follows:

$$Deg(\tilde{f} \succsim_l \tilde{g}) = \frac{\int_0^1 \alpha\sigma(\alpha)d\alpha}{(u_{\max} - u_{\min})\int_0^1 \alpha\delta(\alpha)d\alpha} \tag{5.2}$$

In other words, $Deg(\tilde{f} \succsim_l \tilde{g})$ is determined as a percentage of a weighted average of differences $U_i^\alpha(\tilde{f}) - U_j^\alpha(\tilde{g}), i, j = 1,2$ with respect to $u_{\max} - u_{\min}$ being maximally possible difference ($u_{\min}$ and $u_{\max}$ are respectively the lower and upper bounds of the universe of discourse for utility $\tilde{u}$, and, therefore, $u_{\max} \leq \tilde{U}(\tilde{f}) \leq u_{\min}$ as $\tilde{U}(\tilde{f})$ is an aggregation of $\tilde{u}$). By other words, the closer the difference $U_i^\alpha(\tilde{f}) - U_j^\alpha(\tilde{g})$ of the equally possible values of precise overall utilities of $\tilde{f}$ and $\tilde{g}$ to $u_{\max} - u_{\min}$ the higher is the extent to which $\tilde{f}$ is better than $\tilde{g}$.

Let us show that the famous existing utility models are special cases of the proposed combined states-based fuzzy utility model. To do this, we simplify our model to its non-fuzzy variant and consider its relation with the existing utility models. Bi-capacity-based aggregation of $u(f(s,h))$ on a space Ω would be a natural generalization of an aggregation of $u(f(s))$ on a space S. We will show this by comparing of EU and CEU applied on space $S=\{s_1,s_2,...s_m\}$ with the same models applied on a combined states space Ω. For obvious illustration let us at first look at a general representation of combined states space $\Omega = S\times H$ given in Table 5.1.

Table 5.1. Combined states space

	s_1	s_i	...	s_N
h_1	(s_1,h_1)	(s_i,h_1)	...	(s_N,h_1)
h_j	(s_1,h_j)	(s_i,h_j)	...	(s_N,h_j)
...	...	...	...	...
h_M	(s_1,h_1)	(s_i,h_j)	...	(s_N,h_M)

EU criterion used for combined states space (Table 5.1) will have the following form:

$$U(f)=\sum_{k=1}^{MN}u(f(w_k))p(w_k)=\sum_{j=1}^{M}\sum_{i=1}^{N}u(f(s_i,h_j))p(s_i,h_j) \tag{5.3}$$

In traditional EU (i.e., EU applied on a space S only) they consider that a DM exhibits the same behavior in any state of nature. In our terminology this means that only one state of a DM can exist. Then, to model a classical EU within (5.3) we should exclude all h_j except one, say h_k. This immediately means that $P(s_i,h_j)=0, \forall j\neq k$ (as we consider that all $h_j, j\neq k$ don't exist) and we have

$$U(f)=\sum_{i=1}^{N}u(f(s_i,h_k))p(s_i,h_k)$$

Now, as a DM is always at a state h_k whatever state s_i takes place, we have $p(s_i,h_k)=p(s_i)$. Furthermore, in common EU only risk attitudes as behavioral aspects are taken into account. A DM is considered as either risk averse or risk seeking or risk neutral. So, h_k can represent one of these behaviors. For example, if h_k represents risk aversion then $u()$ will be concave, if h_k represents risk seeking then $u()$ will be convex etc. So, h_k determines form of $u()$. If we use notation $u^*(f(\cdot))$ for $u(f(\cdot,h_k))$ when h_k represents, for example, risk aversion, we have (5.3) as $U(f)=\sum_{i=1}^{N}u^*(f(s_i))p(s_i)$ which is nothing but the traditional EU. So, the traditional EU is a special case of the EU criterion used for Ω. Combined-states based approach as opposed to classical EU allows to take into account that a DM can exhibit various risk attitudes at various states of nature. This usually takes place in real life and is taken into account in PT and CPT (these models are based on experimental observations demonstrating that people exhibit risk aversion for gains and risk seeking for losses).

Let us now show that CEU used for space S is a special case of the analogous aggregation over Ω. CEU over Ω will have the following form:

$$U(f)=\sum_{l=1}^{N}(u(f(w_{(l)}))-u(f(w_{(l+1)})))\eta(\{w_{(1)},...,w_{(l)}\}) \tag{5.4}$$

$w_{(l)}=(s_j,h_k)$, $u(f(w_{(l)}))\geq u(f(w_{(l+1)}))$. Assuming now that only some h_k exists, we have that $\forall w\in\Omega, w=(s_i,h_k)$, that is $\Omega=S\times\{h_k\}$. Then we will have $u(f(w_{(l)}))-u(f(w_{(l+1)}))=0$ whenever $w_{(l)}=(s_i,h_k), w_{(l+1)}=(s_i,h_k)$. Only differences $u(f(w_{(l)}))-u(f(w_{(l+1)}))$ for which $w_{(l)}=(s_i,h_k), w_{(l+1)}=(s_j,h_k), i\neq j$ may not be equal to zero. As a result, making simple transformations, we will have:

$$U(f)=\sum_{i=1}^{n}(u(f(s_{(i)},h_k))-u(f((s_{(i+1)},h_k))))\eta(\{(s_{(1)},h_k),...,(s_{(i)},h_k)\})$$

Now, using notations $u^*(f(\cdot))$ for $u(f(\cdot,h_k))$ and $\eta^*(\{s_{(1)},...,s_{(j)}\})=$ $=\eta(\{(s_{(1)},h_k),...,(s_{(j)},h_k)\})$ we can write

$$U(f)=\sum_{i=1}^{N}(u^*(f(s_{(i)}))-u^*(f(s_{(i+1)})))\eta^*(\{s_{(1)},...,s_{(i)}\})$$

This is nothing but a traditional CEU. Traditional CEU is often used to represent uncertainty attitude as an important behavioral aspect. So, if h_k represents uncertainty aversion (uncertainty seeking) then $\eta(\{(s_{(1)},h_k),...,(s_{(j)},h_k)\})$ can be chosen as lower prevision (upper prevision).

It can also be shown that the utility model used in the CPT is also a special case of the combined states approach. This follows from the fact that representation used in CPT is a sum of two Choquet integrals.

The solution of the problem consists in determination of an optimal action $\tilde{f}^*\in\mathcal{A}$ with $\tilde{U}(\tilde{f}^*)=\max_{\tilde{f}\in\mathcal{A}}\left\{\int_{\Omega}\tilde{u}(\tilde{f}(\tilde{S},\tilde{h}))d\tilde{\eta}\right\}$. The problem is solved as follows. At the first stage it becomes necessary to assign linguistic utility values $\tilde{u}(\tilde{f}(\tilde{S}_i,\tilde{h}_j))$ to every action $\tilde{f}\in\mathcal{A}$ taken at a state of nature $\tilde{S}_i\in S$ when a

DM's state is $\tilde{h}_j$. The second stage consists in construction of a FJP distribution $\tilde{P}^l$ on Ω proceeding from partial information on marginal distributions over $\mathcal{S}$ and $\mathcal{H}$ which is represented by given fuzzy probabilities for all states except one. This requires constructing unknown fuzzy probability for each space [8,9]. Given marginal distribution of fuzzy probabilities for all the states, it is needed to verify consistency, completeness and redundancy of this distribution [7]. Finally, on the base of fuzzy marginal distributions (for $\mathcal{S}$ and H) and information on dependence between states of nature $\tilde{S} \in \mathcal{S}$ and a DM's states $\tilde{h} \in \mathrm{H}$ it is needed to construct FJP distribution $\tilde{P}^l$ on Ω.

At the third stage it is necessary to construct a fuzzy-valued bi-capacity $\tilde{\eta}(\cdot,\cdot)$ based on FJP $\tilde{P}^l$ on Ω. For simplicity one can determine a fuzzy-valued bi-capacity as the difference of two fuzzy-valued capacities.

Next the problem of calculation of an overall utility $\tilde{U}(\tilde{f})$ for every action $\tilde{f} \in \mathcal{A}$ is solved by using formula (5.1). In (5.1) differences between fuzzy utilities $\tilde{u}(\tilde{f}(\tilde{S},\tilde{h}))$ assigned at the first stage are multiplied on the base of the Zadeh's extension principle by the values of the fuzzy valued bi-capacity $\tilde{\eta}(\cdot,\cdot)$ constructed at the third stage.

Finally, an optimal action $\tilde{f}^* \in \mathcal{A}$ as the action with the maximal fuzzy valued utility $\tilde{U}(\tilde{f}^*) = \max_{\tilde{f} \in \mathcal{A}} \left\{ \int_{\Omega} \tilde{u}(\tilde{f}(\tilde{S},\tilde{h}))d\tilde{\eta} \right\}$ is determined by comparing fuzzy overall utilities $\tilde{U}(\tilde{f})$ for all $\tilde{f} \in \mathcal{A}$ (see formula (5.2)).

As we mentioned above, in order to solve the considered problem of behavioral decision making we need to construct a fuzzy-valued bi-capacity over a space of combined states Ω. This fuzzy-valued bi-capacity is used to model relations between combined states under imperfect relevant information. One natural informational basis to construct a fuzzy-valued bi-capacity is a FJP over combined states. The FJP distribution describes dependence of states of a DM on states of nature, that is, dependence of a human behavior on objective conditions that is a quite natural phenomenon. In order to proceed to construction of FJP we need to consider some preliminary concepts that are given below.

There exist mainly two approaches to construction of a joint probability distribution: approaches for modeling dependence among events (e.g. the chance that it will be cloudy and it will rain) and approaches modeling dependence among random variables (e.g. the chance that an air temperature is in-between 20C and 30C and an air humidity is in-between 90%-95%). In modeling dependence of states of a DM on states of nature we will follow dependence of events framework. This framework is more suitable as states of a DM and states of nature are not numerical but are rather qualitative.

To measure a joint probability of two events H and G we need two kinds of information: marginal probabilities for H and G and information on a type of dependence between H and G referred to as a *sign of dependence*. There exist three types of dependence: positive dependence, independence and negative dependence. Positive dependence implies that H and G have tendency to occur together, e.g. one favors occurrence of another. For example, cloudiness and rain are positively dependent. Negative dependence implies they don't commonly occur together, e.g. one precludes occurrence of another. For example, sunny day and raining are negatively dependent. Independence implies that occurrence of one does not affect an occurrence of another. The extreme case of a positive dependence is referred to as a perfect dependence. The extreme case of a negative dependence is referred to as opposite dependence. It is well known that given numerical probabilities $P(H)$ and $P(G)$ of independent events H and G, the joint probability $P(H,G)$ is determined as

$$P(H,G) = P(H)P(G) \tag{5.5}$$

The perfect dependence is determined as [46,48]

$$P(H,G) = \min(P(H), P(G)) \tag{5.6}$$

For explanation of this fact one may refer to [46,48]. It is clear that $P(H)P(G) \le \min(P(H), P(G))$. Positive dependence among H and G is modeled as [46,48]

$$P(H,G) \in [P_1(H,G), P_2(H,G)] = \left[P(H)P(G), \min(P(H), P(G))\right] \tag{5.7}$$

Indeed, positively dependent events occur together more often that independent ones.

Opposite dependence among H and G is determined as

$$P(H,G) = \max(P(H) + P(G) - 1, 0) \tag{5.8}$$

For explanation of this fact one may refer to [8,46]. It is known that $\max(P(H) + P(G) - 1, 0) \le P(H)P(G)$. Negative dependence among H and G is modeled as [46,48]

$$\begin{aligned} P(H,G) &\in [P_1(H,G), P_2(H,G)] = \\ &= [\max(P(H) + P(G) - 1, 0), P(H)P(G)] \end{aligned} \tag{5.9}$$

Indeed, negatively dependent events occur together less often that independent ones.

Unknown dependence is modeled as [46,48]

$$\begin{aligned} P(H,G) &\in [P_1(H,G), P_2(H,G)] = \\ &= [\max(P(H) + P(G) - 1, 0), \min(P(H), P(G))] \end{aligned} \tag{5.10}$$

For the case of interval-valued probabilities of H and G, i.e. when $P(H)\in[P_1(H),P_2(H)]$ and $P(G)\in[P_1(G),P_2(G)]$ the formulas (5.5)-(5.10) are generalized as follows:

$$P(H,G)\in[P_1(H,G),P_2(H,G)]=[P_1(H)P_1(G),P_2(H)P_2(G)] \tag{5.11}$$

$$\begin{aligned}&P(H,G)\in[P_1(H,G),P_2(H,G)]=\\&=[\min(P_1(H),P_1(G)),\min(P_2(H),P_2(G))]\end{aligned} \tag{5.12}$$

$$P(H,G)\in[P_1(H,G),P_2(H,G)]=\left[P_1(H)P_1(G),\min(P_2(H),P_2(G))\right] \tag{5.13}$$

$$\begin{aligned}&P(H,G)\in[P_1(H,G),P_2(H,G)]=\\&=[\max(P_1(H)+P_1(G)-1,0),\max(P_2(H)+P_2(G)-1,0)]\end{aligned} \tag{5.14}$$

$$\begin{aligned}&P(H,G)\in[P_1(H,G),P_2(H,G)]=\\&=[\max(P_1(H)+P_1(G)-1,0),P_2(H)P_2(G)]\end{aligned} \tag{5.15}$$

$$\begin{aligned}&P(H,G)\in[P_1(H,G),P_2(H,G)]=\\&=[\max(P_1(H)+P_1(G)-1,0),\min(P_2(H),P_2(G))]\end{aligned} \tag{5.16}$$

For more details about dependence one may refer to [46,48].

The above mentioned formulas may be extended for the case of fuzzy probabilities $\tilde{P}(H)$ and $\tilde{P}(G)$ as follows. The fuzzy joint probability $\tilde{P}(H,G)$ may be defined as

$$\tilde{P}(H,G)=\bigcup_{\alpha\in[0,1]}\alpha[P_1^{\alpha}(H,G),P_2^{\alpha}(H,G)]$$

where endpoints of an interval $[P_1^{\alpha}(H,G),P_2^{\alpha}(H,G)]$ are determined from endpoints $P_1^{\alpha}(H)$, $P_1^{\alpha}(G)$, $P_2^{\alpha}(H)$ and $P_2^{\alpha}(G)$ on the base of one of formulas (5.11)-(5.16) depending on a sign of dependence. For example, positive dependence is modeled as

$$\tilde{P}(H,G)=\bigcup_{\alpha\in[0,1]}\alpha\left[P_1^{\alpha}(H)P_1^{\alpha}(G),\min(P_2^{\alpha}(H),P_2^{\alpha}(G))\right]$$

and negative dependence as

$$\tilde{P}(H,G)=\bigcup_{\alpha\in[0,1]}\alpha\left[\max(P_1^{\alpha}(H)+P_1^{\alpha}(G)-1,0),P_2^{\alpha}(H)P_2^{\alpha}(G)\right].$$

5.2 Behavioral Modeling of an Economic Agent

We suggest to model an economic agent, a DM, by a set of states. An important issue that arises here is a determination of a state of a DM h. As far as this concept is used to model human behavior which is conditioned by psychological, mental and other behavioral factors, in general, it should not have an abstract or atomic content but should have substantial basis. One approach is a consideration of h as a 'personal quality' of a DM which is formalized as a value of a multivariable function. Each input variable of this function is to be used for measuring one of behavioral factors like risk attitude, ambiguity attitude, altruism, trust, fairness, social responsibility. Thus, a personal quality will have different 'levels' h_i each determined by a vector of measured behavioral factors that describes a behavioral condition of a DM.

Another approach is to consider a state of a DM h_i as a vector of variables describing behavioral factors without converting it into a single generalized value.

The first approach is simpler, i.e. more convenient as a state of a DM will enter decision model as a single value. However a question arises on how to convert a vector into a single value. Anyway, this will lead to a loss of information. The second approach is more adequate, however it is more complex. Consider a small example. Let a DM consider three possible alternatives for investment: to buy stocks, to buy bonds of enterprise or to deposit money in a bank. The results of the alternatives are subject to a state of nature as one of the three possible economic conditions: growth, stagnation, inflation. As the factors underlying behavioral condition of a DM, that is, a state of a DM, we will consider attitudes to risk and ambiguity which are main issues and are especially important for investment problems. The first approach to model a state of a DM is a convolution of the values of the considered factors into a personal quality as a single resulting value. For example, a personal quality of a risk averse and an ambiguity averse investor may be characterized by the typical term "conservative investor", the other combinations will be described by other personal qualities like "aggressive investor" etc. Given such a 'single-valued' reference, it is needed to determine both a joint probability and a utility of any alternative for this state of a DM and any state of nature (economy) – growth, stagnation, inflation. As the state of a DM and a state of nature are both 'single-valued' it will be rather easy to do this. However, this easiness is conditioned by simplistic approach which deprives us of useful information. Influence of each factor will be substantially driven out by convolution to a single value. The second approach models a state of a DM 'as is' – as a pair of risk and ambiguity attitudes without a convolution. Such modeling is more intuitive and transparent. The determination of joint probability and a utility of any alternative as measures of relations between state of nature and state of a DM will then be more adequate because greater useful information is considered. Indeed, even in the first approach, a researcher may have to 'return back' to the behavioral factors in order to more substantially model the relations between the corresponding single-valued personal quality and states of nature. However, the second approach is more complex in terms of mathematical realization – the number of the variables in a model is larger – and this is a price for more adequate modeling.

In this sections we consider two kinds of the first approach to modeling a DM.

Let us consider an approach to agent behavior modeling under second order uncertainty [4,5,50]. It is very difficult to precisely define a term like agent [23,32,33,38,49]. There are tens definitions of agent. A definition similar to [27,38,49] was suggested by us in 1986 [10] and we will use this definition in this work which embraces the following features: autonomy; interaction with an environment and other agents; perception capability; learning; reasoning capability. In [10] an agent with the mentioned characteristics was called a smart agent.

The architecture of a smart agent in accordance with this definition is given in Figure 5.1.

The mathematical description of knowledge in the knowledge base (KB) of agent is based on fuzzy interpretation of antecedents and consequents in production rules.

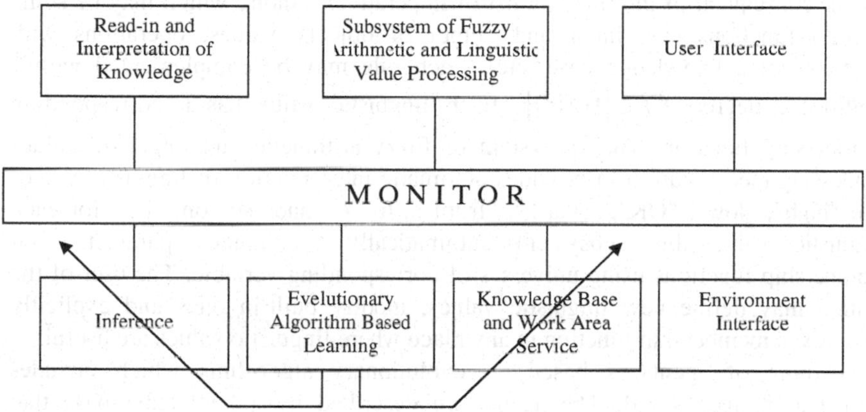

Fig. 5.1 Structure of an intelligent agent

For the knowledge representation the antecedent of each rule contains a conjunction of logical connectives like (Figure 5.2): <name of object> $\left\{\begin{matrix} = \\ \neq \end{matrix}\right\}$ <linguistic value> named elementary antecedent .

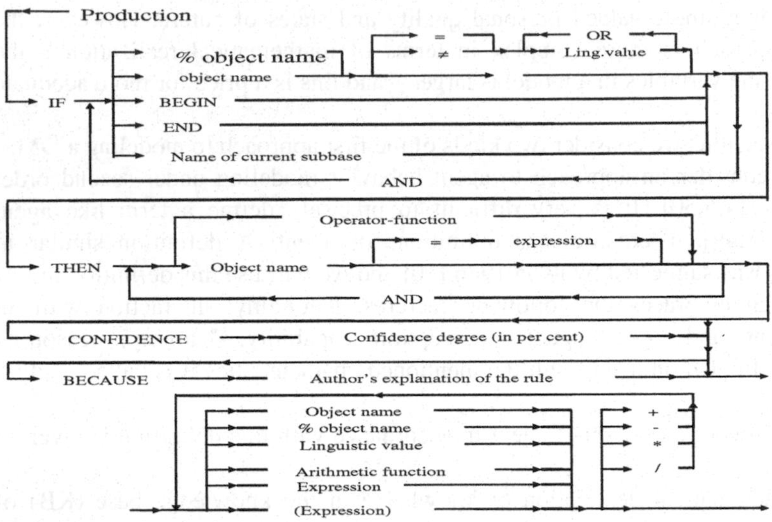

Fig. 5.2 Production rules

The consequent of the rule is a list of imperatives, among which may be some operator-functions (i.e. input and output of objects' values, operations with segments of a knowledge base, etc). Each rule may be complemented with a confidence degree $Cf \in [0,100]$. Each linguistic value has a corresponding membership function. The subsystem of fuzzy arithmetic and linguistic values processing (see Figure 5.1) provides automatic interpretation of linguistic values like "high", "low", "OK", "near...", "from ... to..." and so on; i.e. for each linguistic value this subsystem automatically computes parameters of membership functions using universes of corresponding variable. The user of the system may define new linguistic values, modify built-in ones and explicitly prescribe a membership function in any place where linguistic values are useful.

Learning of agents is based on evolutionary algorithms which includes adjusting of agent's KB. The agents are described by a knowledge-base that consists of a certain number of fuzzy rules related through "ALSO":

$$R^k : IF\ x_1\ is\ \tilde{A}_{k1}\ and\ x_2\ is\ \tilde{A}_{k2}\ and\ ...\ and\ x_m\ is\ \tilde{A}_{km}\ THEN$$
$$u_{k1}\ is\ \tilde{B}_{k1}\ and\ u_{k2}\ is\ \tilde{B}_{k2}\ and\ ...\ and\ u_{kl}\ is\ \tilde{B}_{kl},\quad k=\overline{1,K}$$

where $x_i, i=\overline{1,m}$ and $u_j,\ j=\overline{1,l}$ are total input and local output variables, $\tilde{A}_{ki}, \tilde{B}_{kj}$ are fuzzy sets, and k is the number of rules. Note, that inputs $x_1, x_2, ..., x_m$ may be crisp or fuzzy variables.

Efficiency of inference engine considerably depends on the knowledge base internal organization. That is why an agent's model implements paradigm of "network of production rules" similar to semantic network. Here the nodes are rules and vertexes are objects. Inference mechanism acts as follows. First, some objects take some values (initial data). Then, all production rules, containing each of these objects in antecedent, are chosen from the knowledge base. For these rules the truth degree is computed (in other words, the system estimates the truth degree of the fact that current values of objects correspond to values fixed in antecedents). If the truth degree exceeds some threshold then imperatives from consequent are executed. At that time the same objects as well as a new one take new values and the process continues till work area contains "active" objects ("active" object means untested one).

The assigned value of the object is also complemented by a number, named confidence degree, which is equal to the truth degree of the rule.

A truth degree of a rule's antecedent is calculated according to the following algorithm [4,5].

Let us consider an antecedent of a rule in the form:

$$\text{IF ... AND } \tilde{w}_i \begin{Bmatrix} = \\ \neq \end{Bmatrix} \tilde{a}_{ij} \text{ AND ... AND } \tilde{w}_k \begin{Bmatrix} = \\ \neq \end{Bmatrix} \tilde{a}_{jk} \text{ AND ...}$$

Confidence degree of the rule is $Cf \in [0,100]$.

Objects $\tilde{w}_i$, $\tilde{w}_k$ etc have current values of the form ($\tilde{v}$, *cf*) in the work area (here $\tilde{v}$ is linguistic value with its membership function, $cf \in [0,100]$ is confidence degree of the value $\tilde{v}$). Truth value of *k*-th elementary antecedent is:

$r_k = Poss\left(\tilde{v}_k \middle| \tilde{a}_{jk}\right) cf_k$, if the sign is "=" and $r_k = \left(1 - Poss\left(\tilde{v}_k \middle| \tilde{a}_{jk}\right)\right) cf_k$, if the sign is "≠". *Poss* is defined as

$$Poss\left(\tilde{v} \middle| \tilde{a}\right) = \max_u \min(\mu_{\tilde{v}}(u), \mu_{\tilde{a}}(u)) \in [0,1].$$

The truth degree of the rule:

$$R_j = (\min_k r_k) \frac{Cf_j}{100}$$

After the inference is over, the user may obtain for each object the list of its values with confidence degree which are accumulated in the work area. The desirable value of the object may be obtained using one of the developed algorithms:

$$\tilde{w}_i : (\tilde{v}_i^n, cf_i^n),\ n = \overline{1, N},$$

N is total number of values

Calculation of resulting value:

I. Last - $\tilde{v}_i^N$

II. The value with maximum confidence degree - $\tilde{v}_i^m / cf_i^m = \max_n cf_i^n$

III. The value $\tilde{v}_i = \Lambda_n(\tilde{v}_i^n cf_i^n)$, or $\tilde{v}_i = V_n(\tilde{v}_i^n cf_i^n)$

IV. The average value $\bar{\tilde{v}}_i = \dfrac{\sum_n \tilde{v}_i^n cf_i^n}{\sum_n cf_i^n}$

IF $x_1 = \tilde{a}_1^j$ AND $x_2 = \tilde{a}_2^j$ AND ... THEN $y_1 = \tilde{b}_1^j$ AND $y_2 = \tilde{b}_2^j$ AND ...

IF ... THEN $Y_1 = AVRG(y_1)$ AND $Y_2 = AVRG(y_2)$ AND ...

This model has a built-in function AVRG which calculates the average value. This function simplifies the organization of compositional inference with possibility measures. As a possibility measure here a confidence degree is used. So, the compositional relation is given as a set of production rules like:

$$\text{IF } x_1 = \tilde{A}_1^j \text{ AND } x_2 = \tilde{A}_2^j \text{ AND ... THEN } y_1 = \tilde{B}_1^j \text{ AND } y_2 = \tilde{B}_2^j \text{ AND},$$

where j is a number of a rule (similar to the row of the compositional relation matrix). After all these rules have been executed (with different truth degrees) the next rule (rules) ought to be executed:

$$\text{IF ... THEN } Y_1 = AVRG(y_1) \text{ AND } Y_2 = AVRG(y_2) \text{ AND ...}$$

Fuzzy Hypotheses Generating and Accounting Systems. *Using this model one may construct hypotheses generating and accounting systems. Such system contains the rules:*

$$\text{IF <condition}_j\text{> THEN } X = \tilde{A}_j \text{ CONFIDENCE } cf_j$$

Here "$X = \tilde{A}_j$"is a hypothesis that the object X takes the value $\tilde{A}_j$.Using some preliminary information, this system generates elements $X = (\tilde{A}_j, R_j)$, where R_j is a truth degree of j-th rule. In order to account the hypothesis (i.e. to estimate the truth degree that X takes the value $\tilde{A}_j$) the recurrent Bayes-Shortliffe formula, generalized for the case of fuzzy hypotheses, is used [43]:

$$P_0 = 0$$

$$P_j = P_{j-1} + cf_j Poss(\tilde{A}_0 / \tilde{A})\left(1 - \frac{P_{j-1}}{100}\right)$$

This formula is realized as a built-in function *BS* :

$$\text{IF END THEN } P = BS(X, \tilde{A}_0).$$

Let us consider example. Let us describe the model taking into account the private characteristic features of a DM by using the following rules (inputs and outputs vary within [0,100] range):

Rule 1:
IF altruism level of a DM is *about 45* and emotion level of a DM is *about 40*
THEN personal quality of a DM ($\tilde{D}_i$) is *about 35* and CF is *90*
Rule 2:
IF altruism level of a DM is *about 45* and emotion level of a DM is *about 60*
THEN personal quality of a DM ($\tilde{D}_i$) is *about 45* and CF is *55*
……………
Rule 15:
IF altruism level of a DM is *about 65* and emotion level of a DM is *about 20*
THEN personal quality of a DM ($\tilde{D}_i$) is *about 75* and CF is *60.*

It is required to determine the DM performance (output).

Suppose that emotion level of a DM is *about 65* and altruism level of a DM is *about 60.*

The values of linguistic variable are trapezoidal fuzzy numbers. For example:

$$about\,45 = \begin{cases} \dfrac{x-30}{12}, 30 \le x \le 42 \\ 1, 42 \le x \le 48 \\ \dfrac{50-x}{2}, 48 \le x \le 50 \\ 0, otherwise \end{cases} \qquad about\,60 = \begin{cases} \dfrac{x-50}{5}, 50 \le x \le 55 \\ 1, 55 \le x \le 65 \\ \dfrac{70-x}{5}, 65 \le x \le 70 \\ 0, otherwise \end{cases}$$

$$about\,75 = \begin{cases} \dfrac{x-50}{15}, 50 \le x \le 65 \\ 1, 65 \le x \le 80 \\ \dfrac{85-x}{5}, 80 \le x \le 85 \\ 0, otherwise \end{cases}$$

The above described model is realized by using the ESPLAN expert system shell [4]. For example, for *altruism level* being *about* 65 and *emotion level* being 60 the *personal quality* is computed as *about* 45.

In addition to the imprecision of human conceptualization reflected in natural language many situations that arise in human behavioral modeling entail aspects of probabilistic uncertainty [50]. Now we consider an agent behavioral modeling using fuzzy and Demster-Shater theories suggested in [50].

The Dempster-Shafer approach fits nicely into the fuzzy logic since both techniques use sets as their primary data structure and are important components of the emerging field of granular computing [13,31]. In [50] the behavioral model is represented by partitioning the input space. We can represent relationship between input and output variables by a collection of n "IF-THEN" rules of the form:

$$\text{If } X_1 \text{ is } \tilde{A}_{i1} \text{ and } X_2 \text{ is } \tilde{A}_{i2}, \ldots \text{ and } X_r \text{ is } \tilde{A}_{ir} \text{ then } Y \text{ is } D_i \tag{5.17}$$

Here each $\tilde{A}_{ij}$ typically indicates a linguistic term corresponding to a value of its associated variable, furthermore each $\tilde{A}_{ij}$ is formally represented as a fuzzy subset defined over the domain of the associated variable X_j. Similarly $\tilde{D}_i$ is a value associated with the consequent variable Y that is formally defined as a fuzzy subset of the domain of Y. To find the output of an agent described by (5.17) for given values of the input variables the Mamdani-Zadeh reasoning paradigm is used [51].

It is needed now to add further modeling capacity to model (5.17) by allowing for probabilistic uncertainty in the consequent. For this we consider the consequent to be a fuzzy Dempster-Shafer granule. Thus we shall now consider the output of each rule to be of the form Y is m_i where m_i is a belief structure with focal elements $\tilde{D}_{ij}$ which are fuzzy subsets of the universe Y and associated weights $m_i(\tilde{D}_{ij})$. Thus a typical rule is now of the form

$$\text{If } X_1 \text{ is } \tilde{A}_{i1} \text{ and } X_2 \text{ is } \tilde{A}_{i2}, \ldots \text{ and } X_r \text{ is } \tilde{A}_{ir} \text{ then } Y \text{ is } m_i \tag{5.18}$$

Using a belief structure to model the consequent of a rule is essentially saying that $m_i(\tilde{D}_{ij})$ is the probability that the output of the ith rule lies in the set $\tilde{D}_{ij}$. So rather than being certain as to the output set of a rule we have some randomness in the rule. We note that with $m_i(\tilde{D}_{ij}) = 1$ for some $\tilde{D}_{ij}$ we get the (5.17).

Let us describe the reasoning process in this situation with belief structure consequents. Assume the inputs to the system are the values for the antecedent variables, $X_j = x_j$. For each rule we obtain the firing level, $\tau_i = \mathrm{Min}[A_{ij}(x_j)]$.

The output of each rule is a belief structure $\hat{m}_i = \tau_i \wedge m$. The focal elements of $\hat{m}_i$ are $\tilde{F}_{ij}$, a fuzzy subset of Y where $F_{ij}(y) = \mathrm{Min}[\tau_i, D_{ij}(y)]$, here $\tilde{D}_{ij}$ is a focal element of m_i. The weights associated with these new focal elements are simply $\hat{m}_i(\tilde{F}_{ij}) = \hat{m}_i(\tilde{D}_{ij})$. The overall output of the system m is obtained by taking a union of the individual rule outputs, $m = \bigcup_{i=1}^{n} \hat{m}_i$.

For every a collection $< \tilde{F}_{1j_1}, \ldots \tilde{F}_{nj_1} >$ where $\tilde{F}_{ij_1}$ is a focal element of m_i we obtain a focal element of m, $\tilde{E} = \bigcup_i \tilde{F}_{ij_1}$ and the associated weight is

$$m(\tilde{E}) = \prod_{i=1}^{n} \hat{m}_i(\tilde{F}_{ij_1}).$$

As a result of this step it is obtained a fuzzy D-S belief structure *V* is m as output of the agent. We denote the focal elements of m as the fuzzy subsets $\tilde{E}_j$, j = 1 to q, with weights $m(\tilde{E}_j)$.

Let us describe the model taking into account the characteristic features of economic agent (DM). Here the basic problem is to evaluate personal quality of a DM by using its psychological determinants. For determining psychological determinants as basic factors (inputs of a model) influencing to DM performance, total index of DM (output of a model), we used the fuzzy Delphi method. We have obtained that main psychological determinants (inputs) are following factors: trust, altruism, reciprocity, emotion, risk, social responsibility, tolerance to ambiguity.

For a total index (resulting dimension) of a DM as an overall evaluation to be determined on the base of the determinants we obtained personal quality or power of decision or DM's performance. So, DM's behavioral model can be described as (for simplicity we use 2 inputs):

Rule 1: IF trust level of a DM is *about 76* and altruism level of a DM *about 45* THEN personal quality of a DM (V) is m_1.

Rule 2: IF trust level of a DM is *about 35* and altruism level of a DM *about 77* THEN personal quality of a DM (V) is m_2.

Let us determine the output (personal quality of a DM) if trust level of a DM is *about 70* and altruism level of a DM is *about 70.*

m_1 has focal elements $\tilde{D}_{11} = \tilde{46}$ with $m(\tilde{D}_{11}) = 0.7$ and $\tilde{D}_{12} = \tilde{48}$ with $m(\tilde{D}_{11}) = 0.3$;

m_2 has focal elements $\tilde{D}_{21} = \tilde{76}$ with $m(\tilde{D}_{21}) = 0.2$ and $\tilde{D}_{22} = \tilde{81}$ with $m(\tilde{D}_{22}) = 0.8$

The values of linguistic variables are triangle fuzzy numbers:

$$\tilde{46}=\begin{cases}\frac{x-40}{6}, 40\le x\le 46\\ 1, x=46\\ \frac{65-x}{19}, 46\le x\le 65\\ 0, otherwise\end{cases}\qquad \tilde{48}=\begin{cases}\frac{x-40}{8}, 40\le x\le 48\\ 1, x=48\\ \frac{65-x}{17}, 48\le x\le 65\\ 0, otherwise\end{cases}$$

$$\tilde{76}=\begin{cases}\frac{x-61}{15}, 61\le x\le 76\\ 1, x=76\\ \frac{95-x}{19}, 76\le x\le 95\\ 0, otherwise\end{cases}\qquad \tilde{81}=\begin{cases}\frac{x-61}{20}, 61\le x\le 81\\ 1, x=81\\ \frac{95-x}{14}, 81\le x\le 95\\ 0, otherwise\end{cases}$$

Let us calculate the belief values for each rule. By using [4] in this example the empty set takes the value 0.09. But in accordance with Dempster-Shafer theory m-value of the empty set should be zero. In order to achieve this, m values of the focal elements should be normalized and m value of the empty set made equal to zero. The normalization process is as follows:

1) Determine $T=\sum_{A_i\cap B_i=\varnothing} m_1(\tilde{A}_i)\cdot m_2(\tilde{B}_i)$

2) For all $\tilde{A}_i\cap\tilde{B}_i=\varnothing$ weights are

$m(\tilde{E}_k)=\frac{1}{1-T}m_1(\tilde{A}_i)\cdot m_2(\tilde{B}_j)$

3) For all $\tilde{E}_k=\varnothing$ sets $m(\tilde{E}_k)=0$

In accordance with the procedures described above:

$$m_3=(\{\tilde{46}\})=0.230769,$$

$$m_3=(\{\tilde{46},y\})=0.384615,$$

$$Bel(\{\tilde{46},y\})=0.615385.$$

For the second rule: $Bel(\{\tilde{76},y\})=0.753425$. Firing level of the *i*-th rule is equal to the minimum among all degrees of membership of a system input to antecedent fuzzy sets of this rule: $\tau_i = \min_{j=1}^{n}[\max_{X_j}(A'(x_j) \wedge A_{ij}(x_j))]$. The firing levels of each rule are $\tau_1 = 0.26$ and $\tau_2 = 0.28$. The defuzzified values of focal elements obtained by using the center of gravity method are the following: $Defuz(\tilde{E}_1) = \overline{y}_1 = 61.56$; $Defuz(\tilde{E}_2) = \overline{y}_2 = 64.15$; $Defuz(\tilde{E}_3) = \overline{y}_3 = 62.52$; $Defuz(\tilde{E}_4) = \overline{y}_4 = 65.11$. The defuzzified value of m is $\overline{y} = 63.92$.

By using the framework described above we arrive at the following Dempster-Shafer structure:

IF trust level of a DM is *about 70* and altruism level of a DM *about 70* THEN personal quality of a DM (V) is equal to *about* 63.92.

References

1. Alexeyev, A.V., Borisov, A.N., Glushkov, V.I., Krumberg, O.A., Merkuryeva, G.V., Popov, V.A., Slyadz, N.N.: A linguistic approach to decision-making problems. Fuzzy Sets and Systems 22, 25–41 (1987)
2. Aliev, R.A.: Theory of decision making under second-order uncertainty and combined states. In: Proceedings of the Ninth International Conference on Application of Fuzzy Systems and Soft Computing (ICAFS 2010), Prague, pp. 5–6. b-Quadrat Verlag, Czech Republic (2010)
3. Aliev, R.A.: Decision making with combined states under imperfect information. In: Proceedings of the Sixth International Conference on Soft Computing and Computing with Words in System Analysis, Decision and Control (ICSCCW 2011), pp. 3–4. b-Quadrat Verlag, Antalya (2011)
4. Aliev, R.A., Aliev, R.R.: Soft Computing and its Application. World Scientific, New Jersey (2001)
5. Aliev, R.A., Fazlollahi, B., Aliev, R.R.: Soft Computing and its Application in Business and Economics. Springer, Heidelberg (2004)
6. Aliev, R.A., Huseynov, O.H.: A new approach to behavioral decision making with imperfect information. In: Proceedings of the Sixth International Conference on Soft Computing and, Computing with Words in System Analysis, Decision and Control, ICSCCW 2011, pp. 227–237. b-Quadrat Verlag, Antalya (2011)
7. Aliev, R.A., Huseynov, O.H.: Decision making under imperfect information with combined states. In: Proceedings of the Ninth International Conference on Application of Fuzzy Systems and Soft Computing (ICAFS 2010), Prague, pp. 400–406. b-Quadrat Verlag, Czech Republic (2010)
8. Aliev, R.A., Pedrycz, W., Fazlollahi, B., Huseynov, O.H., Alizadeh, A.V., Guirimov, B.G.: Fuzzy logic-based generalized decision theory with imperfect information. Information Sciences 189, 18–42 (2012)
9. Aliev, R.A., Pedrycz, W., Huseynov, O.H.: Decision theory with imprecise probabilities. International Journal of Information Technology & Decision Making 11(2), 271–306 (2012), doi:10.1142/S0219622012400032
10. Aliev, R.A., Tserokvny, A.E.: "Smart" manufacturing systems. News of Academy of Sciences of USSR. Tech. Cybernetics 6, 99–108 (1988) (in English and Russian)

11. Andreoni, J., Miller, J.: Giving according to garp: an experimental test of the consistency of preferences for altruism. Econometrita 70, 737–753 (2002)
12. Asto Buditjahjanto, I.G.P., Miyauchi, H.: An intelligent decision support based on a subtractive clustering and fuzzy inference system for multiobjective optimization problem in serious game. International Journal of Information Technology and Decision Making 10(5), 793–810 (2011)
13. Bargiela, A., Pedrycz, W.: Granular Computing: An Introduction. Kluwer Academic Publishers, Amsterdam (2003)
14. Berg, J., Dickhaut, J., McCabe, K.: Trust, reciprocity, and social history. Games and Economic Behavior 10, 122–142 (1995)
15. Belohlavek, R., Sigmund, E., Zacpal, J.: Evaluation of IPAQ questionnaires supported by formal concept analysis. Information Science 181(10), 1774–1786 (2011)
16. Brafman, R.I., Tennenholz, M.: Modeling agents as qualitative decision makers. Artificial Intelligence 94(1-2), 217–268 (1997)
17. Cao, L.B., Dai, R.W.: Agent-oriented metasynthetic engineering for decision making. International Journal of Information Technology and Decision Making 2(2), 197–215 (2003)
18. Compte, O., Postlewaite, A.: Mental processes and decision making. Working paper. Yale University, New Haven (2009)
19. Cover, T., Hellman, M.: Learning with finite memory. Annals of Mathematical Statistics 41, 765–782 (1970)
20. Cox, J.C.: How to identify trust and reciprocity. Games and Economic Behavior 46, 260–281 (2004)
21. Cox, J.C., Friedman, D., Sadiraj, V.: Revealed altruism. Econometrica 76(1), 31–69 (2008)
22. Falk, A., Fischbacher, U.: A theory of reciprocity. Games and Economic Behavior 54, 293–315 (2006)
23. Franklin, S., Graesser, A.: Is it an Agent, or Just a Program?: A Taxonomy for Autonomus Agents. In: Jennings, N.R., Wooldridge, M.J., Müller, J.P. (eds.) ECAI-WS 1996 and ATAL 1996. LNCS, vol. 1193, pp. 21–36. Springer, Heidelberg (1997)
24. Ghirardato, P.: Coping with ignorance: unforeseen contingencies and non-additive uncertainty. Economic Theory 17, 247–276 (2001)
25. Ghirardato, P., Maccheroni, F., Marinacci, M.: Differentiating Ambiguity and Ambiguity Attitude. Journal of Economic Theory 118, 133–173 (2004)
26. Gilboa, I., Schmeidler, D.: Maximin Expected utility with a non-unique prior. Journal of Mathematical Economics 18, 141–153 (1989)
27. Hayes Roth, B.: An Architecture for Adaptive Intelligent Systems. Artificial Intelligence 72, 329–365 (1995)
28. Kahneman, D., Tversky, A.: Prospect theory: an analysis of decision under uncertainty. Econometrica 47, 263–291 (1979)
29. Klibanoff, P., Marinacci, M., Mukerji, S.: A smooth model of decision making under ambiguity. Econometrica 73(6), 1849–1892 (2005)
30. Labreuche, C., Grabisch, M.: Generalized Choquet-like aggregation functions for handling bipolar scales. European Journal of Operational Research 172, 931–955 (2006)
31. Lin, T.S., Yao, Y.Y., Zadeh, L.A.: Data Mining, Rough Sets and Granular Computing. Physica-Verlag, Heidelberg (2002)

32. Luck, M., D'lnverno, M.: A formal Framework for Agency and Autonomy. In: Proc. of the 1st International Conference on Multi-Agent Systems, pp. 254–260. AAAI Press/MIT Press, San-Francisko, CA (1995)
33. Luck, M., D'lnverno, M.: Engagement and Cooperation in Motivated Agent Modeling. In: Distibuted Artifical Intelligence. Architecture and Modeling. First Australian Workshop on Distributed Artificial Intelligence, Canberra, ACT, Australia, vol. 70 (1995a)
34. Magni, C.A., Malagoli, S., Mastroleo, G.: An alternative approach to firms' evaluation: expert systems and fuzzy logic. International Journal of Information Technology and Decision Making 6, 195–225 (2006)
35. McCarthy, J.: Ascribing mental qualities to machines. In: Ringle, M. (ed.) Philosophical Perspectives in Artificial Intelligence, pp. 161–195. Humanities Press, Atlantic Highlands (1979)
36. Neumann, J., Morgenstern, O.: Theory of games and economic behaviour. Princeton University Press (1944)
37. Newell, A.: The knowledge level. AI Magazine 2(2), 1–20 (1981)
38. Nwana, H.S., Ndumu, D.T.: An Introduction to Agent Technology. In: Nwana, H.S., Azarmi, N. (eds.) Software Agents and Soft Computing: Towards Enhancing Machine Intelligence. LNCS, vol. 1198, pp. 3–26. Springer, Heidelberg (1997)
39. Rao, A.S., Georgeff, M.P.: Trader species with different decision strategies and price dynamics in financial markets: an agent-based modeling perspective. International Journal of Information Technology and Decision Making 9(2), 327–344 (2010)
40. Schmeidler, D.: Subjective probability and expected utility without additivity. Econometrita 57(3), 571–587 (1989)
41. Shoham, Y.: BSV investors versus rational investors: an agent-based computational finance model. International Journal of Information Technology and Decision Making 5(3), 455–466 (2006)
42. Shoham, Y., Cousins, S.B.: Logics of Mental Attitudes in AI: a Very Preliminary Survey. In: Lakemeyer, G., Nebel, B. (eds.) ECAI-WS 1992. LNCS, vol. 810, pp. 296–309. Springer, Heidelberg (1994)
43. Shortliffe, E.H.: Computer-based medical consultations: MYCIN. American Elsevier, NY (1976)
44. Simon, H.: Models of Bounded Rationality: Empirically Grounded Economic Reason, vol. 3. MIT Press, Cambridge (1997)
45. Tversky, A., Kahneman, D.: Advances in Prospect theory: Cumulative Representation of Uncertainty. Journal of Risk and Uncertainty 5(4), 297–323 (1992)
46. Williamson, R.C.: Probabilistic Arithmeti. Ph.D dissertation, University of Queensland, Austrialia (1989), `http://theorem.anu.edu.au/~williams/papers/thesis300dpi.ps`
47. Wilson, A.: Bounded memory and biases in information processing. NAJ Economics, 5 (2004)
48. Wise, B.P., Henrion, M.: A framework for comparing uncertain inference systems to probability. In: Kanal, L.N., Lemmer, J.F. (eds.) Uncertainty in Artificial Intelligence, pp. 69–83. Elsevier Science Publishers, Amsterdam (1986)
49. Wooldridge, M., Jennings, N.: Agents Theories, Architectures, and Languages: a Survey. In: Wooldridge, M., Jennings, N.R. (eds.) Intelligent Agents, pp. 1–22. Springer, Berlin (1995)

50. Yager, R.R.: Human behavioral modeling using fuzzy and Dempster–Shafer theory. In: Liu, H., Salerno, J.J., Young, M.J. (eds.) Social Computing, Behavioral Modeling, and Prediction, pp. 89–99. Springer, New York (2008)
51. Yager, R.R., Filev, D.P.: Essentials of Fuzzy Modeling and Control. John Wiley, New York (1994)
52. Zadeh, L.A.: Fuzzy Sets. Information and Control 8, 338–353 (1965)
53. Zadeh, L.A.: Fuzzy logic = Computing with Words. IEEE Transactions on Fuzzy Systems 4(2), 103–111 (1996)
54. Zadeh, L.A.: A note on web intelligence, world knowledge and fuzzy logic. Data and Knowledge Engineering 50, 291–304 (2004)

Chapter 6
Decision Making on the Basis of Fuzzy Geometry

6.1 Motivation

Decision making is conditioned by relevant information. This information very seldom has reliable numerical representation. Usually, decision relevant information is perception-based. A question arises of how to proceed from perception-based information to a corresponding mathematical formalism. When perception-based information is expressed in NL, the fuzzy set theory can be used as a corresponding mathematical formalism and then the theories presented in Chapters 3,4,5 can be applied for decision analysis. However, sometimes perception-based information is not sufficiently clear to be modeled by means of membership functions. In contrast, it remains at a level of some cloud images which are difficult to be caught by words. This imperfect information caught in perceptions cannot be precisiated by numbers or fuzzy sets and is referred to as *unprecisiated* information. In order to better understand a spectrum of decision relevant information ranging from numbers to unprecisiated information, let us consider a benchmark problem of decision making under imperfect information suggested Prof. Lotfi Zadeh. The problem is as follows.

Assume that we have two open boxes, A and B, each containing twenty black and white balls. A ball is picked at random. If I pick a white ball from A, I win a1 dollars; if I pick a black ball, I lose a2 dollars. Similarly, if I pick a white ball from B, I win b1 dollars; and if I pick a black ball, I lose b2 dollars. Then, we can formulate the five problems dependent on the reliability of the available information:

Case 1. I can count the number of white balls and black balls in each box. Which box should I choose?

Case 2. I am shown the boxes for a few seconds, not enough to count the balls. I form a perception of the number of white and black balls in each box. These perceptions lead to perception-based imprecise probabilities which allow to be described as fuzzy probabilities. The question is the same: which box should I choose.

Case 3. I am given enough time to be able to count the number of white and black balls, but it is the gains and losses that are perception-based and can be described as fuzzy numbers. The question remains the same.

Case 4. Probabilities, gains and losses are perception-based and can be described as fuzzy probabilities and fuzzy numbers. The question remains the same.

R.A. Aliev: *Fuzzy Logic-Based Generalized Theory of Decisions*, STUDFUZZ 293, pp. 217–230.
DOI: 10.1007/ 978-3-642-34895-2_6

Case 5. The numbers of balls of each color in each box cannot be counted. All a I have are visual perceptions which cannot be precisiated by fuzzy probabilities.

Let us discuss these cases. Case 1 can be successfully solved by the existing theories because it is stated in numerical information. Cases 2-4 are characterized by linguistic decision-relevant information, and therefore, can be solved by the decision theory suggested in Chapter 4. No theory can be used to solve Case 5 as it is stated, including the theory suggested in Chapter 4, because this case is initially stated in informational framework of visual perceptions for which no formal decision theory is developed. However, humans are able to make decisions based on visual perceptions. Modeling of this outstanding capability of humans, even to some limited extent, becomes a difficult yet a highly promising research area. This arises as a motivation of the research suggested in this chapter. In this chapter we use Fuzzy Geometry and the extended fuzzy logic [15] to cope with uncertain situations coming with unprecisiated information. In this approach, the objects of computing and reasoning are geometric primitives, which model human perceptions when the latter cannot be defined in terms of membership functions. The fuzzified axioms of Euclidean geometry are used and the main operations over fuzzy geometric primitives are introduced. A decision making method with outcomes and probabilities described by geometrical primitives is developed. In this method, geometrical primitives like fuzzy points and fuzzy lines represent the basic elements of the decision problem as information granules consisting of an imprecise value of a variable and the confidence degree for this value. The decision model considers a knowledge base with fuzzy geometric "if-then" rules.

All works on decision analysis assume availability of numeric or measurement-based information. In other words, the available imperfect information is always considered to admit required precision. The fundamental question remains: what if the information is not only imperfect and perception-based, but also unprecisiated?

As stated in [15] while fuzzy logic delivers an important capability to reason precisely in presence of imperfect information, the extended (or unprecisiated) fuzzy logic delivers a unique ability to reason imprecisely with imperfect information. The capability to reason imprecisely is used by human being when precise reasoning is infeasible, excessively costly or not required at all. A typical real-life case example is a case when the only available information is perception-based and no trustworthy precision or fuzzy numeric models (e.g. articulated through membership functions) are possible to obtain. As a model of unprecisiated fuzzy logic we consider fuzzy geometry [15].

The concept of fuzzy geometry is not new. Many authors suggest various versions of fuzzy geometry. Some of well-known ones are the Poston's fuzzy geometry [8], coarse geometry [9], fuzzy geometry of Rosenfeld [10] , fuzzy geometry of Buckley and Eslami [2], fuzzy geometry of Mayburov [8], fuzzy geometry of Tzafestas [13], and fuzzy incidence geometry of Wilke [14]. Along this line of thought, many works are devoted to model spatial objects with fuzzy boundaries [3,4,12].

The study reported in [12] proposes a general framework to represent ill-defined information regarding boundaries of geographical regions by using the concept of relatedness measures for fuzzy sets. Regions are represented as fuzzy sets in a two–dimensional Euclidean space, and the notions of nearness and relative orientation are expressed as fuzzy relations. To support fuzzy spatial reasoning, the authors derive transitivity rules and provide efficient techniques to deal with the complex interactions between nearness and cardinal directions.

The work presented in [3] introduces a geometric model for uncertain lines that is capable of describing all the sources of uncertainty in spatial objects of linear type. Uncertain lines are defined as lines that incorporate uncertainty description both in the boundary and interior and can model all the uncertainty by which spatial data are commonly affected and allow computations in presence of uncertainty without oversimplification of the reality.

Qualitative techniques for spatial reasoning are adopted in [4]. The author formulates a computational model for defining spatial constraints on geographic regions, given a set of imperfect quantitative and qualitative constraints.

What is common in all currently known fuzzy geometries is that the underlying logic is the fuzzy logic. Fuzzy logic implies existence of valid numerical information (qualitative or quantitative) regarding the geometric objects under consideration. In situations, when source information is very unreliable to benefit from application of computationally-intensive mathematical computations of traditional fuzzy logic, some new method is needed. The new fuzzy geometry, the concept of which is proposed by Zadeh and referred to as F-Geometry, could be regarded as a highly suitable vehicle to model unprecisiated or extended fuzzy logic [15].

Of the geometries mentioned above, the fuzzy incidence geometry of Wilke [14] can form a starting point for developing the new F-Geometry. Thus fuzzy incidence geometry extends the Euclidean geometry by providing concepts of extended points and lines as subsets of coordinate space, providing fuzzy version of incidence axioms, and reasoning mechanism by taking into account the positional tolerance and truth degree of relations among primitives. To allow for partially true conclusions from partially true conditions, the graduated reasoning with Rational Pavelka Logic (RPL) is used [7].

The purpose of this chapter is to develop a concept and a technique that can be used to more adequately reflect the human ability to formally describe perceptions for which he/she could hardly suggest acceptable linguistic approximations due to their highly uncertain nature or for which such precision, if provided, would lead to a loss or degradation of available information. Such unprecisiatable perceptions many times form an underlying basis for everyday human reasoning as well as decision making in economics and business.

It is suggested that Fuzzy Geometry or F-geometry (or geometry for extended primitives) can be used to more adequately reflect the human ability to describe decision-relevant information by means of geometric primitives. Classical geometry is not useful in this case. As it was mentioned in [5], classical geometry fails to acknowledge that visual space is not an abstract one but its properties are defined by perceptions.

The main idea is to describe uncertain data (which are perceptions of human observer, researcher, or a decision maker) in geometric language using extended primitives: points, lines, bars, stripes, curves etc. to prevent possible loss of information due to the precision of such data to classical fuzzy sets based models (e.g. when using membership functions etc.).

6.2 Fuzzy Geometry Primitives and Operations

F-geometry is a simple and natural approach that can be used to express human perceptions in a visual form so that they can be used in further processing with minimal distortion and loss of information. In F-geometry, we use different primitive geometric concepts such as f-points, f-intervals, f-lines etc. as well as more complex f-transform concepts such as f-parallel, f-similar, f-convex, f-stable, etc. to express the underlying information. The primitive concepts can be entered by hand using simple graphic interface tool such as spray-pen or Z-mouse [15].

Pieces of information, describing the properties required for decision-making, are represented in forms of 2D geometric objects. For one-dimensional properties the second dimension can be used for expressing additional information.

For entering the information regarding a certain property, for example, to define a range of probabilities, the decision-maker (DM), instead of entering numbers, manually draws strips using a spray can. By doing so, the DM could also implicitly express his/her confidence degree about the entered information by drawing physically greater objects or thick lines for less confident information granules and more compact sized marks, e.g., points or thinner lines with strictly defined boundaries, to express more reliable information granules.

Generally F-geometry primitives can be defined as two-dimensional sets, which are subsets of R^2. F-marks are primitives of F-geometry that can be used in arithmetic, comparison, and set-theoretic operations. Therefore, we require that F-marks (but not necessarily their F-transforms) are convex sets. A concept named Z-number has been suggested in [16], which also could be used to approximately represent F-marks (see Section 1.1). A Z-number consists of a pair of fuzzy sets, basically trapezoidal fuzzy numbers, entered by using a specialized graphical interface tool. The fundamental difference between a Z-number and an F-mark suggested here is that a Z-number is still explicitly based on membership functions whereas an F-mark is not.

Definition 6.1 [6,14]. **F-mark.** An ***F-mark*** is a bounded subset of R^2, representing a graphical hand-mark drawn by human being to indicate visually a value of a perception-based information granule.

So, formally an F-mark *A* can be represented as a bounded subset of $R^2 : A \subset R^2$. But an F-mark is more than just a physical area as it is meant to hold a perception of a measurable value. Therefore, we will use two notations: *A* (when meaning a perception, an unprecisiated value) and A (when meaning an area or a variable to hold a measurable value). Usually, an area A, representing an F-mark is assumed to be a convex set [14].

If required, we should be able to approximately represent, i.e. precisiate, F-marks by using two-dimensional membership functions (e.g. of truncated pyramidal or con form) based on density, intensity, or width of the spray pen (Z-mouse) used for the drawing [15].

Let us define some basic primitives that we will use in context of decision making.

Any F-mark, which represents a convex subset A of R^2, can be approximately defined by its center $c = (x_c, y_c)$ (which is a Euclidean point) and two diameters ($\phi_{\min}, \phi_{\max}$) [14]:

$$A = P(c, \phi_{\min}, \phi_{\max}) \tag{6.1}$$

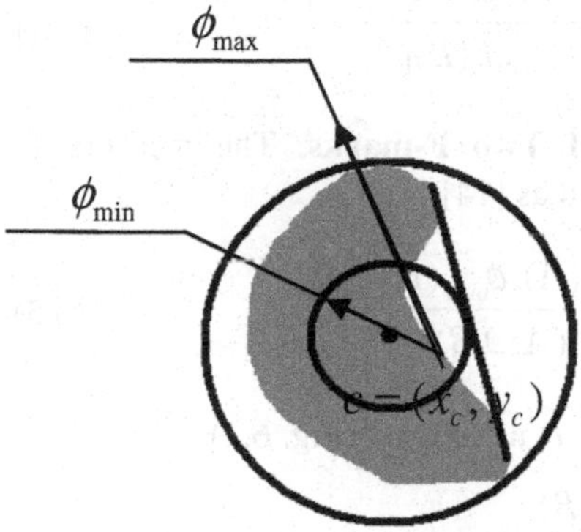

Fig. 6.1 An F-mark with two diameters and its convex hull

The center c can be computed as a center of gravity of convex hull: $c=C(ch(A))$ while the two diameters are [14]:

$$\phi_{\min} = \min_t \left| ch(A) \cap \{c + t \cdot R_\alpha (0,1)^T\} \right|$$

$$\phi_{\max} = \max_t \left| ch(A) \cap \{c + t \cdot R_\alpha (0,1)^T\} \right|$$

where $t \in R$ and R_α is the rotation matrix describing rotation by angle α.

The illustration of the concept is presented in Fig. 6.1.

Definition 6.2. F-point. The degree to which an F-mark $A = P(c, \phi_{\min}, \phi_{\max})$ is an ***F-point*** is determined as follows [14]:

$$p(A) = \phi_{\min} / \phi_{\max} \tag{6.2}$$

Definition 6.3. The degree to which an F-mark A is an ***F-line*** is determined as:

$$l(A)=1-p(A).$$

Definition 6.4. Truth Degree of an Incidence of Two F-marks. The truth degree of predicate for the incidence of F-marks A and B is determined as [14]:

$$inc(A,B) = \max\left(\frac{|ch(A) \cap ch(B)|}{|ch(A)|}, \frac{|ch(A) \cap ch(B)|}{|ch(B)|}\right), \tag{6.3}$$

here $ch(A)$ is a convex hull of an f-mark A, $|ch(A)|$ is the area covered by $ch(A)$.

Definition 6.5. Truth Degree of an Equality of Two F-marks. The truth degree of a predicate defining the equality of F-marks A and B, is determined as follows [14]:

$$eq(A,B) = \min\left(\frac{|ch(A) \cap ch(B)|}{|ch(A)|}, \frac{|ch(A) \cap ch(B)|}{|ch(B)|}\right) \tag{6.4}$$

Definition 6.6. Measure of Distinctness of Two F-marks. The measure of distinctness of f-marks A and B is determined as [14]:

$$dp(A,B) = \max\left(0, 1 - \frac{\max\left(\phi_{\max}(A), \phi_{\max}(B)\right)}{\phi_{\max}(ch(A \cup B))}\right) \tag{6.5}$$

Two f-points A and B can generate an f-line L as follows (Fig. 6.2):

$$L = ch(A \cup B)$$

Axioms of the Fuzzy Incidence Geometry

The following axioms formalize the behavior of points and lines in incident geometry [14]:

(A1) For every two distinct points *p* and *q*, at least one line *l* exists that is incident with *p* and *q*.

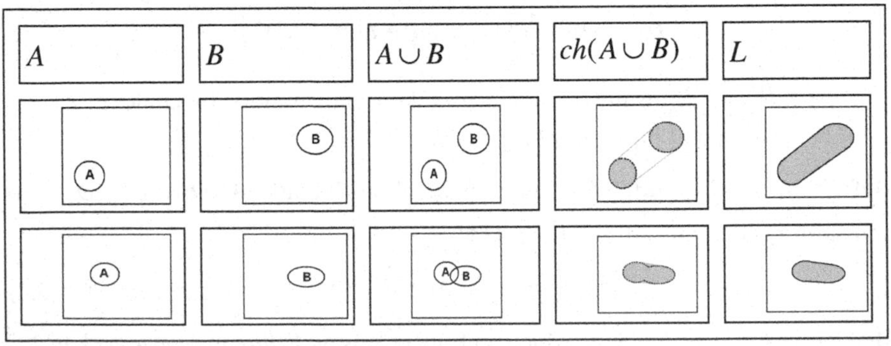

Fig. 6.2 Generation of an f-line from two f-points

(A2) Such a line is unique.

(A3) Every line is incident with at least two points.

(A4) At least three points exist that are not incident with the same line.

For fuzzy version of incident geometry each of the above axioms may not evaluate to absolute truth for all possible inputs.

A fuzzy version of the incident geometry, which is suitable to work with f-marks can be axiomatized as follows [14]:

$$\text{(A1')} \left(dp(x,y) \to \sup_z \left[l(z) \otimes inc(x,z) \otimes inc(y,z) \right], r_1 \right)$$

$$\text{(A2')} \left(dp(x,y) \to \right.$$

$$\left. \to \left[l(z) \to \left[inc(x,z) \to \left[\begin{array}{l} inc(y,z) \to l(z') \to \\ \to \left[inc(x,z') \to \left[inc(y,z') \to eq(z,z') \right] \right] \end{array} \right] \right] \right], r_3 \right)$$

$$\text{(A3')} \left(l(z) \to \sup_{x,y} \left\{ \begin{array}{l} p(x) \otimes p(y) \otimes \neg \\ \neg eq(x,y) \otimes inc(x,z) \otimes inc(y,z) \end{array} \right\}, r_3 \right) \tag{6.6}$$

$$\text{(A4')} \left(\sup_{u,v,w,z} \left[\begin{array}{l} p(u) \otimes p(v) \otimes p(w) \otimes l(z) \to \\ \to \neg \left(inc(u,z) \otimes inc(v,z) \otimes inc(w,z) \right) \end{array} \right], r_4 \right),$$

where x, y, z, z', u, v, and w are measurable variables to hold F-marks, $\otimes$ denotes Lukasiewicz t-norm, r_1, r_2, r_3, and r_4 are truth values of the associated axioms.

In this study, we consider two basic types of F-marks: F-points and F-lines.

When it is needed for a concise representation or fast computation, any convex F-mark can be approximately represented as (6.1) [14]. Instead of it, we suggest an approximation illustrated in Fig. 6.3.

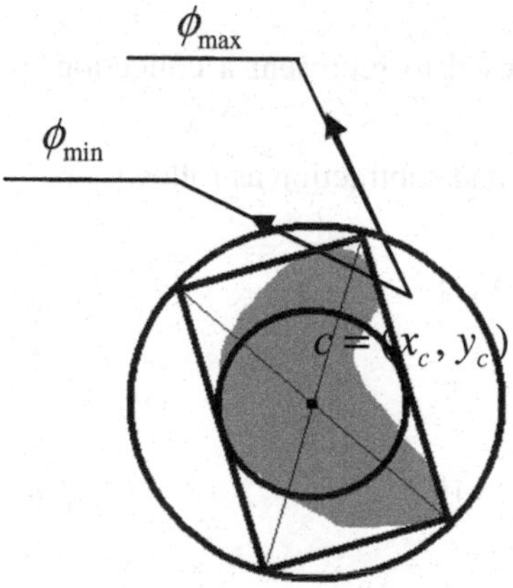

Fig 6.3 The suggested approximation of an F-mark

The idea behind this method is to find a parallelogram with a minimum square, the intersection with f-mark of which is the same f-mark. Then the sides of the parallelogram are the two diameters, the shorter one is $\phi_{\min}$ and the longer one is $\phi_{\max}$.

If a parallelogram can be specified as $S\left((c_x, c_y), \alpha, h, w\right)$, where (c_x, c_y) is its center (centroid), α is rotation angle (e.g. counter-clockwise vs. Y axis, which is 0), and h and w are its sides.

Then $\phi_{\min} = \min(h, w)$ and $\phi_{\max} = \max(h, w)$, where h and w are found by solving the optimization task:

$$\begin{gathered} h \cdot w \rightarrow \min \\ \text{s.t. } S\left((c_x, c_y), \alpha, h, w\right) \cap A = A . \end{gathered} \tag{6.7}$$

As it can be seen $A_{\min} \subseteq A \subseteq A_{\max}$, where $A_{\min}$ and $A_{\max}$ are 2D disks with diameters $\phi_{\min} = \min(h, w)$ and $\phi_{\max} = \max(h, w)$, respectively.

As we pointed above, all F-marks are 2D sets, which are bounded subsets of R^2.

$$A \equiv \int_{(x,y) \in A} \{(x, y)\},$$

where $A \subset R^2$ and the integral sign $\int$ does not mean integration but denotes a collection of all points $(x, y) \in A$. With any desired accuracy an F-mark can be represented as a discrete set:

$$A \equiv \sum_{(x_i, y_i) \in A} \{(x_i, y_i)\},$$

where $A \subset R^2$, and a summation sign $\sum$ is used to represent a collection of elements of a discrete set.

We define arithmetic operations of summation and subtraction as follows:

$$A + B = \int_{\substack{(x_1, y_1) \in A \\ (x_2, y_2) \in B}} \{(x_1 + x_2, y_1 + y_2)\} \tag{6.8}$$

$$A - B = \int_{\substack{(x_1, y_1) \in A \\ (x_2, y_2) \in B}} \{(x_1 - x_2, y_1 - y_2)\} \tag{6.9}$$

Also we define the above arithmetic operations with respect to one of the axes *X* or *Y*. As we see it below, the operation is done with respect to axis *X*. For example, when the entered data is a scalar value (i.e. 1D) and the axis *Y* is used to represent the user's confidence degree (or vice versa):

$$A \underset{X}{+} B = \int\limits_{\substack{(x_1,y_1)\in A\\(x_2,y_2)\in B}} \{(x_1+x_2,y_1)\} \cup \int\limits_{\substack{(x_1,y_1)\in A\\(x_2,y_2)\in B}} \{(x_1+x_2,y_2)\} \tag{6.10}$$

$$A \underset{Y}{+} B = \int\limits_{\substack{(x_1,y_1)\in A\\(x_2,y_2)\in B}} \{(x_1,y_1+y_2)\} \cup \int\limits_{\substack{(x_1,y_1)\in A\\(x_2,y_2)\in B}} \{(x_2,y_1+y_2)\} \tag{6.11}$$

$$A \underset{X}{-} B = \int\limits_{\substack{(x_1,y_1)\in A\\(x_2,y_2)\in B}} \{(x_1-x_2,y_1)\} \cup \int\limits_{\substack{(x_1,y_1)\in A\\(x_2,y_2)\in B}} \{(x_1-x_2,y_2)\} \tag{6.12}$$

$$A \underset{Y}{-} B = \int\limits_{\substack{(x_1,y_1)\in A\\(x_2,y_2)\in B}} \{(x_1,y_1-y_2)\} \cup \int\limits_{\substack{(x_1,y_1)\in A\\(x_2,y_2)\in B}} \{(x_2,y_1-y_2)\} \tag{6.13}$$

Let us define an operation of multiplication of an F-mark A by a numeric value k :

$$A \underset{X}{\cdot} k = \int\limits_{(x,y)\in A} \{(x\cdot k, y)\} \tag{6.14}$$

We define the Max and Min operations:

$$\underset{X}{Max}(A,B) = \int\limits_{\substack{(x_1,y_1)\in A\\(x_2,y_2)\in B}} \{\max(x_1,x_2), y_1\} \cup \int\limits_{\substack{(x_1,y_1)\in A\\(x_2,y_2)\in B}} \{\max(x_1,x_2), y_2\} \tag{6.15}$$

$$\underset{X}{Min}(A,B) = \int\limits_{\substack{(x_1,y_1)\in A\\(x_2,y_2)\in B}} \{\min(x_1,x_2), y_1\} \cup \int\limits_{\substack{(x_1,y_1)\in A\\(x_2,y_2)\in B}} \{\min(x_1,x_2), y_2\} \tag{6.16}$$

To compare f-marks A and B, we use the method adopted from the Jaccard compatibility measure to compare degree to which A exceeds B: $g_{\geq}(A,B)$. We assume that $A > B$ if $g_{\geq}(A,B) > g_{\geq}(B,A)$, $A < B$ if $g_{\geq}(A,B) < g_{\geq}(B,A)$ and $A = B$ otherwise.

For the two extended points A and B one has [1,6]

$$g_{\geq}(A,B) = \frac{1}{2}\left(\frac{\|\mathrm{Max}(A,B)\cap A\|}{\|\mathrm{Max}(A,B)\cup A\|} + \frac{\|\mathrm{Min}(A,B)\cap B\|}{\|\mathrm{Min}(A,B)\cup B\|} \right) \tag{6.17}$$

An F-point A can approximately be represented parametrically as $A = M((c_x, c_y), h, w)$. We use the notation $M()$ to denote parametrically a general

F-mark, while the set of parameters in parentheses depends on chosen approximation model. Without any loss of generality, we assume that $0 < h \le w$, and, hence, h and w are convenient replacements for $\phi_{\min}$ and $\phi_{\max}$, respectively. An F-line L can be (approximately) produced from a convex hull of two F-points $A_1 = M((c_{1x}, c_{1y}), h_1, w_1)$ and $A_2 = M((c_{2x}, c_{2y}), h_2, w_2)$:

$$L = ch\left(M((c_{1x}, c_{1y}), h_1, w_1), M((c_{2x}, c_{2y}), h_2, w_2)\right).$$

Let $h = \max\left(h_1, w_1, h_2, w_2\right)$, then an F-line can be represented approximately as an F-point:

$$\tilde{M}\left(\left((c_{1x} + c_{2x})/2, (c_{1y} + c_{2y})/2\right), h, w\right),$$

where $w \sim h + \sqrt{(c_{2x} - c_{1x})^2 + (c_{2y} - c_{1y})^2}$.

Therefore, we can parametrically represent both F-line and F-point either as $M((c_x, c_y), h, w)$ or as $M((c_{1x}, c_{1y}), (c_{2x}, c_{2y}), h)$.

The parameter h could visually be interpreted as the height or thickness of an F-mark. Likewise the parameter w can be regarded as the width or length of an F-mark. When F-marks representing information regarding a value of a scalar (1D) uncertain variable (e.g. a probability of an event or expected profit) are accepted from the user (e.g. a decision maker, a DM), the second dimension is assumed to express the degree of confidence (or belief or trust) of the user in the entered data. For example, thicker (or long) lines would mean less trustworthy data than the one associated with the thinner (or short) lines.

Without loss of generality, we relate the parameter h with the degree of confidence of DM in the value of specified F-mark (either F-point or F-line).

To do so we define a decreasing function $\sigma(h)$ expressing a relationship between the height and the associated confidence degree for which the following conditions hold true:

$$\lim_{h \to +0} \sigma(h) = 1$$

$$M(c_1, c_2, \sigma^{-1}(0)) = \varnothing,$$

where $\sigma^{-1}(d)$ is the reciprocal function producing a confidence degree d and associated value of h. A suitable function could be $\sigma(h) - 1 - h / h_{\max}$ for which $\sigma^{-1}(d) = h_{\max}(1 - d)$, where $h \in (0, h_{\max}], h_{\max} > 0$.

Let us also introduce a function that for any F-mark $A = M((c_{1x}, c_{1y}), (c_{2x}, c_{2y}), h)$ returns its parameter h ($\phi_{\min}$):

$$H(A) = H\left(M((c_{1x}, c_{1y}), (c_{2x}, c_{2y}), h)\right) = h$$

F-geometry can be effectively used in decision-making. The decision-making "if-then" rules for an uncertain environment can be composed on the basis of F-geometry concepts used to more adequately reflect the perceived information granules and relationships. F-geometry based decision-making allows for better modeling of the knowledge of human observer, researcher, or a DM, thereby making the inference system's output more realistic (through minimizing losses of meaning and distortion of source information).

Let us start with a formal problem statement.

In an unprecisiated perception-based information setting, we consider a decision making problem as a 4-tuple $(\mathcal{S}, \mathcal{Y}, \mathcal{A}, \succsim)$ where the set of states of nature $\mathcal{S} = \{S_1, S_2, ..., S_n\}$, corresponding probability distribution P and set of outcomes $\mathcal{Y}$ are generally considered as spaces of F-marks. The set of actions $\mathcal{A}$ is considered as a set of mappings from $\mathcal{S}$ to $\mathcal{Y}$. In turn, preferences $\succsim$ are to be implicit in some knowledge base described as some "if-then" rules, which include $\mathcal{S}, \mathcal{Y}, \mathcal{A}$- based description of various decision making situations faced before and a DM's or experts' opinion-based evaluations of actions' assessment U (combined outcome) which are also to be described by f-marks.

A typical knowledge base may look as follows:

If $S_1, S_2, ..., S_n$ and (P_2 is P_{i2}) and…and (P_n is P_{in})

Then $\mathrm{U}_1 = U_{i1}$ and $\mathrm{U}_2 = U_{i2}$ and ... and $\mathrm{U}_\mathrm{m} = U_{im}$, α_i) ... , $(i = \overline{1,q})$.

Here P_i is the variable describing user entered F-mark for probability of the state of nature S_j and P_{ij} is an F-mark describing the probability of the state of nature S_j used in rule i $(i = \overline{1,q})$, $\mathrm{S}_j \in \mathcal{S}$ $(j = \overline{1,n})$, n is the number of states, U_{ik} is an F-mark describing the assessment of k-th action $(k = \overline{1,m})$ in rule i $(i = \overline{1,q})$, m is the number of considered alternative actions, α_i is the degree of confidence of the expert (designer of the knowledge base) in the rule i $(i = \overline{1,q})$.

The purpose of reasoning is to produce the vector of aggregated assessments $U_1, U_2, ..., U_m$ for different actions f_k $(k = \overline{1,m})$.

The best action then can be selected by ranking of F-marks describing the respective integrated assessments. For integrated assessments' f-marks $U_{f_{k1}}$ and $U_{f_{k2}}$ (corresponding to actions f_{k_1} and f_{k_1} respectively), f_{k_1} is the better action if $U_{f_{k1}} > U_{f_{k2}}$.

6.3 Fuzzy Geometry Gased If-Then Rules and the Reasoning Method

For simplicity, let us consider that the states of nature ($S_i, i=\overline{1,n}$) remain unchanged and thus could be removed from consideration in the rules. Then the above mentioned knowledge base takes on the following form:

If $(\mathrm{P}_1 \text{ is } P_{11})$ and $(\mathrm{P}_2 \text{ is } P_{12})$ and … and $(\mathrm{P}_n \text{ is } P_{1n})$
Then $\mathrm{U}_1 = U_{11}, \mathrm{U}_2 = U_{12}, \ldots, \mathrm{U}_\mathrm{m} = U_{1m}, \alpha_1$

If $(\mathrm{P}_1 \text{ is } P_{i1})$ and $(\mathrm{P}_2 \text{ is } P_{i2})$ and … and $(\mathrm{P}_\mathrm{n} \text{ is } P_{in})$
Then $\mathrm{U}_1 = U_{i1}, \mathrm{U}_2 = U_{i2}, \ldots, \mathrm{U}_\mathrm{m} = U_{im}, \alpha_i$

If and $(\mathrm{P}_1 \text{ is } P_{q1})$ and $(\mathrm{P}_2 \text{ is } P_{q2})$ and … and $(\mathrm{P}_\mathrm{n} \text{ is } P_{qn})$
Then $\mathrm{U}_1 = U_{q1}, \mathrm{U}_2 = U_{q2}, \ldots, \mathrm{U}_\mathrm{m} = U_{qm}, \alpha_q$

Assume that based on available cases or expert data, the knowledge base in form of F-geometry based "If-Then" rules shown above has been formulated.

The following steps describe the essence of the underlying methodology and reasoning procedure of decision-making using the suggested F-geometry based approach.

1. Obtain the F-lines P_j from the user and apply them for P_j, $j=\overline{1,n}$.
2. Obtain the minimum value of satisfaction of the fuzzy incidence axioms (A1')-(A4') for all the F-lines P_j generated by the user: $r(j)$, $j=\overline{1,n}$. If $r=\min_j(r(j))$ is lower than a predefined minimum threshold value (e.g., 0.3), ask the user to resubmit the primitives (go to step 1).
3. For each rule compute $\theta_i = \left(\wedge_j(\theta_{ij})\right)$, where $\theta_{ij} = \min\left(r, inc(\mathrm{P}_j, P_{ij})\right)$, $i=\overline{1,q}$
4. For each rule compute $R_i = \theta_i \cdot \alpha_i$
5. Find the indexes i of the rules for which $R_i \geq R_{\min}$, where $R_{\min}$ is the minimum creditability value that a rule must exhibit to be activated. For all such rules $R_{i'} \geq R_{\min}, i'=\overline{1,q'}, q' \leq q$, where i' are new indexes for the rules after removing those for which the above condition fails. If there are no such rules repeat the process starting from step 1.

6. Compute aggregated output components from all rules:

$$\mathrm{U}_{\mathrm{f}_k} = U_{f_k} = ch\left(\frac{\sum_{i'=1,q'} \left(U_{i'k} \cdot R_{i'} \right)}{\sum_{i'=1,q'} R_{i'}} \right), (k = \overline{1,m}).$$

7. Do ranking of the output F-marks U_k, $(k = \overline{1,m})$, and choose the best action depending on the index i_{best} such that $\mathrm{U}_{i_{best}} \geq \mathrm{U}_k, (k = \overline{1,m})$.

References

1. Aliev, R.A., Alizadeh, A.V., Guirimov, B.G.: Unprecisiated information-based approach to decision making with imperfect information. In: Proc. of the 9th International Conference on Application of Fuzzy Systems and Soft Computing (ICAFS 2010), Prague, Czech Republic, pp. 387–397 (2010)
2. Buckley, J.J., Eslami, E.: Fuzzy plane geometry I: Points and lines. Fuzzy Sets and Systems 86(2), 179–187 (1997)
3. Clementini, E.: A model for uncertain lines. Journal of Visual Languages and Computing 16, 271–288 (2005)
4. Dutta, S.: Qualitative Spatial Reasoning: A Semi-Quantitative Approach Using Fuzzy Logic. In: Buchmann, A., Smith, T.R., Wang, Y.-F., Günther, O. (eds.) SSD 1989. LNCS, vol. 409, pp. 345–364. Springer, Heidelberg (1990)
5. Ferrato, M., Foster, D.H.: Elements of a fuzzy geometry for visual space. In: O, Y.L., Toet, A., Foster, D.H., Heijmans, H.J.A.M., Meer, P. (eds.) Shape in Picture. Mathematical Description of Shape in Grey-level Images, pp. 333–342. Springer, Berlin (1994)
6. Guirimov, B.G., Gurbanov Ramiz, S., Aliev Rafik, A.: Application of fuzzy geometry in decision making. In: Proc. of the Sixth International Conference on Soft Computing and Computing with Words in System Analysis, Decision and Control (ICSCCW 2011), Antalya, pp. 308–316 (2011)
7. Hajek, P., Paris, J., Shepherdson: Rational Pavelka Predicate Logic is a Conservative Extension of Lukasiewicz Predicate Logic. The Journal of Syumbolic Logi. Association for Syumbolic Logic 65(2), 669–682 (2000), `http://www.jstor.org/stable/258660`
8. Mayburov, S.: Fuzzy geometry of phase space and quantization of massive fields. Journal of Physics A: Mathematical and Theoretical 41, 1–10 (2008)
9. Poston, T.: Fuzzy geometry. Ph.D Thesis, University of Warwick (1971)
10. Roe, J.: Index theory, coarse geometry, and topology of manifolds. In: CBMS: Regional Conf. Ser. in Mathematics. The American Mathematical Society, Rhode Island (1996)
11. Rosenfeld, A.: Fuzzy geometry: an updated overview. Information Science 110(3-4), 127–133 (1998)
12. Schockaert, S., De Cock, M., Kerre, E.: Modelling nearness and cardinal directions between fuzzy regions. In: Proceedings of the IEEE World Congress on Computational Intelligence (FUZZ-IEEE), pp. 1548–1555 (2008)

13. Tzafestas, S.G., Chen, C.S., Fokuda, T., Harashima, F., Schmidt, G., Sinha, N.K., Tabak, D., Valavanis, K. (eds.): Fuzzy logic applications in engineering science. Microprocessor Based and Intelligent Systems Engineering, vol. 29, pp. 11–30. Springer, Netherlands (2006)
14. Wilke, G.: Approximate Geometric Reasoning with Extended Geographic Objects. In: Proceedings of the Workshop on Quality, Scale and Analysis Aspects of City Models, Lund, Sweden, December 3-4 (2009), http://www.isprs.org/proceedings/XXXVIII/2-W11/Wilke.pdf
15. Zadeh, L.A.: Toward extended fuzzy logic. A first step. Fuzzy Sets and Systems 160, 3175–3181 (2009)
16. Zadeh, L.A.: A Note on Z-numbers. Information Sciences 181, 2923–2932 (2010)

Chapter 7
Fuzzy Logic Based Generalized Theory of Stability

7.1 Underlying Motivation

Stability is one of the most essential properties of complex dynamical systems, no matter whether technical or human-oriented (social, economical, etc.). In classical terms, the stability property of a dynamical system is usually quantified in a binary fashion. This quantification states whether the system under consideration reaches equilibrium state after being affected by disturbances. Even if we define a region of stability, in every point of operation of the system we can only conclude that "*the system is stable*" or "*the system is unstable*". No particular quantification as to a degree of stability could be offered. In many cases when such a standard bivalent two-valued definition of stability is being used, we may end up with counterintuitive conclusions.

In contrast, human-generated statements would involve degrees of stability which are articulated linguistically and expressed by some fuzzy numbers which link with some quantification of stability positioned somewhere in-between states of being *absolutely* stable and *absolutely* unstable i.e., a stability degree could be expressed by a fuzzy number defined over the unit interval in which 0 is treated as absolutely unstable and 1 corresponds to that state that is absolutely stable. It is then advantageous to introduce linguistic interpretation of degrees of stability, i.e., a degree of stability becomes a linguistic variable assuming terms such as "*unstable*", "*weakly* stable", "*more or less* stable", "*strongly* stable", "*completely* stable" each of them being described by the corresponding fuzzy numbers defined over [0,1] [2] .

We can conclude that the concept of stability is a fuzzy concept in the sense that it is a matter of degree. In general, fuzzy concepts cannot be defined within the conceptual framework of bivalent logic [49]. If the concept of stability cannot be defined within the conceptual structure of bivalent logic, then how can it be defined? What is needed for this purpose is PNL (Precisiated Natural Language) - a language that is based on fuzzy logic – on logic in which everything is or is allowed to be a matter of degree [47]. To define a concept through the use of PNL, with PNL serving as a definition language, the concept is defined in a natural language; second, the natural language is precisiated. Since fuzzy sets regarded as basic information granules are human-centric, fuzzy stability concept meets

R.A. Aliev: *Fuzzy Logic-Based Generalized Theory of Decisions*, STUDFUZZ 293, pp. 231–264.
DOI: 10.1007/ 978-3-642-34895-2_7

user-defined objectives and is not counterintuitive. Unlike bivalent-logic-based definitions of stability, PNL-based definitions of stability are context-dependent rather than context-free. Human-centricity of stability has become even more essential.

Accuracy has been a dominant facet of mathematics. However, as the systems under study become more and more complex, nonlinear or uncertain, the use of well-positioned tools of fully deterministic analysis tends to exhibit some limitations and show a lack of rapport with the real world problem under consideration. As a matter of the fact, this form of limitation has been emphasized by the principle of incompatibility [48].

In many cases, information about a behavior of a dynamical system becomes uncertain. In order to obtain a more realistic model of reality, we have to take into account existing components of uncertainty. Furthermore, uncertainties might not be of probabilistic type. The generalized theory of uncertainty (GTU), outlined by Prof. L.A. Zadeh in [51], breaks with the tradition of viewing uncertainty as a province of probability and puts it in a much broader perspective. In this setting, the language and formalism of the dynamic fuzzy "if-then" rules and fuzzy differential equations (FDE) [1,3,5,7,12-14,17,18,21,24,31] become natural ways to model dynamical systems.

In this study, we propose a setting of the Generalized Theory of Stability (GTS) for complex dynamical systems, described by fuzzy differential equations (FDEs). Different PNL-based definitions of stability of dynamical systems are introduced. Also fuzzy stability (FS) of systems, binary stability of fuzzy systems (BFS), binary stability of systems (BS) which are special cases of GTS are considered. The introduced definitions offer a continuous classification of stability solutions by admitting different degrees of stability. We also show that under some conditions, the fuzzy stability coincides with classical notion of stability of a Boolean character.

7.2 Classical Stability Theory

In general, in classical stability theory there are two fundamental approaches for investigating stability of dynamical systems. The first approach is related to Lyapunov stability theory [9], while the second one is based on the Lipschitz stability theory [11]. For linear systems, the concepts of stability by Lipschitz and Lyapunov are the same, while for nonlinear systems these concepts differ. It can be shown that the system identified to be Lipschitz stable, is also Lyapunov stable, but not vice-versa [11]. In what follows, we briefly recall the concepts and offer some comparative analysis.

Consider a nonlinear differential system

$$x' = f(t,x), \qquad x(t_0) = x_0, \qquad x \in R^n,\ f(t,0) \equiv 0. \tag{7.1}$$

where $f(t,x)$ is Lipschitz continuous with respect to x uniformly in t and piecewise continuous in t. Its associated variational systems are

$$y' = f_x'(t,0)\,y, \tag{7.2}$$

$$z' = f'_x(t, x(t, t_0, x_0))z, \tag{7.3}$$

where $f \in C\left[R_+ \times R^n, R^n\right]$ and f'_x denotes $\frac{\partial f}{\partial x}$ which exists and continuous on $R_+ \times R^n$, $t_0 \geq 0$, $f(t,0) \equiv 0$, $x(t, t_0, x_0)$ is the solution of (7.1), $x(t_0, t_0, x_0) = x_0$.

The zero solution $x(t) = 0$ of the system (7.1) is called Lyapunov stable, if given any $\varepsilon > 0$, $t_0 \in R_+$ there exists $\delta = \delta(\varepsilon, t_0) > 0$ that is continuous in $t_0 \in R_+$, such that $\|x_0\| < \delta$ implies $\|x(t, t_0, x_0)\| < \varepsilon$ for $t \geq t_0$. If δ is independent of t_0, then the zero solution $x(t) = 0$ of the system (7.1) is called uniformly Lyapunov stable.

The basic theorem of Lyapunov gives sufficient conditions for the stability of the origin of a system. Here the problem of ensuring stability is related to the presence of a so-called special function of Lyapunov *V*, and the satisfaction of the inequality $\frac{dV}{dt} \leq 0$. The indirect method of Lyapunov uses the linearization of the original dynamical system (7.1) in order to determine the local stability of the original system.

Let us now consider the Lipschitz stability [11]. The zero solution $x(t) = 0$ of (7.1) is said to be uniformly Lipschitz stable if $\exists M > 0$ and $\delta > 0$ such that $\|x(t, t_0, x_0)\| \leq M \|x_0\|$, whenever $\|x_0\| \leq \delta$ and $t \geq t_0 \geq 0$. Also, if we consider its associated variational systems [9] then the zero solution $x(t) = 0$ of (7.1) is said to be uniformly Lipschitz stable in variation, if $\exists M > 0$ and $\delta > 0$ such that $\|\Phi(t, t_0, x_0)\| \leq M$, for $\|x_0\| \leq \delta$ and $t \geq t_0 \geq 0$, where $\Phi(t, t_0, x_0)$ is the fundamental matrix solution of the variational equation (7.3).

So far we have reviewed the fundamental concepts pertinent to the two-valued (binary) stability of systems (BS). Now let us consider binary stability of fuzzy systems (BFS stability). A huge number of books and papers is devoted to stability analysis of fuzzy controllers [4,6,10,15,16,19,20,22,25,26,29,30,32,34,36,37,38, 39,41,44,46,53]. In most of works the describing function, the Popov criterion, the circle criterion [15,25,26,34,36] and the criterion of hyperstability are used for stability analysis of fuzzy controllers. Often with the purpose of stability analysis of fuzzy control systems Lyapunov functions are used [6,20,41,44], especially for systems with TSK representation. But the existence of Lyapunov function is normally the critical point of this method. Tanaka and Sugeno [39] demonstrated that the stability of a Takagi-Sugeno (T-S) model could be focused on finding a common positive definite matrix *P* to the quadratic Lyapunov equation. In [37] a new approach for the stability analysis of continuous Sugeno Type II and III fuzzy systems is proposed. It is based on the use of positive definite and fuzzy negative definite systems under arguments similar to those of standard Lyapunov stability theory. Several refinements following the central issue arose in [8,23,28,40,43,45,]. Main essential defect of these methods is that they need the separation of the dinamical system into a linear and nonlinear parts. However, for fuzzy dinamical systems quite often the required separation would not be possible.

The investigation of stability of fuzzy systems described by FDE is considered in [18,31]. Considered is a fuzzy differential system

$$\tilde{x}' = f(t,\tilde{x}), \tag{7.4}$$

where $f \in C^1\left[R_+ \times E^n, E^n\right], \tilde{x}(t_0) = \tilde{x}_0 \in E^n,\ t \geq t_0,\ t_0 \in R_+, f(t,\hat{0}) \equiv \hat{0}$

The stability analysis of (7.4) is reduced to an investigation of a binary stability of a scalar differential equation

$$w' = g(t,w), \quad w(t_0) = w_0 \geq 0.$$

This equation is called a scalar comparative differential equation and is obtained from (7.4) by means of Lyapunov-like functions. Stability of the zero solution of the scalar comparative differential equation implies stability of the zero solution of (7.4).

We mention that the existing methods of stability analysis of a fuzzy system (not completely fuzzified system) is reduced to an investigation of a crisp system.

Summarizing, in all these investigations, the stability analysis of a fuzzy system is reduced to the use of the binary stability technique (the "yes – no" stability outcome) without any further refinement that could involve a possible quantification of stability degrees of systems under consideration. The outcome of the analysis also stresses that the classical binary stability concepts could become quite counterintuitive by ignoring possible gradation of the concept of stability. As an example, let us consider the equilibrium states of mechanical systems shown in Fig. 7.1, Fig. 7.2, and Fig. 7.3. Fig. 7.1 depicts an equilibrium state of a system consisting of a rod and a bar placed on it.

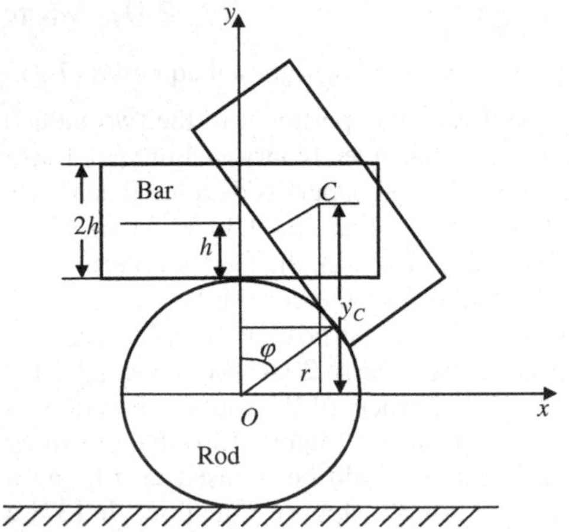

Fig. 7.1 The equilibrium state of a mechanical system

In the Fig. 7.1 we used the following notation pertaining to the parameters of the system:

h – the half of the bar's thickness,
r – the radius of the rod,
φ – an angle of bar's turn,
C – the center of the bar,
O – the center of the rod, where we place the origin of the coordinate system,
y_C – the distance between the center of the bar and the x axis.

Potential energy of a gravity of the bar is expressed as $\Pi = mg(y_C - h - r)$ (Fig. 7.2).

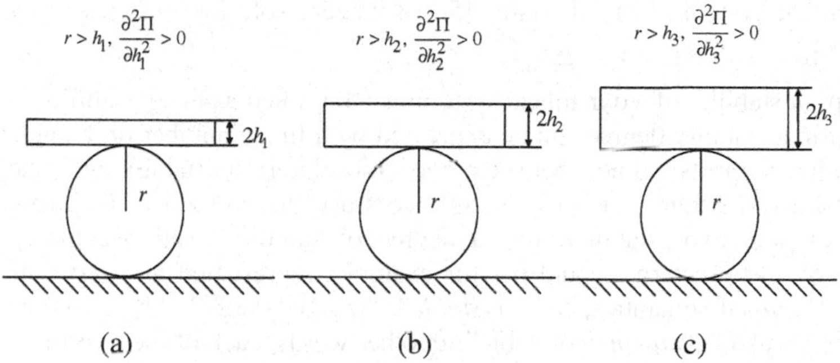

Fig. 7.2 Examples of different equilibrium states

In terms of Lyapunov stability, in all the three cases shown above the system is stable no matter how large the thickness of the bar is. But using intuitive or qualitative assessment of stability in each case we arrive to different conclusions. By reviewing the system more carefully, we note that in Fig. 7.2 (a) the state of the system is *strongly* stable, in Fig. 7.2 (b) the state of the system is *more or less* stable, in Fig. 7.2 (c) the state of the system is *weakly* stable. The classical theory of stability does not offer any classification with respect to the varying degrees of stability.

Fig. 7.3 brings another set of examples involving different equilibrium states of the system which consist of a platform and a ball placed on it. Again, in terms of

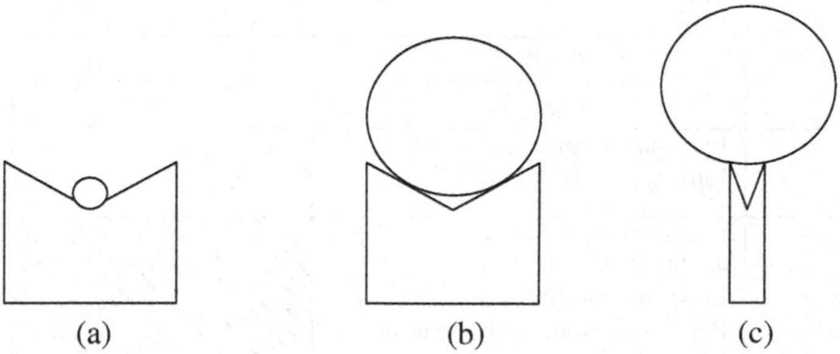

Fig. 7.3 Different equilibrium states

Lyapunov stability, the system is stable no matter how large the diameter of the ball is. Our intuition suggests a certain gradation of the concept of stability. The system shown in Fig. 7.3 (a) is *strongly* stable, in Fig. 7.3 (b) the system is *more or less* stable, and in Fig. 7.3 (c) the system is *weakly* stable.

7.3 Fuzzy Stability Concept

As it has been shown above, we need to introduce the concept of fuzzy stability. Prof. L. Zadeh firstly addressed the problem of fuzzy stability of the dynamical systems (FS) [50]. He considered a differential system

$$x' = f(x), \quad x \in R^n, \quad x_0 = 0 \tag{7.5}$$

and concluded that system (7.5) is F-stable [50] if its zero solution satisfies fuzzy Lipschitz condition with respect to Δx_0.

This concept of stability of a dynamical system may be called as fuzzy stability of a system. In turn, a stability degree can be expressed by a fuzzy number or a single numeric quantity with its values between zero (absolutely unstable) and one (absolutely stable). In light of this, it is necessary to introduce linguistic interpretation of degree of stability, i.e., a degree of stability itself becomes a linguistic variable, whose terms could be, for example, represented as fuzzy sets with some well-defined semantics, say "*unstable*", "*weakly* stable", "*more or less* stable", "*strongly* stable", "*completely* stable". In other words, each of these terms is expressed by a fuzzy number defined in the closed interval [0,1]. Let us note that we sometimes use linguistic terms to evaluate a state of a system under consideration. For example, in modeling macro economical processes economists often verbally evaluate a stability of an equilibrium point. In particular, for economical inflation dynamics described by nonlinear equation, they use such terms as "*weakly* stable", "*semi-stable*", "*weakly unstable*" to evaluate stability of an equilibrium point.

Given some concepts of stability discussed so far, we can arrive at certain taxonomy, Table 1, in which we involve two axes, namely a type of stability and a form of the system under study. This brings forward four general cases.

Table 7.1 Overview of studies devoted to stability theory versus the classes of systems and the concept of stability

		Stability	
		binary	fuzzy
Systems	system	Lyapunov stability [9] Lipschitz stability [11]	Fuzzy Lipschitz stability [51]
	fuzzy system	Lyapunov stability [6,18,20,31,35, 41,44,52] Lipschitz stability [2,3] Popov criterion, circle criterion etc [15,25,26,34,36]	

It becomes apparent that there are no studies devoted to fuzzy stability of fuzzy dynamical systems (shaded region of the table). Our intent is to make this table complete by considering GTS in the setting of complex dynamical systems described by FDE.

As opposed to the existing technique of stability analysis of fuzzy systems in the approach suggested by us an original fuzzy system described by FDE is investigated directly. But main advantage of the methodology proposed is to consider and investigate linguistic degree of stability. Linguistic stability (fuzzy stability) corresponds to human intuition as opposed to binary view of stability (stable-unstable) which is counterintuitive. We propose methodology that allows constructing systems with the predefined linguistic degrees of stability. This provides to design both high-stable control systems when there are high requirements for reliability and safety (e.g. for control of atomic and chemical reactors, planes etc), and also to save expenses in construction of systems without redundant stability factors. Different PNL-based definitions of stability of dynamical systems are introduced. Also fuzzy stability of systems (FS), binary stability of fuzzy system (BFS), binary stability of systems (BS), which are particular cases of GTS, are discussed and contrasted. The introduced definitions give a continuous classification of stability solutions into different degrees of stability and coincide with classical stability under some conditions.

7.4 The Statement of the Generalized Stability Problem

Let us consider a fuzzy differential system

$$\tilde{x}' = f(t,\tilde{x}) \tag{7.6}$$

where f in (11) is continuous and has continuous partial derivatives $\dfrac{\partial f}{\partial x}$ on $R_+ \times E^n$,i.e. $f \in C^1\left[R_+ \times E^n, E^n\right]$, and $\tilde{x}(t_0) = \tilde{y}_0 \in E^n$, $t \geq t_0$, $t_0 \in R_+$.

Definition 7.1. The solution $\tilde{x}(t,t_0,\tilde{y}_0)$ of the system (7.6) is said to be fuzzy Lipschitz stable with respect to the solution $\tilde{x}(t,t_0,\tilde{x}_0)$ of the system (7.6) for $t \geq t_0$, where $\tilde{x}(t,t_0,\tilde{x}_0)$ is any solution of the system (7.6), if there exists a fuzzy number $\tilde{M} = \tilde{M}(t_0) > \tilde{0}$, such that

$$\left\|\tilde{x}(t,t_0,\tilde{y}_0) -_h \tilde{x}(t,t_0,\tilde{x}_0)\right\|_{fH} \leq \tilde{M}(t_0)\left\|\tilde{y}_0 -_h \tilde{x}_0\right\|_{fH} \tag{7.7}$$

If $\tilde{M}$ is independent on t_0, then the solution $\tilde{x}(t,t_0,\tilde{y}_0)$ of the system (7.6) is said to be uniformly fuzzy Lipschitz stable with respect to the solution $\tilde{x}(t,t_0,\tilde{x}_0)$.

Let $\tilde{x}(t,t_0,\tilde{x}_0)$ be the solution to (7.6) for $t \geq t_0$. Then $\tilde{\Phi}(t,t_0,\tilde{x}_0) = \dfrac{\partial \tilde{x}(t,t_0,\tilde{x}_0)}{\partial \tilde{x}_0}$ exists and is the fundamental matrix solution of the variational equation

$$\tilde{z}' = \frac{\partial f(t, \tilde{x}(t,t_0,\tilde{x}_0))}{\partial \tilde{x}} \tilde{z}, \tag{7.8}$$

and $\dfrac{\partial \tilde{x}(t,t_0,\tilde{x}_0)}{\partial t_0}$ exists, is a solution of (7.8), and satisfies the relation:

$$\frac{\partial \tilde{x}(t,t_0,\tilde{x}_0)}{\partial t_0} + \tilde{\Phi}(t,t_0,\tilde{x}_0) f(t_0,\tilde{x}_0) = \tilde{0}, \text{ for } t \geq t_0.$$

Definition 7.2. The solution $\tilde{x}(t,t_0,\tilde{y}_0)$ of the system (7.6) through $(t_0,\tilde{y}_0)$ for $t \geq t_0$ is said to be ***fuzzy Lipschitz stable*** with respect to the solution $\tilde{x}(t,t_0,\tilde{x}_0)$ of (7.6) for $t \geq t_0$, where $\tilde{x}(t,t_0,\tilde{x}_0)$ is any solution of the system (7.6) if and only if there exist $\tilde{M} = \tilde{M}(t_0) > \tilde{0}$ and $\tilde{\delta} > \tilde{0}$ such that $\left\| \tilde{x}(t,t_0,\tilde{y}_0) -_h \tilde{x}(t,t_0,\tilde{x}_0) \right\|_f \leq \tilde{M}(t_0) \left\| \tilde{y}_0 -_h \tilde{x}_0 \right\|_f$ for $t \geq t_0$, provided $\left\| \tilde{y}_0 - \tilde{x}_0 \right\|_f \leq \tilde{\delta}$. If $\tilde{M}$ is independent of t_0, then the solution $\tilde{x}(t,t_0,\tilde{y}_0)$ of the system (7.6) is ***uniformly fuzzy Lipschitz stable*** with respect to the solution $\tilde{x}(t,t_0,\tilde{x}_0)$.

Definition 7.3. The solution $\tilde{x}(t,t_0,\tilde{y}_0)$ of system (7.6) through $(t_0,\tilde{y}_0)$ for $t \geq t_0$ is said to be ***fuzzy Lipschitz stable in variation*** with respect to the solution $\tilde{x}(t,t_0,\tilde{x}_0)$ of (7.6) for $t \geq t_0$, where $\tilde{x}(t,t_0,\tilde{x}_0)$ is any solution of the system (7.6) if and only if there exist $\tilde{M} = \tilde{M}(t_0) > \tilde{0}$ and $\tilde{\delta} > \tilde{0}$ such that $\left\| \tilde{\Phi}(t,t_0,\tilde{y}_0) \right\|_f \leq \tilde{M}(t_0)$ for $t \geq t_0$, provided $\left\| \tilde{y}_0 \right\|_f \leq \tilde{\delta}$, where $\tilde{\Phi}(t,t_0,\tilde{y}_0)$ is the fundamental matrix solution of (7.8) such that $\tilde{\Phi}(t_0,t_0,\tilde{y}_0) = \tilde{I}$.

If $\tilde{M}$ is independent of t_0, then the solution $\tilde{x}(t,t_0,\tilde{y}_0)$ of the system (7.6) is ***uniformly fuzzy Lipschitz stable in variation*** with respect to the solution $\tilde{x}(t,t_0,\tilde{x}_0)$.

Definition 7.4. The solution $\tilde{x}(t,t_0,\tilde{x}_0)$ of system (7.6) through $(t_0,\tilde{x}_0)$ for $t \geq t_0$ is said to be ***asymptotically fuzzy Lipschitz stable in variation*** if $\int_{t_0}^{t} \left\| \tilde{\Phi}(t,s) \right\|_f ds \leq \tilde{M}$ for every $t_0 \geq 0$ and all $t \geq t_0$, where $\tilde{\Phi}(t,t_0)$ is the fundamental matrix solution of (7.8) such that $\tilde{\Phi}(t_0,t_0) = \tilde{I}$.

Remark. The theory of FDE which utilizes the Hukuhara derivative (H-derivative) has a certain disadvantage that the $diam\left((x(t))^{\alpha} \right)$ of the solution $\tilde{x}(t)$ of FDE is a nondecreasing function of time [17,24]. As it was mentioned in [17], this formulation of FDE cannot reflect any rich behavior of solutions of ODE, such as stability, periodicity, bifurcation and others, is not well suited for modeling purposes. In view of this it is useful to utilize b)-type of strongly generalized differentiability. For example, let us consider the following FDE:

$$\tilde{x}' = -\tilde{x},$$

The solution of the numeric analogon of this equation, that is $y' = -y$ comes in the form $y = y_0 e^{-t}$. This solution is stable. If we use H-derivative (a) - type of strongly generalized differentiability for the above FDE, then α-cut of its solution is $x_l^{\alpha} = \frac{1}{2}(x_{l0}^{\alpha} - x_{r0}^{\alpha})e^{t} + \frac{1}{2}(x_{l0}^{\alpha} + x_{r0}^{\alpha})e^{-t}, x_r^{\alpha} = \frac{1}{2}(x_{r0}^{\alpha} - x_{l0}^{\alpha})e^{t} + \frac{1}{2}(x_{l0}^{\alpha} + x_{r0}^{\alpha})e^{-t}$. In general it is not stable, and so, we lose the stability property. But if we use b)-type of strongly generalized differentiability, then α-cut of its solution is $x_l^{\alpha} = x_{l0}^{\alpha} e^{-t}, x_r^{\alpha} = x_{r0}^{\alpha} e^{-t}$. It is stable and coincides with the solution obtained before for the numeric case.

7.5 Stability Criteria

In stability theory an object to be investigated is a given solution of some equations system. In the case when the general solution of the system is known, it is possible to investigate a given solution by a direct analysis of the general solution. But as it is difficult to obtain general solutions of nonlinear differential equations, it is needed to develop some criteria which allow investigating stability of a given solution when a general solution is unknown. Here we present the Direct and Indirect methods for Lipschitz stability analysis of the systems described by FDE.

The Direct Method

Theorem 7.1. Let us consider the solutions $\tilde{x}(t,t_0,\tilde{y}_0)$, $t \geq t_0$, and $\tilde{x}(t,t_0,\tilde{x}_0)$, $t \geq t_0$ of the system (7.6). Let us assume the following

1) There exists a fuzzy number $\tilde{L}(t_0)$, such that $\int_{t_0}^{\infty}\tilde{\lambda}(s)ds = \tilde{L}(t_0)$, where $\tilde{\lambda}(s) \in C\left[[0,\infty), E_+ \subset E\right], E_+ = \left\{\tilde{\lambda} \in E, \mathrm{supp}(\tilde{\lambda}) \geq 0\right\}$.

2) f satisfies the following fuzzy Lipschitz condition with respect to $\tilde{x}$: $\left\| f(t,\tilde{v}(t,t_0,\tilde{v}_0) + \tilde{x}(t,t_0,\tilde{x}_0)) -_h f(t,\tilde{x}(t,t_0,\tilde{x}_0)) \right\|_{fH} \leq \tilde{\lambda}(t)\left\|\tilde{v}(t)\right\|_{fH}$, where $\tilde{v}(t,t_0,\tilde{v}_0) = \tilde{x}(t,t_0,\tilde{y}_0) -_h \tilde{x}(t,t_0,\tilde{x}_0)$.

Then the solution $\tilde{x}(t,t_0,\tilde{y}_0)$ of the system (7.6) is fuzzy Lipschitz stable with respect to $\tilde{x}(t,t_0,\tilde{x}_0)$.

Proof. Let us consider the solutions $\tilde{x}(t,t_0,\tilde{y}_0)$, $t \geq t_0$, and $\tilde{x}(t,t_0,\tilde{x}_0)$, $t \geq t_0$, of the system (7.6). Then

$$\tilde{v}(t,t_0,\tilde{v}_0) = \tilde{x}(t,t_0,\tilde{y}_0) -_h \tilde{x}(t,t_0,\tilde{x}_0), \tag{7.9}$$

$$\tilde{v}'(t,t_0,\tilde{v}_0) = \tilde{F}(t,\tilde{v}) = f(t,\tilde{v}(t,t_0,\tilde{v}_0) + \tilde{x}(t,t_0,\tilde{x}_0)) -_h f(t,\tilde{x}(t,t_0,\tilde{x}_0)), \tag{7.10}$$

$$\left\|\tilde{F}(t,\tilde{v})\right\|_{fH} = \left\| f(t,\tilde{v}(t,t_0,\tilde{v}_0) + \tilde{x}(t,t_0,\tilde{x}_0)) -_h f(t,\tilde{x}(t,t_0,\tilde{x}_0)) \right\|_{fH}. \tag{7.11}$$

Integrating (7.10) with respect to s from t_0 to t and taking the norm yields:

$$\tilde{v}(t,t_0,\tilde{v}_0) = \tilde{v}(t_0) + \int_{t_0}^{t}\tilde{F}(s,\tilde{v}(s))ds,$$

$$\left\|\tilde{v}(t,t_0,\tilde{v}_0)\right\|_{fH} \leq \left\|\tilde{y}_0 -_h \tilde{x}_0\right\|_{fH} + \left\|\int_{t_0}^{t}\tilde{F}(s,\tilde{v}(s))ds\right\|_{fH} \leq \left\|\tilde{y}_0 -_h \tilde{x}_0\right\|_{fH} + \int_{t_0}^{t}\left\|\tilde{F}(s,\tilde{v}(s))\right\|_{fH} ds$$

According to (7.11) and condition 2),

$$\left\|\tilde{v}(t,t_0,\tilde{v}_0)\right\|_{fH} \leq \left\|\tilde{y}_0 -_h \tilde{x}_0\right\|_{fH} + \int_{t_0}^{t}\tilde{\lambda}(s)\left\|\tilde{v}(s)\right\|_{fH} ds.$$

Let us denote $\tilde{z}(t) = \left\|\tilde{v}(t,t_0,\tilde{v}_0)\right\|_{fH}$. Then

$$\tilde{z}(t) \leq \tilde{z}(t_0) + \int_{t_0}^{t}\tilde{\lambda}(s)\tilde{z}(s)ds \tag{7.12}$$

But as $z_l^\alpha(t),\ z_r^\alpha(t),\ \lambda_l^\alpha(t),\ \lambda_r^\alpha(t) \geq 0$, then

$$z_l^\alpha(t) \leq z_l^\alpha(t_0) + \int_{t_0}^{t} \lambda_l^\alpha(s) z_l^\alpha(s) ds,$$

$$z_r^\alpha(t) \leq z_r^\alpha(t_0) + \int_{t_0}^{t} \lambda_r^\alpha(s) z_r^\alpha(s) ds.$$

Using Gronwall's inequality, from (7.12), we get

$$z_l^\alpha(t) \leq z_l^\alpha(t_0) e^{\int_{t_0}^{t} \lambda_l^\alpha(s) ds}, \quad z_r^\alpha(t) \leq z_r^\alpha(t_0) e^{\int_{t_0}^{t} \lambda_r^\alpha(s) ds}.$$

So,

$$\tilde{z}(t) \leq \tilde{z}(t_0) e^{\int_{t_0}^{t} \tilde{\lambda}(s) ds}.$$

Using condition 1) of the Theorem for $\tilde{\lambda}(s)$ yields:

$$\tilde{z}(t) \leq \tilde{z}(t_0) e^{\tilde{L}(t_0)}.$$

This implies

$$\left\| \tilde{x}(t, t_0, \tilde{y}_0) -_h \tilde{x}(t, t_0, \tilde{x}_0) \right\|_{fH} \leq e^{\tilde{L}(t_0)} \left\| \tilde{y}_0 -_h \tilde{x}_0 \right\|_{fH}.$$

But as $\tilde{L}(t_0)$ is fuzzy, $e^{\tilde{L}(t_0)}$ is also fuzzy, and for some $\tilde{M} \geq e^{\tilde{L}(t_0)}$ ($\tilde{M}$ is fuzzy, for example, when $\tilde{M}$ is $e^{\tilde{L}(t_0)}$) we have

$$\left\| \tilde{x}(t, t_0, \tilde{y}_0) -_h \tilde{x}(t, t_0, \tilde{x}_0) \right\|_{fH} \leq \tilde{M} \left\| \tilde{y}_0 -_h \tilde{x}_0 \right\|_{fH}.$$

The proof is complete.

In other words, the obtained $\tilde{M}$ provides the stability of system (7.6) according to (7.7). $\tilde{M}$ must be equal or greater than $e^{\int_{t_0}^{\infty} \tilde{\lambda}(t) dt}$. Let us note that $\tilde{M}$ can be crisp too. (7.7) is soft constraint and degree of its satisfaction gives degree of stability of a considered dynamical system.

The Indirect Method

Theorem 7.2. Let $\tilde{\Phi}(t, t_0)$ be the fundamental matrix of (7.8). If there exist positive continuous functions $\tilde{k}(t)$ and $\tilde{h}(t) \in E^1$, $t \geq t_0$, such that

$$\int_{t_0}^{t} \tilde{h}(s) \| \tilde{\Phi}(t,s) \|_f \, ds \le \tilde{k}(t) \quad \text{for } t \ge t_0 \ge 0, \tag{7.13}$$

and

$$\tilde{k}(t) \exp\left(-\int_{t_1}^{t} \frac{\tilde{h}(s)}{\tilde{k}(s)} ds \right) \le \tilde{K} \quad \text{for } t \ge t_1 \ge t_0, \tag{7.14}$$

where $\tilde{K} \in E^1$ is a fixed positive constant, then the solution $\tilde{x}(t, t_0, \tilde{y}_0)$ of (7.6) is uniformly fuzzy Lipschitz stable.

Proof. Let $b^{\alpha}(t) = \dfrac{1}{\| \Phi^{\alpha}(t, t_0) \|_f}$.

Furthermore, it satisfies $\Phi^{\alpha}(t, s)\Phi^{\alpha}(s, t_0) = \Phi^{\alpha}(t, t_0)$ and $\Phi^{\alpha}(t, t_0)^{-1} = \Phi^{\alpha}(t_0, t)$.

Then

$$\left(\int_{t_0}^{t} h^{\alpha}(s) b^{\alpha}(s) ds \right) \Phi^{\alpha}(t,s) = \int_{t_0}^{t} \Phi^{\alpha}(t,s) \Phi^{\alpha}(s,t_0) b^{\alpha}(s) h^{\alpha}(s) ds .$$

Hence

$$\frac{1}{b^{\alpha}(t)} \left(\int_{t_0}^{t} h^{\alpha}(s) b^{\alpha}(s) ds \right) \le \int_{t_0}^{t} h^{\alpha}(s) \| \Phi^{\alpha}(t,s) \|_f \, ds \le k^{\alpha}(t) .$$

Let $B^{\alpha}(t) = \int_{t_0}^{t} h^{\alpha}(s) b^{\alpha}(s) ds$. Then $h^{\alpha}(t) B^{\alpha}(t) \le k^{\alpha}(t) (B'(t))^{\alpha}$ or

$$(B'(t))^{\alpha} \ge \frac{h^{\alpha}(t)}{k^{\alpha}(t)} B^{\alpha}(t) .$$

Multiplying both sides by $\exp\left(-\int_{t_0}^{t} \frac{h^{\alpha}(s)}{k^{\alpha}(s)} ds \right)$, for some $t_1 \ge t_0$, we obtain

$$\frac{d}{dt} \left(\exp\left(-\int_{t_1}^{t} \frac{h^{\alpha}(s)}{k^{\alpha}(s)} ds \right) \cdot B^{\alpha}(t) \right) \ge 0 .$$

This implies that $\exp\left(-\int_{t_1}^{t} \frac{h^{\alpha}(s)}{k^{\alpha}(s)} ds \right) \cdot B^{\alpha}(t) \ge B^{\alpha}(t_1)$.

Thus $B^{\alpha}(t) \geq B^{\alpha}(t_1)\exp\left(\int_{t_1}^{t}\frac{h^{\alpha}(s)}{k^{\alpha}(s)}ds\right)$.

It is clear that $k^{\alpha}(t)b^{\alpha}(t) \geq B^{\alpha}(t) \geq B^{\alpha}(t_1)\exp\left(\int_{t_1}^{t}\frac{h^{\alpha}(s)}{k^{\alpha}(s)}ds\right)$.

Hence $\|\Phi^{\alpha}(t,t_0)\|_f = \frac{1}{b^{\alpha}(t)} \leq \frac{k^{\alpha}(t)}{B^{\alpha}(t_1)}\exp\left(-\int_{t_1}^{t}\frac{h^{\alpha}(s)}{k^{\alpha}(s)}ds\right)$. If we now take $N^{\alpha} \geq \frac{K^{\alpha}}{B^{\alpha}(t_1)}$, we obtain $\|\Phi^{\alpha}(t,t_0)\|_f \leq N^{\alpha}$. For $\tilde{M} \geq \tilde{N}$ we have $\left\|\tilde{x}(t,t_0,\tilde{y}_0) -_h \tilde{x}(t,t_0,\tilde{x}_0)\right\|_f \leq \tilde{M}\left\|\tilde{y}_0 -_h \tilde{x}_0\right\|_f$.

The theorems 7.1, 7.2 provide a possibility to find a number $\tilde{M}$ (if it exists), which determines the maximal value of divergence between the investigated and disturbed solutions. The number $\tilde{M}$ gives information on degree of stability of a considered system and we will take into account its value in our measure of degree of stability.

7.6 Examples

Example 7.1. Let an economical inflation process changes as provided by the differential equation:

$$\tilde{y}' = \frac{\tilde{a}}{(b+t)^2}\tilde{y}, \; \tilde{y}(t_0) \in E^1, \quad t_0 = 0, \tag{7.15}$$

where $\tilde{y}$ is a value of inflation. At first let $\tilde{a} = \tilde{1}$, where $\tilde{1}$ is a fuzzy number with a membership function shown in Fig. 7.4.

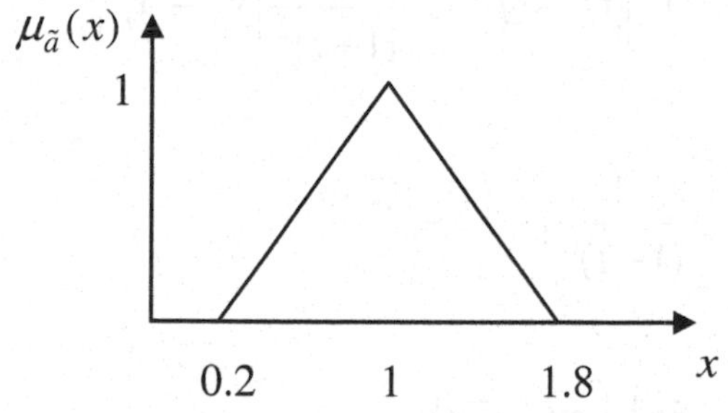

Fig. 7.4. Triangular membership function of the fuzzy number $\tilde{1}$

Using α cuts, we obtain $1^{\alpha} = (1_l^{\alpha}, 1_r^{\alpha})$, where

$$1_l^{\alpha} = 1 - 0.8(1-\alpha),\ \ 1_r^{\alpha} = 1 + 0.8(1-\alpha).$$

Let us verify if the second condition of the Theorem 7.1. is satisfied. Denote $f(t,\tilde{x}) = \overline{f},\ \ f(t,\tilde{y}) = f$. Using α cuts, one has

$$\left\| f(t,\tilde{y}) -_h f(t,\tilde{x}) \right\|_{fH}^{\alpha} = \left(\left| f_l^{\alpha} - \overline{f}_l^{\alpha} \right|, \left| f_r^{\alpha} - \overline{f}_r^{\alpha} \right| \right),$$

where

$$f_l^{\alpha} = \left(\frac{1}{(1+t)^2} \tilde{y} \right)_l^{\alpha},\ f_r^{\alpha} = \left(\frac{1}{(1+t)^2} \tilde{y} \right)_r^{\alpha},\ \overline{f}_l^{\alpha} = \left(\frac{1}{(1+t)^2} \tilde{x} \right)_l^{\alpha},\ \overline{f}_r^{\alpha} = \left(\frac{1}{(1+t)^2} \tilde{x} \right)_r^{\alpha}$$

Considering the cases when

$$x_l^{\alpha} \geq 0, y_l^{\alpha} \geq 0;$$
$$x_l^{\alpha} \geq 0, y_l^{\alpha} \leq 0,\ x_r^{\alpha} \geq 0, y_r^{\alpha} \geq 0;$$
$$x_l^{\alpha} \geq 0, y_l^{\alpha} \leq 0,\ x_r^{\alpha} \geq 0, y_r^{\alpha} \leq 0;$$
$$x_l^{\alpha} \leq 0, y_l^{\alpha} \geq 0,\ x_r^{\alpha} \leq 0, y_r^{\alpha} \geq 0;$$
$$x_l^{\alpha} \leq 0, y_l^{\alpha} \geq 0,\ x_r^{\alpha} \geq 0, y_r^{\alpha} \geq 0;$$
$$x_l^{\alpha} \leq 0, y_l^{\alpha} \leq 0,\ x_r^{\alpha} \leq 0, y_r^{\alpha} \leq 0;$$
$$x_l^{\alpha} \leq 0, y_l^{\alpha} \leq 0,\ x_r^{\alpha} \geq 0, y_r^{\alpha} \geq 0;$$
$$x_l^{\alpha} \leq 0, y_l^{\alpha} \leq 0,\ x_r^{\alpha} \leq 0, y_r^{\alpha} \geq 0;$$
$$x_l^{\alpha} \leq 0, y_l^{\alpha} \leq 0,\ x_r^{\alpha} \geq 0, y_r^{\alpha} \leq 0$$

we found that

$$\left| f_l^{\alpha} - \overline{f}_l^{\alpha} \right| \leq \frac{1_r^{\alpha}}{(1+t)^2} \left| y_l^{\alpha} - x_l^{\alpha} \right|,\ \left| f_r^{\alpha} - \overline{f}_r^{\alpha} \right| \leq \frac{1_r^{\alpha}}{(1+t)^2} \left| y_r^{\alpha} - x_r^{\alpha} \right|$$

always hold. But the inequality

$$\left| f^{\alpha=1} - \overline{f}^{\alpha=1} \right| \leq \frac{1}{(1+t)^2} \left| y^{\alpha=1} - x^{\alpha=1} \right|$$

also always holds. Therefore,

$$\max\left(\left| f_l^{\alpha} - \overline{f}_l^{\alpha} \right|, \left| f_r^{\alpha} - \overline{f}_r^{\alpha} \right| \right) \leq$$

$$\frac{1_r^\alpha}{(1+t)^2}\max\left(\left|y_l^\alpha - x_l^\alpha\right|,\left|y_r^\alpha - x_r^\alpha\right|\right).$$

The fuzzy Haussdorff distance between $f(t,\tilde{y}), f(t,\tilde{x})$ is defined as $d_{fH}\left(f(t,\tilde{y}), f(t,\tilde{x})\right) = \int_\alpha \left(\alpha, d_H\left(f^\alpha(t,\tilde{y}), f^\alpha(t,\tilde{x})\right)\right) = \int_\alpha \max\left(\left|f_l^\alpha - \overline{f}_l^\alpha\right|,\left|f_r^\alpha - \overline{f}_r^\alpha\right|\right)$

Using α cuts

$$\left(d_{fH}\left(f(t,\tilde{y}), f(t,\tilde{x})\right)\right)^\alpha =$$
$$=\left(\min\left(\left|f^{\alpha=1} - \overline{f}^{\alpha=1}\right|, \max\left(\left|f_l^\alpha - \overline{f}_l^\alpha\right|,\left|f_r^\alpha - \overline{f}_r^\alpha\right|\right)\right), \max\left(\left|f^{\alpha=1} - \overline{f}^{\alpha=1}\right|, \max\left(\left|f_l^\alpha - \overline{f}_l^\alpha\right|,\left|f_r^\alpha - \overline{f}_r^\alpha\right|\right)\right)\right).$$

But

$$\left(\min\left(\left|f^{\alpha=1} - \overline{f}^{\alpha=1}\right|, \max\left(\left|f_l^\alpha - \overline{f}_l^\alpha\right|,\left|f_r^\alpha - \overline{f}_r^\alpha\right|\right)\right), \max\left(\left|f^{\alpha=1} - \overline{f}^{\alpha=1}\right|, \max\left(\left|f_l^\alpha - \overline{f}_l^\alpha\right|,\left|f_r^\alpha - \overline{f}_r^\alpha\right|\right)\right)\right) =$$

$=\left(\left|f^{\alpha=1} - \overline{f}^{\alpha=1}\right|, \max\left(\left|f_l^\alpha - \overline{f}_l^\alpha\right|,\left|f_r^\alpha - \overline{f}_r^\alpha\right|\right)\right)$, because there exists the Hukuhara difference between $f(t,\tilde{y}), f(t,\tilde{x})$, i.e.

$$\left(d_{fH}\left(f(t,\tilde{y}), f(t,\tilde{x})\right)\right)_l^\alpha = \left(d_{fH}\left(f(t,\tilde{y}), f(t,\tilde{x})\right)\right)^{\alpha=1}.$$

Similarly,

$$\left(d_{fH}(\tilde{y},\tilde{x})\right)^\alpha = \left(\left|y^{\alpha=1} - x^{\alpha=1}\right|, \max\left(\left|y_l^\alpha - x_l^\alpha\right|,\left|y_r^\alpha - x_r^\alpha\right|\right)\right),$$

i.e. $\left(d_{fH}(\tilde{y},\tilde{x})\right)_l^\alpha = \left(d_{fH}(\tilde{y},\tilde{x})\right)^{\alpha=1}$.

As $\max\left(\left|f_l^\alpha - \overline{f}_l^\alpha\right|,\left|f_r^\alpha - \overline{f}_r^\alpha\right|\right) \le \frac{1_r^\alpha}{(1+t)^2}\max\left(\left|y_l^\alpha - x_l^\alpha\right|,\left|y_r^\alpha - x_r^\alpha\right|\right)$,

and $\left|f^{\alpha=1} - \overline{f}^{\alpha=1}\right| \le \frac{1}{(1+t)^2}\left|y^{\alpha=1} - x^{\alpha=1}\right|$, we can write

$$\left(d_{fH}\left(f(t,\tilde{y}), f(t,\tilde{x})\right)\right)_l^\alpha = \left(d_{fH}\left(f(t,\tilde{y}), f(t,\tilde{x})\right)\right)^{\alpha=1} \le \frac{1}{(1+t)^2}\left|y^{\alpha=1} - x^{\alpha=1}\right|$$

$$\left(d_{fH}\left(f(t,\tilde{y}), f(t,\tilde{x})\right)\right)_r^\alpha \le \frac{1_r^\alpha}{(1+t)^2}\max\left(\left|y_l^\alpha - x_l^\alpha\right|,\left|y_r^\alpha - x_r^\alpha\right|\right).$$

Then

$$\left(\left(d_{fH}\left(f(t,\tilde{y}),f(t,\tilde{x})\right)\right)_l^{\alpha},\left(d_{fH}\left(f(t,\tilde{y}),f(t,\tilde{x})\right)\right)_r^{\alpha}\right)\leq$$

$$\leq\left(\frac{1}{(1+t)^2},\frac{1_r^{\alpha}}{(1+t)^2}\right)\left(\left|y^{\alpha=1}-x^{\alpha=1}\right|,\max\left(\left|y_l^{\alpha}-x_l^{\alpha}\right|,\left|y_r^{\alpha}-x_r^{\alpha}\right|\right)\right).$$

Let us denote $\left(A_l^{\alpha},A_r^{\alpha}\right)=\left(\frac{1}{(1+t)^2},\frac{1_r^{\alpha}}{(1+t)^2}\right)$. Then we can write $d_{fH}\left(f\left(t,\tilde{y}\right),f\left(t,\tilde{x}\right)\right)\leq\frac{\tilde{A}}{(1+t)^2}d_{fH}\left(\tilde{y},\tilde{x}\right)$, where the obtained fuzzy number $\tilde{A}$ is shown in Fig. 7.5.

Then $\tilde{\lambda}(t)=\frac{\tilde{A}}{(1+t)^2}$, and $\tilde{L}(0)=\int_0^{\infty}\frac{\tilde{A}}{(1+t)^2}dt=\tilde{A}$.

Hence, if

$$\tilde{M}\geq e^{\tilde{L}}=e^{\tilde{A}},\text{ i.e } M^{\alpha}=\left(e^{\tilde{A}}\right)^{\alpha}=\left[\left(e^{A}\right)_l^{\alpha},\left(e^{A}\right)_r^{\alpha}\right]=\left[e^{A_l^{\alpha}},e^{A_r^{\alpha}}\right].$$

Then $d_{fH}\left(\tilde{y}(t),\tilde{x}(t)\right)\leq\tilde{M}d_{fH}\left(\tilde{y}_0,\tilde{x}_0\right)$, i.e. the solution $\tilde{y}\left(t,t_0,\tilde{y}\left(t_0\right)\right)$ of (7.15) is Lipschitz stable with respect to the solution $\tilde{y}\left(t,t_0,\tilde{x}\left(t_0\right)\right)$.

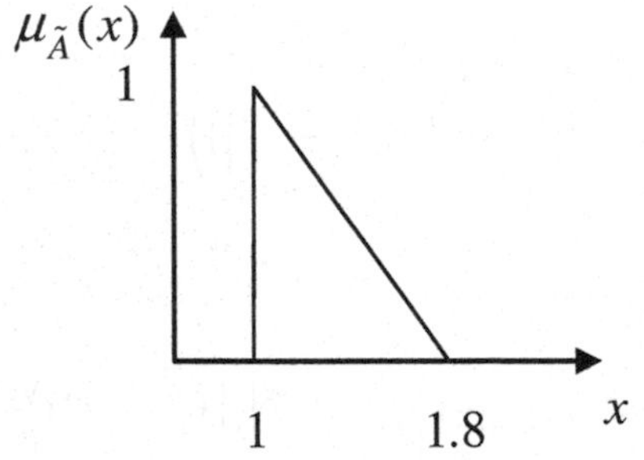

Fig.7.5 Fuzzy number $\tilde{A}$ bounding the Hukuhara difference

Let us at first introduce generic terms of the linguistic variable of the "degree of stability": "*Unstable*", "*Weakly* stable", "*More or less* stable", "*Strongly* stable", "*Completely* stable". They are shown in Fig. 7.6 and reflect our perception of qualitative (linguistic) stability of the system. Obviously one could use different membership functions, however those presented here have been selected because of their underlying semantics and usefulness in further processing.

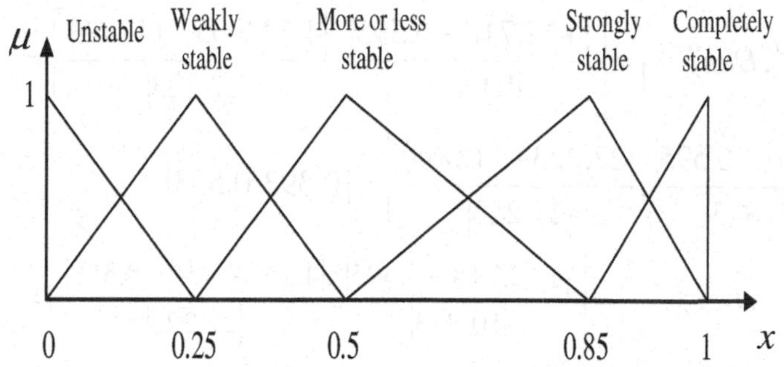

Fig. 7.6 Linguistic terms – fuzzy sets of degree of stability

The degree of stability will be determined by considering the scaled distance between the left and the right parts of the inequality $\left\| \tilde{y}(t,t_0,\tilde{y}_0) -_h \tilde{y}(t,t_0,\tilde{x}_0) \right\|_{fH} \le \tilde{M} \left\| \tilde{y}_0 -_h \tilde{x} \right\|_{fH}$. Formally, to define the degree of stability we use the following formula:

$$\widetilde{Deg} = \frac{\left\| \left\| \tilde{y}(t,t_0,\tilde{y}_0) - \tilde{y}(t,t_0,\tilde{x}_0) \right\|_{fH} -_h \tilde{M} \left\| \tilde{y}_0 - \tilde{x}_0 \right\|_{fH} \right\|_{fH}}{\tilde{M} \left\| \tilde{y}_0 -_h \tilde{x}_0 \right\|_{fH}}, \tag{7.16}$$

$$Deg^{\alpha} = \left[Deg_l^{\alpha}, Deg_r^{\alpha} \right] =$$

$$= \left[\frac{\left(\left\| \tilde{M} \left\| \tilde{y}_0 - \tilde{x}_0 \right\|_{fH} -_h \left\| \tilde{y}(t,t_0,\tilde{y}_0) - \tilde{x}(t,t_0,\tilde{x}_0) \right\|_{fH} \right\|_{fH} \right)_l^{\alpha}}{\left(\tilde{M} \left\| \tilde{y}_0 -_h \tilde{x}_0 \right\|_{fH} \right)_l^{\alpha}}, \frac{\left(\left\| \tilde{M} \left\| \tilde{y}_0 - \tilde{x}_0 \right\|_{fH} -_h \left\| \tilde{y}(t,t_0,\tilde{y}_0) - \tilde{x}(t,t_0,\tilde{x}_0) \right\|_{fH} \right\|_{fH} \right)_r^{\alpha}}{\left(\tilde{M} \left\| \tilde{y}_0 -_h \tilde{x}_0 \right\|_{fH} \right)_r^{\alpha}} \right]$$

$$= \left[\frac{\left\| M_l^{\alpha} \left(\left\| \tilde{y}_0 -_h \tilde{x}_0 \right\|_{fH} \right)_l^{\alpha} - \left(\left\| \tilde{y}(t,t_0,\tilde{y}_0) -_h \tilde{x}(t,t_0,\tilde{x}_0) \right\|_{fH} \right)_l^{\alpha} \right\|}{M_l^{\alpha} \left(\left\| \tilde{y}_0 -_h \tilde{x}_0 \right\|_{fH} \right)_l^{\alpha}}, \frac{\left\| M_r^{\alpha} \left(\left\| \tilde{y}_0 -_h \tilde{x}_0 \right\|_{fH} \right)_r^{\alpha} - \left(\left\| \tilde{y}(t,t_0,\tilde{y}_0) - \tilde{x}(t,t_0,\tilde{x}_0) \right\|_{fH} \right)_r^{\alpha} \right\|}{M_r^{\alpha} \left(\left\| \tilde{y}_0 -_h \tilde{x}_0 \right\|_{fH} \right)_r^{\alpha}} \right].$$

The initial conditions $\tilde{y}_0$ and $\tilde{x}_0$ of the solutions $\tilde{y}(t,t_0,\tilde{y}_0)$ (investigated solution), and $\tilde{y}(t,t_0,\tilde{x}_0)$ (an arbitrary solution) we choose as triangular fuzzy numbers. Let us evaluate the distance at $t = 1$. To present a procedure of Deg^{α} calculation, let us consider the cases $\alpha = 0$, $\alpha = 0.5$, $\alpha = 1$:

$$Deg^{\alpha=0} = \left[Deg_l^{\alpha=0}, Deg_r^{\alpha=0}\right] = \left[\frac{|4*2.718-6.595|}{|10.873|}, \frac{|4.5*6.05-11.068|}{|27.223|}\right] =$$

$$= \left[\frac{|10.873-6.595|}{|10.873|}, \frac{|27.223-11.068|}{|27.223|}\right] \approx [0.393, 0.593];$$

$$Deg^{\alpha=0.5} = \left[Deg_l^{\alpha=0.5}, Deg_r^{\alpha=0.5}\right] = \left[\frac{|4*2.718-6.595|}{|10.873|}, \frac{|4.25*4.055-8.558|}{|27.223|}\right] =$$

$$= \left[\frac{|10.873-6.595|}{|10.873|}, \frac{|17.235-8.558|}{|17.235|}\right] \approx [0.393, 0.499];$$

$$Deg^{\alpha=1} = \left[Deg_l^{\alpha=1}, Deg_r^{\alpha=1}\right] = \left[\frac{|4*2.718-6.595|}{|10.873|}, \frac{|4*2.718-6.595|}{|10.873|}\right] =$$

$$= \left[\frac{|10.873-6.595|}{|10.873|}, \frac{|10.873-6.595|}{|10.873|}\right] \approx 0.393.$$

Fig. 7.7 displays the degree of stability being the result of the computing shown above.

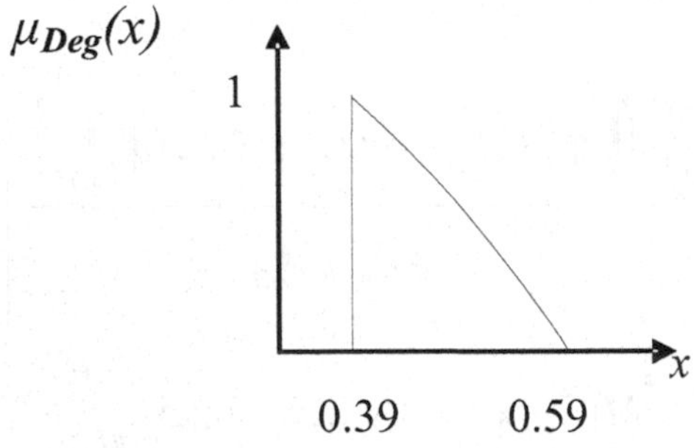

Fig. 7.7 The degree of stability

The core of this fuzzy number is approximately equal to 0.39. Approximating this fuzzy number by a triangular membership function and using (1.57) (section 1.4) we calculate the possibility measure of the similarity between the latter and the terms present in the codebook (Fig. 7.6). The possibility measure of this fuzzy number and the term "*weakly* stable" is approximately 0.43, whereas that of the possibility of this fuzzy number and the term "*more or less* stable" is approximately 0.76. Based on this computing, we conclude that the system is *more or less* stable.

Let now consider $\tilde{a} = \tilde{1}$, where $\tilde{1}$ is defined as shown in Fig. 7.8. The degree of stability is calculated following the same procedure as before and the result becomes expressed by the fuzzy number, see Fig. 7.9.

The core of the resulting fuzzy number is approximately 0.39. The possibility measure of this fuzzy number and the term "*weakly* stable" is approximately 0.43, whereas that of this fuzzy number and the term "*more or less* stable" is approximately 0.7. Hence we can conclude that the system is *more or less* stable.

When $\tilde{a} = 1$, i.e., it is a single numeric value, the degree of stability is shown in Fig. 7.10. In other words, the degree of stability is numeric and equal to 0.39. The membership degree of this number vis-a-vis the linguistic term "*weakly* stable" is approximately 0.43, whereas that of this number to the term "*more or less* stable" is approximately 0.57. In conclusion, we note that the linguistic degree of stability is somewhere in-between "*weakly* stable" and "*more or less* stable", but with a tendency of being closer to the linguistic term "*more or less* stable".

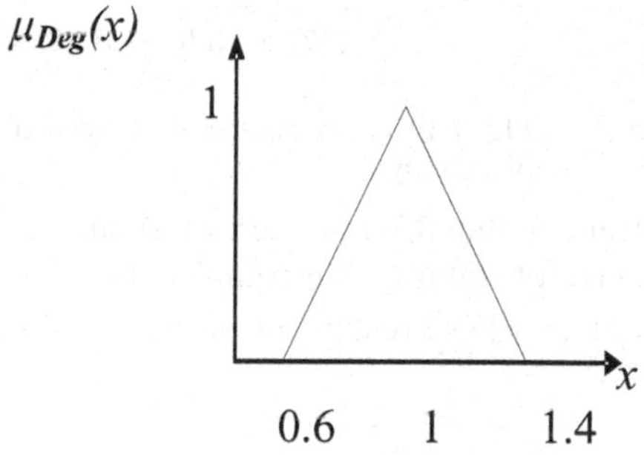

Fig. 7.8 Membership function of fuzzy number $\tilde{1}$

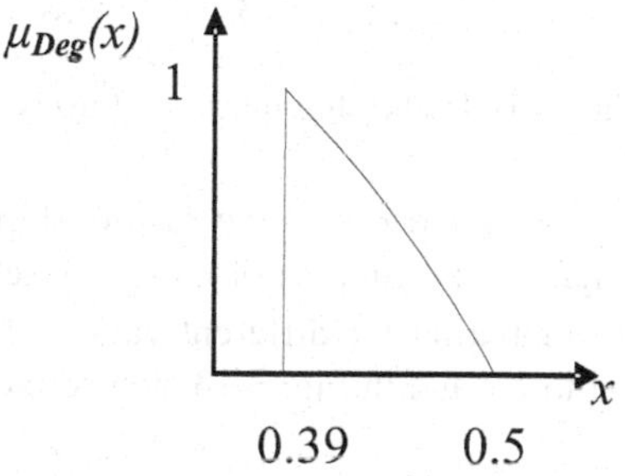

Fig. 7.9 The degree of stability

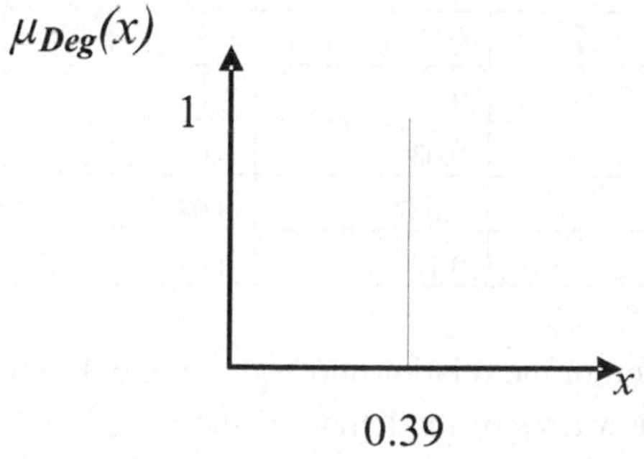

Fig. 7.10 The numeric degree of stability

If $\tilde{a} = \tilde{3}$, where $\tilde{3}$ is shown in Fig. 11, we obtain the degree of stability illustrated in Fig. 7.12.

The core of this fuzzy number is approximately 0.77. The resulting possibility measure of this fuzzy number and the term "*more or less* stable" is 0.23, whereas for the term "*strongly* stable" we obtain the possibility measure of approximately 0.83. Subsequently, we can conclude that the degree of stability is somewhere in-between "*strongly* stable" and "*more or less* stable" terms, but is closer to the linguistic term "*strongly* stable".

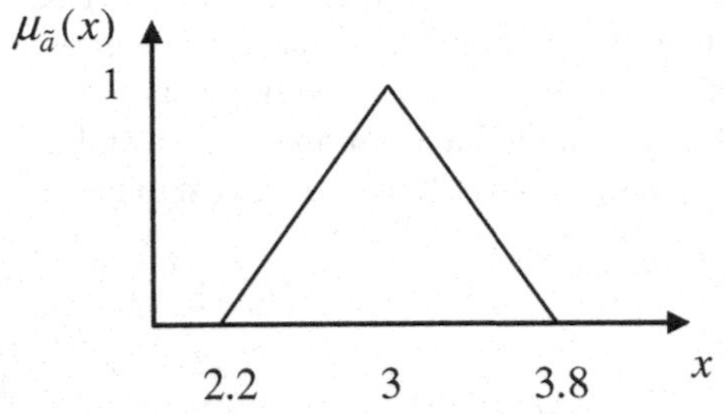

Fig. 7.11 Membership function of fuzzy number $\tilde{3}$

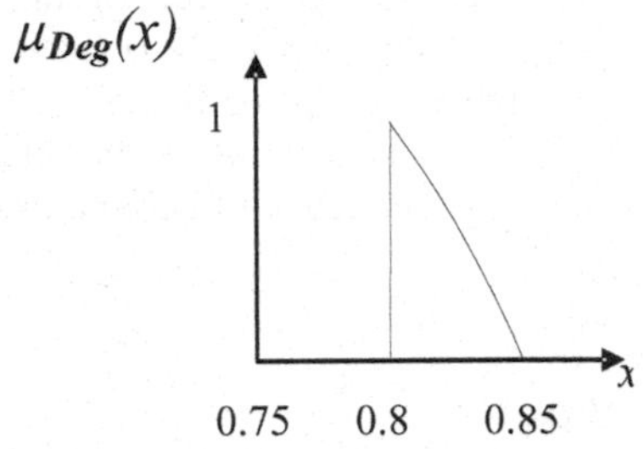

Fig. 7.12 Fuzzy number of the degree of stability

In order to reveal the relationship between the length of support of $\tilde{a}$ and the length of the support of the fuzzy set of degree of stability, we completed a series of computing for different values of cores of $\tilde{a}$ whose results are summarized in Table 7.2 and in Fig. 7.13, respectively.

Table 7.2 Relationships between the length of support of $\tilde{a}$ and the length of support of the degree of stability for selected values of the core of $\tilde{a}$

		The length of the support of the degree of stability		
		Core of $\tilde{a}$ is 1	Core of $\tilde{a}$ is 2	Core of $\tilde{a}$ is 3
The length of the support of $\tilde{a}$	0	0	0	0
	0.4	0.06	0.035	0.02
	0.8	0.11	0.067	0.04
	1.6	0.2	0.12	0.07

Furthermore the graphical interpretation of the relationship between the degree of stability and the core of $\tilde{a}$ for various values of the length of support of $\tilde{a}$ is included in Fig. 7.13.

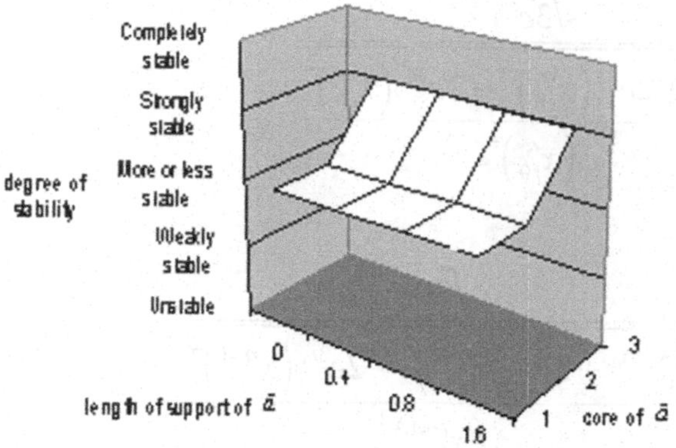

Fig. 7.13 The relationship between the length of the support of $\tilde{a}$ and the degree for different values of the cores of $\tilde{a}$

The obtained result coincides with our intuition: the level of uncertainty of the degree of stability increases as the uncertainty of $\tilde{a}$ (the length of the support) increases. What we obtained is a numeric quantification of this relationship.

Example 7.2. Let us consider a fuzzy dynamical system described by the differential equation:

$$\tilde{x}'(t) = \tilde{a}\tilde{x}(t) -_h e^t \tilde{x}^3(t)\,,\ \tilde{x}(t_0) = \tilde{x}_0\,, \tag{7.17}$$

where $\tilde{x}(t)$, $\tilde{x}_0 \in E^1$. We express the α-level set of $\tilde{x}(t)$ as the compact interval $x^\alpha(t) = [x_l^\alpha(t), x_r^\alpha(t)]$, and $x^\alpha(t_0) = [x_{l0}^\alpha, x_{r0}^\alpha]$, and in this way obtain the ordinary initial value problem

$$\begin{cases} \left(x'(t)\right)_l^\alpha = a_l^\alpha x_l^\alpha(t) - e^t\left(x^3(t)\right)_l^\alpha, \ x_l^\alpha(t_0) = x_{l0}^\alpha \\ \left(x'(t)\right)^{\alpha=1} = a^{\alpha=1}x^{\alpha=1}(t) - e^t\left(x^3(t)\right)^{\alpha=1}, \ x^{\alpha=1}(t_0) = x_0^{\alpha=1} \\ \left(x'(t)\right)_r^\alpha = a_r^\alpha x_r^\alpha(t) - e^t\left(x^3(t)\right)_r^\alpha, \ x_r^\alpha(t_0) = x_{r0}^\alpha \end{cases} \tag{7.18}$$

For $\tilde{a} = (1,1,1)$ the solution to (7.18) is

$$x_l^\alpha(t) = \frac{\sqrt{3}e^t}{\sqrt{\dfrac{3e^{2t_0} + 2e^{3t}\left(x_{l0}^\alpha\right)^2 - 2e^{3t_0}\left(x_{l0}^\alpha\right)^2}{\left(x_{l0}^\alpha\right)^2}}},$$

$$x_r^{\alpha}(t)=\frac{\sqrt{3}e^{t}}{\sqrt{\dfrac{3e^{2t_0}+2e^{3t}\left(x_{r0}^{\alpha}\right)^2-2e^{3t_0}\left(x_{r0}^{\alpha}\right)^2}{\left(x_{r0}^{\alpha}\right)^2}}}$$

For $\alpha=1$ we have

$$x^{\alpha=1}(t)=\frac{\sqrt{3}e^{t}}{\sqrt{\dfrac{3e^{2t_0}+2e^{3t}\left(x_{0}^{\alpha=1}\right)^2-2e^{3t_0}\left(x_{0}^{\alpha=1}\right)^2}{\left(x_{0}^{\alpha=1}\right)^2}}}$$

The α - level of the fundamental matrix solution of the variational system of (7.17) is

$$\Phi_l^{\alpha}=-\frac{3e^{t+2t_0}}{\sqrt{\dfrac{2e^{3t}}{3}-\dfrac{2e^{3t_0}}{3}+\dfrac{e^{2t_0}}{\left(x_{l0}^{\alpha}\right)^2}}\left(-2e^{3t}\left(x_{l0}^{\alpha}\right)^3+e^{2t_0}x_{l0}^{\alpha}\left(-3+2e^{t_0}\left(x_{l0}^{\alpha}\right)^2\right)\right)}$$

$$\Phi_r^{\alpha}=-\frac{3e^{t+2t_0}}{\sqrt{\dfrac{2e^{3t}}{3}-\dfrac{2e^{3t_0}}{3}+\dfrac{e^{2t_0}}{\left(x_{r0}^{\alpha}\right)^2}}\left(-2e^{3t}\left(x_{r0}^{\alpha}\right)^3+e^{2t_0}x_{r0}^{\alpha}\left(-3+2e^{t_0}\left(x_{r0}^{\alpha}\right)^2\right)\right)}$$

For $\alpha=1$, the fundamental matrix solution is

$$\Phi^{\alpha=1}=-\frac{3e^{t+2t_0}}{\sqrt{\dfrac{2e^{3t}}{3}-\dfrac{2e^{3t_0}}{3}+\dfrac{e^{2t_0}}{\left(x_{0}^{\alpha=1}\right)^2}}\left(-2e^{3t}\left(x_{0}^{\alpha=1}\right)^3+e^{2t_0}x_{0}^{\alpha=1}\left(-3+2e^{t_0}\left(x_{0}^{\alpha=1}\right)^2\right)\right)}$$

The graphs of the two fuzzy solutions of (7.17), with the initial conditions $x_{l0}^{\alpha=0}=0.1$, $x_0^{\alpha=1}=0.2$, $x_{r0}^{\alpha=0}=0.3$, $y_{l0}^{\alpha=0}=0.8$, $y_0^{\alpha=1}=0.9$, $y_{r0}^{\alpha=0}=1$ respectively, are shown in Fig. 14.

From Fig. 7.14 one can see that the two solutions converge to each other that confirms stability of solution to (7.18).

If $x_{l0}^{\alpha=0}\geq 0.1$, and $t_0=0$, then the inequality

$$\left\|\tilde{x}(t,t_0,\tilde{y}_0)-_h\tilde{x}(t,t_0,\tilde{x}_0)\right\|_f\leq\tilde{M}(t_0)\left\|\tilde{y}_0-_h\tilde{x}_0\right\|_f$$

is satisfied with the triangular fuzzy number $\tilde{M}\ (t_0)=(1,\ 1.56063,\ 2.41925)$. Thus the solution of (7.17) is fuzzy Lipschitz stable, but it is not uniformly fuzzy Lipschitz stable.

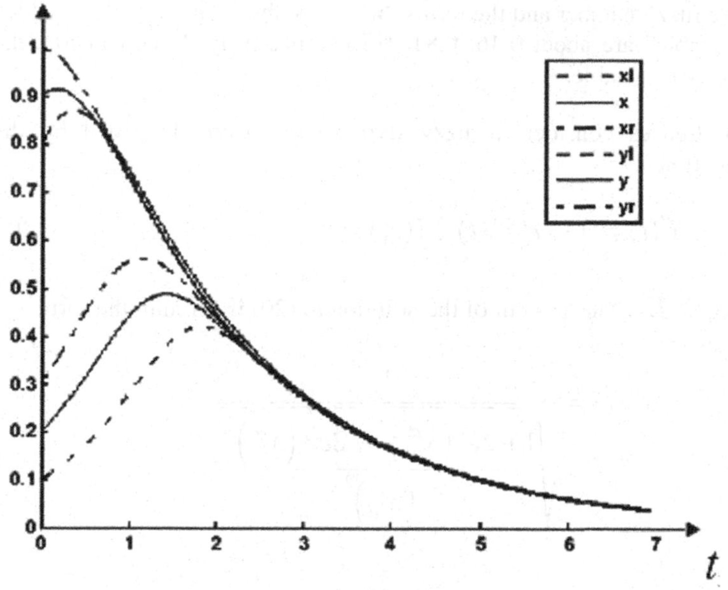

Fig. 7.14 Fuzzy solutions to (7.17)

The degree of stability is defined as the degree to which the inequality $\left\|\tilde{x}(t,t_0,\tilde{y}_0)-_h \tilde{x}(t,t_0,\tilde{x}_0)\right\|_f \leq \tilde{M}(t_0)\left\|\tilde{y}_0 -_h \tilde{x}_0\right\|_f$ is satisfied. This degree in its turn depends on the distance between the initial conditions of the given solution and arbitrary solutions. Consequently, as the degree of stability of the investigated dynamical system we can use the ratio (7.19):

$$\widetilde{Deg}=\frac{\int_0^{\delta}\left(\tilde{M}\,\|\Delta\tilde{x}_0\|_f-\|\Delta\tilde{x}\|_f\right)d\left\|\Delta\tilde{x}_0\right\|_f}{\int_0^{\delta}\left(\tilde{M}\,\|\Delta\tilde{x}_0\|_f\right)d\left\|\Delta\tilde{x}_0\right\|_f} \tag{7.19}$$

The resulting degree of stability is a triangular fuzzy number $\widetilde{Deg}=(0.312813,0.423109,0.461868)$. The possibility measures of this fuzzy number and the terms "*weakly* stable" and "*more or less* stable" are about 0.51 and 0.73 respectively. Given this quantification, we can conclude that the system is rather *more or less* stable than *weakly* stable.

If $\tilde{a}=(0.5,1,1.5)$ then $\widetilde{M}(t_0)=(1,\ 1.56063,\ 3.15505)$ and the degree of stability is represented in the form of the triangular fuzzy number $\widetilde{Deg}=(0.35125,0.423109,0.575086)$. Here the computed possibility measures of the fuzzy number and the terms "*weakly* stable", "*more or less* stable" and "*strongly* stable" are about 0.46, 0.81, 0.15 respectively. In conclusion, the system is *more or less* stable.

Example 7.3. Let us consider a fuzzy dynamical system described by the differential equation

$$\tilde{x}'(t)=\tilde{0}-_h e^t\tilde{x}^3(t)\ ,\tilde{x}(t_0)=\tilde{x}_0\,. \tag{7.20}$$

where $\tilde{x}(t)$, $\tilde{x}_0\in E^1$. The α - cut of the solution to (20) is given in the form

$$x_l^\alpha(t)=\frac{1}{\sqrt{\dfrac{1+2e^t\left(x_{l0}^\alpha\right)^2-2e^{t_0}\left(x_{l0}^\alpha\right)^2}{\left(x_{l0}^\alpha\right)^2}}},$$

$$x_r^\alpha(t)=\frac{1}{\sqrt{\dfrac{1+2e^t\left(x_{r0}^\alpha\right)^2-2e^{t_0}\left(x_{r0}^\alpha\right)^2}{\left(x_{r0}^\alpha\right)^2}}}$$

If $\alpha=1$ then the α - cut of the solution is

$$x^{\alpha=1}(t)=\frac{1}{\sqrt{\dfrac{1+2e^t\left(x_0^{\alpha=1}\right)^2-2e^{t_0}\left(x_0^{\alpha=1}\right)^2}{\left(x_0^{\alpha=1}\right)^2}}}$$

The variational equation of (7.20) is

$$\tilde{z}'(t)=\tilde{A}(t,t_0,\tilde{x}_0)\tilde{z}(t)\,, \tag{7.21}$$

where the α -cut of (7.21) is

$$\begin{cases} (z'(t))_l^\alpha = \left(-3e^t \left(\dfrac{1}{\sqrt{\dfrac{1+2e^t\left(x_{l0}^\alpha\right)^2 - 2e^{t_0}\left(x_{l0}^\alpha\right)^2}{\left(x_{l0}^\alpha\right)^2}}} \right)^2 \right) z_l^\alpha(t) \\ (z'(t))^{\alpha=1} = \left(-3e^t \left(\dfrac{1}{\sqrt{\dfrac{1+2e^t\left(x^{\alpha=1}\right)^2 - 2e^{t_0}\left(x^{\alpha=1}\right)^2}{\left(x^{\alpha=1}\right)^2}}} \right)^2 \right) z^{\alpha=1}(t) \\ (z'(t))_r^\alpha = \left(-3e^t \left(\dfrac{1}{\sqrt{\dfrac{1+2e^t\left(x_{r0}^\alpha\right)^2 - 2e^{t_0}\left(x_{r0}^\alpha\right)^2}{\left(x_{r0}^\alpha\right)^2}}} \right)^2 \right) z_r^\alpha(t) \end{cases} \quad (7.22)$$

The fundamental matrix solution of (7.22) is

$$\Phi_l^\alpha = \frac{1}{\left(x_{l0}^\alpha + 2\left(e^t - e^{t_0}\right)\left(x_{l0}^\alpha\right)^3\right)\sqrt{2e^t - 2e^{t_0} + \dfrac{1}{\left(x_{l0}^\alpha\right)^2}}},$$

$$\Phi_l^\alpha = \frac{1}{\left(x_{r0}^\alpha + 2\left(e^t - e^{t_0}\right)\left(x_{r0}^\alpha\right)^3\right)\sqrt{2e^t - 2e^{t_0} + \dfrac{1}{\left(x_{r0}^\alpha\right)^2}}}$$

For $\alpha = 1$, the fundamental matrix solution is

$$\Phi^{\alpha=1} = \frac{1}{\left(x^{\alpha=1} + 2\left(e^{t} - e^{t_0}\right)\left(x^{\alpha=1}\right)^{3}\right)\sqrt{2e^{t} - 2e^{t_0} + \dfrac{1}{\left(x^{\alpha=1}\right)^{2}}}}$$

The graphs of the two fuzzy solutions to (7.20) with the initial conditions $x_{l0}^{\alpha=0} = 0.1$, $x_0^{\alpha=1} = 0.2$, $x_{r0}^{\alpha=0} = 0.3$, $y_{l0}^{\alpha=0} = 0.8$, $y_0^{\alpha=1} = 0.9$, $y_{r0}^{\alpha=0} = 1$ are shown in Fig. 7.15.

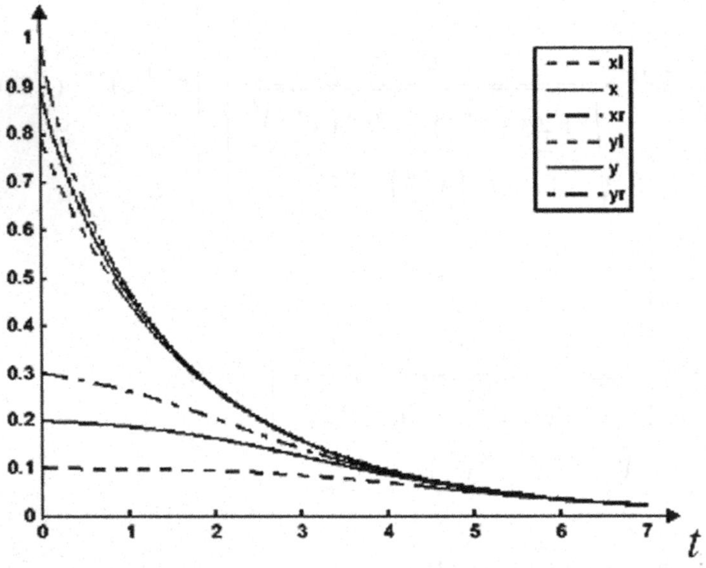

Fig.7.15 Plots of the fuzzy solutions to (7.20)

If $x_{l0}^{\alpha=0} \geq 0.0$, and $t_0 = 0$, then the relationship

$$\left\|\tilde{x}(t,t_0,\tilde{y}_0) -_h \tilde{x}(t,t_0,\tilde{x}_0)\right\|_f \leq \tilde{M}\left\|\tilde{y}_0 -_h \tilde{x}_0\right\|_f$$

is satisfied with $\tilde{M} = (1,1,1)$. Thus, the solution of (7.20) is uniformly fuzzy Lipschitz stable. The degree of stability is the triangular fuzzy number $\widetilde{Deg} = (0.39875, 0.472187, 0.635312)$. Here the possibility measures of this fuzzy number and the terms "*weakly* stable", "*more or less* stable" and "*strongly* stable" are about 0.31, 0.93, 0.26 respectively, which gives rise to the conclusion that the system is *more or less* stable.

Economical System Application

Nonlinear Model of a Manufacture Dynamics. A simple nonlinear model describing the dynamics of a manufacture represents the relation between the level of relative production output and its rate of change [42]:

$$\frac{d\tilde{q}(t)}{dt} = \tilde{q}(t) -_h \tilde{q}^2(t),\ \tilde{q}(0) = \tilde{q}_0,\ \tilde{q}(t),\ \tilde{q}_0 \in E^1, \tag{7.23}$$

where $\tilde{q}(t)$ is a relative production output. $\tilde{q}(t) = Q/Q^*$, where Q is production output and Q^* is its equilibrium value [42]. Production output is one of the main indices of macroeconomics.

The α-cut and the core of (7.23) can be written as follows:

$$\begin{cases} \left(\dfrac{dq(t)}{dt}\right)_l^{\alpha} = q_l^{\alpha}(t) - \left(q^2(t)\right)_l^{\alpha}, \\ \left(\dfrac{dq(t)}{dt}\right)^{\alpha=1} = q^{\alpha=1}(t) - \left(q^2(t)\right)^{\alpha=1}, \\ \left(\dfrac{dq(t)}{dt}\right)_r^{\alpha} = q_r^{\alpha}(t) - \left(q^2(t)\right)_r^{\alpha}, \\ q_l^{\alpha=0}(0) = q_{l0}^{\alpha=0}, \\ q^{\alpha=1}(0) = q_0^{\alpha=1}, \\ q_r^{\alpha=0}(0) = q_{r0}^{\alpha=0}. \end{cases} \tag{7.24}$$

The associated variation system for (7.24) is

$$\begin{cases} \left(\dfrac{dv(t)}{dt}\right)_l^{\alpha} = \left(1 - \dfrac{2e^t q_{l0}^{\alpha}}{e^{t_0} + e^t q_{l0}^{\alpha} - e^{t_0} q_{l0}^{\alpha}}\right) v_l^{\alpha}(t), \\ \left(\dfrac{dv(t)}{dt}\right)^{\alpha=1} = \left(1 - \dfrac{2e^t q_0^{\alpha=1}}{e^{t_0} + e^t q_0^{\alpha=1} - e^{t_0} q_0^{\alpha=1}}\right) v^{\alpha=1}(t), \\ \left(\dfrac{dv(t)}{dt}\right)_r^{\alpha} = \left(1 - \dfrac{2e^t q_{r0}^{\alpha}}{e^{t_0} + e^t q_{r0}^{\alpha} - e^{t_0} q_{r0}^{\alpha}}\right) v_r^{\alpha}(t). \end{cases} \tag{7.25}$$

The fundamental matrix solution for (7.25) comes in the form

$$\Phi_l^\alpha = \frac{e^{t+t_0}}{\left(e^{t_0}\left(q_{0l}^\alpha - 1\right) - e^t q_{0l}^\alpha\right)^2},\ \Phi_r^\alpha = \frac{e^{t+t_0}}{\left(e^{t_0}\left(q_{0r}^\alpha - 1\right) - e^t q_{0r}^\alpha\right)^2}.$$

For $\alpha = 1$, the fundamental matrix solution is

$$\Phi^{\alpha=1} = \frac{e^{t+t_0}}{\left(e^{t_0}\left(q_0^{\alpha=1} - 1\right) - e^t q_0^{\alpha=1}\right)^2}$$

The graphs of the two fuzzy solutions of (7.25) with the initial conditions $q_{l0}^{\alpha=0} = 0.3$, $q_0^{\alpha=1} = 0.4$, $q_{r0}^{\alpha=0} = 0.5$, $qq_{l0}^{\alpha=0} = 0.8$, $qq_0^{\alpha=1} = 0.9$, $qq_{r0}^{\alpha=0} = 1$ are shown in the Fig. 7.16.

If $q_{l0}^{\alpha=0} \ge 0.1$ and $t_0 = 0$ then $\| \tilde{\Phi}(t, t_0, \tilde{q}_0) \|_f \le \tilde{M}$, where $\tilde{M} = (1, 1, 2.8)$, and thus

$$\left\| \tilde{q}(t, t_0, \bar{\tilde{q}}_0) -_h \tilde{q}(t, t_0, \tilde{q}_0) \right\|_f \le \tilde{M} \left\| \bar{\tilde{q}}_0 -_h \tilde{q}_0 \right\|_f.$$

$\tilde{M}$ is independent of t_0, and thus solution of (7.25) is uniformly fuzzy Lipschitz stable in variation. The degree of stability is $\widetilde{Deg} = (0.396562, 0.5404, 0.679825)$. The possibility measures computed for this fuzzy number and the terms "*weakly* stable", "*more or less* stable" and "*strongly* stable" are about 0.26, 0.918, 0.37 respectively. In essence, the system is *more or less* stable.

For the system

$$\frac{d\tilde{q}(t)}{dt} = \tilde{a}\tilde{q}(t) -_h \tilde{a}\tilde{q}^2(t),\ \tilde{q}(0) = \tilde{q}_0,\ \tilde{q}(t),\ \tilde{q}_0 \in E^1,$$

where $\tilde{a} = (0.5, 1, 1.5)$, and the initial conditions

$$q_{l0}^{\alpha=0} = 0.3,\ q_0^{\alpha=1} = 0.4,\ q_{r0}^{\alpha=0} = 0.5,\ qq_{l0}^{\alpha=0} = 0.8,\ qq_0^{\alpha=1} = 0.9,\ qq_{r0}^{\alpha=0} = 1,$$

the degree of stability is $\widetilde{Deg} = (0.369775, 0.5404, 0.5975)$. In this case the possibility measures of this fuzzy number and the terms "*weakly* stable", "*more or less* stable" and "*strongly* stable" are about 0.31, 0.922, 0.24 respectively. Again the system is *more or less* stable.

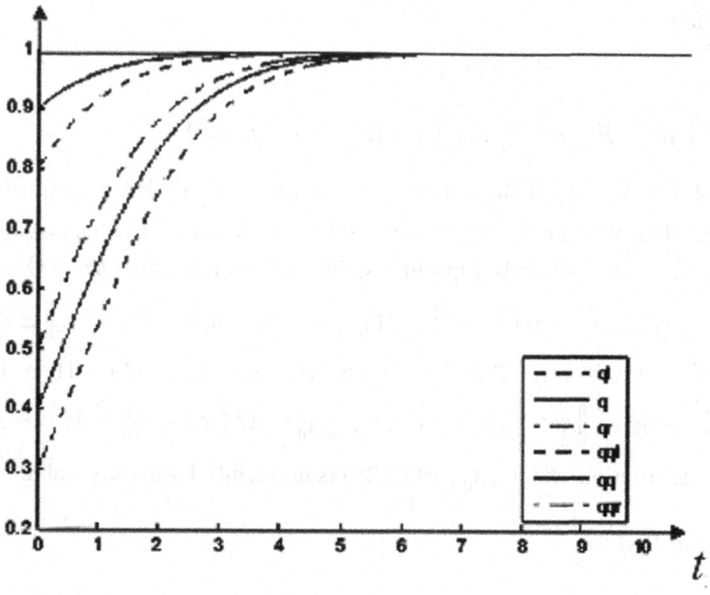

Fig. 7.16 Plot of the fuzzy solutions to (7.25)

If $\tilde{a} = (2, 2.1, 2.2)$, and the initial conditions $q_{l0}^{\alpha=0} = 0.3$, $q_0^{\alpha=1} = 0.4$, $q_{r0}^{\alpha=0} = 0.5$, $qq_{l0}^{\alpha=0} = 0.8$, $qq_0^{\alpha=1} = 0.9$, $qq_{r0}^{\alpha=0} = 1$, the degree of stability is $\widetilde{Deg} = (0.7256, 0.7812, 0.8399)$. In this case the possibility measures of the similarities between this fuzzy number and the terms "*more or less* stable" and "*strongly* stable" are about 0.306, 0.831 respectively, i.e. the system becomes *strongly* stable.

7.7 Comparison of Existing Stability Approaches with the Generalized Theory of Stability (GTS)

It is worth to compare the main approaches to stability as far the definitions and the ensuring methods are concerned. This type of comparison could offer us a certain perspective as to the different points of views and a way in which these concepts of stability relate to each other.

Classical Notions of Stability

If f in the system (7.6) is a function of the variable x and $\tilde{M}(t_0)$ in Definition 7.1 is also a numeric entity, then we encounter the problem of binary Lipschitz stability [31] of a system (BS). To show the relationship between the notion of Lipschitz stability and that of Lyapunov stability, let us refer to the differential system:

$$x' = f(t,x) \tag{7.26}$$

where $f \in C^1\left[R_+ \times R^n, R^n\right]$, $f(t,0) \equiv 0$, $t \geq t_0$, $t_0 \in R_+$.

As it was shown in [31] if the solution $x(t,t_0,x_0)$ of (7.26) is uniformly Lipschitz stable then it is uniformly Lyapunov stable. Assume that the solution $x(t,t_0,x_0)$ of (7.26) is uniformly Lipschitz stable. Then there exist $M > 0$ and $\delta_1 > 0$ such that $\|x(t,t_0,x_0) - x(t,t_0,y_0)\| \leq M\|x_0 - y_0\|$ whenever $\|x_0 - y_0\| \leq \delta_1$, $t \geq t_0 \geq 0$. Now given any $\varepsilon > 0$, choose $\delta = \min(\delta_1, \varepsilon / M)$. Then for $\|x_0 - y_0\| \leq \delta$ we have $\|x(t,t_0,x_0) - x(t,t_0,y_0)\| \leq M\|x_0 - y_0\| \leq M\delta \leq \varepsilon$. It follows that the solution $x(t,t_0,x_0)$ of (7.26) is uniformly Lyapunov stable.

Fuzzy Stability of Systems

If f in the system (7.6) is a function of some variable x and $\tilde{M}(t_0)$ in Definition 7.1 is a fuzzy number then GTS is reduced to the problem of fuzzy Lipschitz stability [50] of a system (FS). Lipschitz constant for f in (7.6) may be determined from the relationship $\left\|\dfrac{\partial f(t,x)}{\partial \tilde{x}}\right\|_f \leq \tilde{N}$. Then $\tilde{M}(t_0)$ is determined on the basis of $\tilde{N}$.

Binary Stability of Fuzzy Systems

If $f(t,\tilde{x})$ in the system (7.6) is fuzzy function of fuzzy variable $\tilde{x}$ and $\tilde{M}(t_0)$ in Definition 7.1 is a numeric entity, then the problem of binary stability of fuzzy system (BFS) could be considered [31]. In this case the problem of stability of fuzzy system (described by fuzzy differential equations or by dynamic "if-then" rules) is reduced to the problem of binary stability viewed in the classical sense.

When the stability of a fuzzy system described by "if-then" rules (TSK or Zadeh model) is considered, the investigation is reduced to the use of the Lyapunov stability theory. The studies covered in [27,33,35,39,52] are devoted to this direction.

When the stability of a fuzzy system described by fuzzy differential equations is considered, then the comparison principle [31] is used. In this method, a fuzzy differential equation is transformed into a scalar comparison differential equation by means of some Lyapunov function. After this, it is enough to investigate a stability of a simpler comparison equation.

Below it is given the theorem on the binary stability of fuzzy differential equation (7.6). Consider the following scalar differential equation:

$$v' = g(t,v) \tag{7.27}$$

where $g(t,v)$ is nondecreasing in v for any t and $g(t,0) \equiv 0$

Theorem 7.3. Assume that

1. The right part of (7.6) satisfies the conditions of existence and uniqueness in some ball $s(\rho)$ of radius ρ (crisp) in E^n with the center in the origin and also $f(t,0) \equiv 0$. The ball is defined in terms of the supremum metric d (see [31]), used for fuzzy sets.
2. For $h > 0, t \in R_+, u \in s(\rho)$,

$$\limsup_{h \to 0^+} \frac{1}{h}\left[d\left[\tilde{u} + hf(t,\tilde{u}), \hat{0} \right] - d\left[\tilde{u}, \hat{0} \right] \right] \le g\left(t, d\left[\tilde{u}, \hat{0} \right]\right); \tag{7.28}$$

3. Function g from R_+^2 to the reals is continuous and $g(t,0) \equiv 0$.

Then from the uniform Lipschitz stability of the zero solution of (7.28) it follows the uniform Lipschitz stability of the zero solution of (7.6).

Example 7.4. Let us consider the fuzzy differential equation [18]:

$$\tilde{x}' = \frac{1}{1+t^2}\tilde{x}$$

$$\lim_{h \to 0^+} \frac{1}{h}\left(d\left[\tilde{x} - h\frac{1}{1+t^2}\tilde{x}, \hat{0} \right] - d\left[\tilde{x}, \hat{0} \right] \right) \le \lim_{h \to 0^+} \frac{1}{h}\left(d\left[\tilde{x}, \hat{0} \right] + d\left[-h\frac{1}{1+t^2}\tilde{x}, \hat{0} \right] - d\left[\tilde{x}, \hat{0} \right] \right) =$$

$$= \lim_{h \to 0^+} \frac{1}{h}\left(d\left[-h\frac{1}{1+t^2}\tilde{x}, \hat{0} \right] \right) = d\left[\frac{1}{1+t^2}\tilde{x}, \hat{0} \right] = \frac{1}{1+t^2} d\left[\tilde{x}, \hat{0} \right] \le g\left(t, d\left[\tilde{x}, \hat{0} \right]\right),$$

where $g\left(t,w\right) = \frac{1}{1+t^2} w$. The solution $w = 0$ is uniformly Lipschitz stable.

Hence the solution $\tilde{x} = \hat{0}$ is uniformly Lipschitz stable too. Thus from the GTS it follows the binary stability of fuzzy systems (BFS).

References

1. Aliev, R.A., Aliev, R.R.: Soft Computing and its Applications. World Scientific, Singapore (2001)
2. Aliev, R.A., Pedrycz, W.: Fundamentals of a fuzzy-logic-based generalized theory of stability. IEEE Transactions on Systems, Man, and Cybernetics, Part B: Cybernetic 39(4), 971–988 (2009)

3. Aliev, R.A., Gurbanov, R.S., Alizadeh, A.V., Aliev, R.R.: Fuzzy Stability Analysis. In: Proc. Seventh International Conference on Applications of Fuzzy Systems and Soft Computing, ICAFS 2006, pp. 87–105. b-Quadrat Verlag, Kaufering (2006)
4. Babuska, R., Verbruggen, H.B.: Fuzzy modeling and model-based control for nonlinear systems. In: Jamshidi, M., Titli, A., Boverie, S., Zadeh, L.A. (eds.) Applications of Fuzzy Logic: Towards High Machine Intelligence Quotient Systems, pp. 49–74. Prentice Hall, New Jersey (1997)
5. Bede, B., Gal, S.G.: Almost periodic fuzzy-number-valued functions. Fuzzy Sets and Systems 147, 385–403 (2004)
6. Boulsama, F., Ichikawa, A.: Application of limit fuzzy controllers to stability analysis. Fuzzy Sets and Systems 49, 103–120 (1992)
7. Buckley James, J., Feuring, T.: Fuzzy differential equations. Fuzzy Sets and Systems 151, 581–599 (2005)
8. Cao, S.G., Rees, N.W., Feng, G.: Analysis and design of fuzzy control systems using dynamic fuzzy global model. Fuzzy Sets and Systems 75, 47–62 (1995)
9. Cezari, L.: Asymptotic behaviour and stability problems in ordinary differential equations. Springer (1959)
10. Chen, Y.Y., Tsao, T.C.: A description of the dynamical behavior of fuzzy systems. IEEE Transactions on Systems, Man and Cybernetics – Part B: Cybernetics 19, 745–755 (1989)
11. Dannan, F.M., Elaydi, S.: Lipschitz stability of nonlinear systems of differential equations. Journal of Mathematical Analysis and Applications 113, 562–577 (1986)
12. Diamond, P.: Stability and periodicity in fuzzy differential equations. IEEE Transactions on Fuzzy Systems 8(5), 583–590 (2000)
13. Diamond, P., Kloeden, P.: Metric Spaces of Fuzzy Sets: Theory And Applications. World Scientific, Singapore (1994)
14. Diamond, P.: Brief note on the variation of constants formula for fuzzy differential equations. Fuzzy Sets and Systems 129, 65–71 (2002)
15. Fuh, C.C., Tung, P.C.: Robust stability analysis of fuzzy control systems. Fuzzy Sets and Systems 88, 289–298 (1997)
16. Gao, S.G., Rees, N.W., Feng, G.: Stability analysis of fuzzy control systems. IEEE Transactions 26, 201–204 (1996)
17. Gnana Bhaskar, T., Lakshmikantham, V., Devi, V.: Revisiting fuzzy differential equations. Nonlinear Analysis 64, 895–900 (2006)
18. Hien, L.V.: A note on the asymptotic stability of fuzzy differential equations. Ukrainian Mathematical Journal 57(7), 904–911 (2005)
19. Hsiao, F.H., Hwang, J.D.: Stability analysis of fuzzy large-scale systems. IEEE Transactions on Systems, Man and Cybernetics – Part B: Cybernetics, vol 32(1), 122–126 (2002)
20. Hwang, G.C., Lin, S.C.: A stability approach to fuzzy control design for nonlinear systems. Fuzzy Sets and Systems 48, 279–287 (1992)
21. Jeong, J.: Stability of a periodic solution for fuzzy differential equations. Journal of Applied Mathematics and Computing 13(1-2), 217–222 (2003)
22. Jianqin, C., Laijiu, C.: Study on stability of fuzzy closed-loop control systems. Fuzzy Sets and Systems 57, 159–168 (1993)
23. Jon, J., Chen, Y.H., Langari, R.: On the stability issues of linear Takagi-Sugeno fuzzy model. IEEE Trans. Fuzzy Syst. 6, 402–410 (1998)
24. Kaleva, O.: A note on fuzzy differential equations. Nonlinear Analysis 64, 895–900 (2006)
25. Kang, H.J., Kwon, C., Lee, H., Park, M.: Robust stability analysis and design method for the fuzzy feedback linearization regulator. IEEE Trans. Fuzzy Syst. 6, 464–472 (1998)

26. Kickert, W.J.M., Mamdani, E.H.: Analysis of a fuzzy logic controller. Fuzzy Sets and Systems 1, 29–114 (1978)
27. Kiriakidis, K.: Robust stabilization of the Takagi-Sugeno fuzzy model via bilinear matrix inequalities. IEEE Transactions on Fuzzy Systems 9(2), 269–277 (2001)
28. Kiriakidis, K., Grivas, A., Tzes, A.: Quadratic stability analysis of the Takagi-Sugeno fuzzy model. Fuzzy Sets and Systems 98, 1–14 (1998)
29. Kubica, E., Madill, D., Wang, D.: Designing Stable MIMO Fuzzy Controllers. IEEE Transactions on Systems, Man and Cybernetics – Part B: Cybernetics 35(2), 372–380 (2005)
30. Klawonn, F.: On a Lukasiewicz logic based controller. In: MEPP 1992 International Seminar on Fuzzy Control through Neural Interpretation of Fuzzy Sets. Reports Computer Science & Mathematics, Ser B, vol. 14, pp. 53–56. Abo Akademi, Turku (1992)
31. Lakshmikantham, V., Mohapatra, R.N.: Theory of fuzzy differential equations and inclusions. Taylor & Francis, London (2003)
32. Lam, H.K., Leung, F.H.F., Tam, P.K.S.: Design and stability analysis of fuzzy model-based nonlinear controller for nonlinear systems using genetic algorithm. IEEE Transactions on Systems, Man and Cybernetics – Part B: Cybernetics 33(2), 250–257 (2003)
33. Lam, H.K., Leung, F.H.F., Tam, P.K.S.: An improved stability analysis and design of fuzzy control systems. In: Proc. Fuzzy Systems Conference, vol. 1, pp. 430–433 (1999)
34. Lin, H.R., Wang, W.J.: L2-stabilization design for fuzzy control systems. Fuzzy Sets and Systems 100, 159–172 (1998)
35. Margaliot, M., Langholz, G.: New approaches to fuzzy modeling and control. Design and analysis. World Scientific (2000)
36. Ray, K.S., Majumder, D.D.: Application of circle criterion for stability analysis of SISO and MIMO systems associated with fuzzy logic controller. IEEE Trans. Syst., Man, Cybern. SMC 14(2), 345–349 (1984)
37. Sonbol, A., Fadali, M.S.: TSK Fuzzy Systems Types II and III Stability Analysis: Continuous Case. IEEE Transactions on Systems, Man, and Cybernetics, Part B 36(1), 2–12 (2006)
38. Sugeno, M., Yasukawa, T.: A fuzzy logic-based approach to qualitative modeling. IEEE Transactions on Fuzzy Systems 1, 7–31 (1993)
39. Tanaka, K., Sugeno, M.: Stability analysis and design of fuzzy control systems. Fuzzy Sets and Systems 45, 135–156 (1992)
40. Tanaka, K., Ikeda, T., Wang, H.O.: Robust stabilization of a class of uncertain nonlinear systems via fuzzy control: Quadratic stabilizability, H^{∞} Control Theory, and Linear Matrix Inequalities. IEEE Trans. Fuzzy Syst. 4, 1–13 (1996)
41. Thathachar, M.A.L., Viswanath, P.: On the stability of fuzzy systems. IEEE Trans. Fuzzy Syst. 5, 145–151 (1997)
42. Turnovsky, S.: Methods of Macroeconomic Dynamics. The MIT Press, Cambridge (1995)
43. Ying, J.: Constructing nonlinear variable gain controllers via Takagi-Sugeno fuzzy control. IEEE Trans. Fuzzy Syst. 6, 226–234 (1998)
44. Wang, L.X.: Adaptive Fuzzy Systems and Control: Design and Stability Analysis. Prentice-Hall (1994)
45. Wang, H.O., Tanaka, K., Griffin, M.F.: An approach to fuzzy control of nonlinear systems: Stability and design issues. IEEE Trans. Fuzzy Syst. 4, 14–23 (1996)
46. Wang, L.X.: Stable adaptive fuzzy controllers with applications to inverted pendulum tracking. IEEE Transactions on Systems, Man and Cybernetics – Part B: Cybernetics 26, 677–691 (1996)

47. Zadeh, L.A.: Fuzzy sets. Information and Control 8, 338–353 (1965)
48. Zadeh, L.A.: The concept of a linguistic variable and its application to approximate reasoning – I. Information Sciences 8, 199–249 (1975)
49. Zadeh, L.A.: Fuzzy logic = Computing with Words. IEEE Transactions on Fuzzy Systems 4(2), 103–111 (1996)
50. Zadeh, L.A.: Protoform Theory and Its Basic Role in Human Intelligence, Deduction, Definition and Search. In: Proc. Second International Conference on Soft Computing and Computing with Words in System Analysis, Decision and Control, ICSCCW 2003, pp. 1–2 (2003)
51. Zadeh, L.A.: Toward the generalized theory of uncertainty (GTU) – an outline. Information Sciences 172, 1–40 (2005)
52. Zhou, C.: Fuzzy-arithmetic-based Lyapunov synthesis in the design of stable fuzzy controllers: a computing-with-words approach. Int. J. Applied Mathematics and Computer Science 12(3), 101–111 (2002)
53. Zhang, J., Morris, A.J.: Recurrent neuro-fuzzy networks for nonlinear process modelling. IEEE Transactions on Neural Networks 10(2), 313–326 (1999)

Chapter 8
Experiments and Applications

In real-world problems, as a rule, perceptions are original sources of decision relevant information on environment and a DM's behavior. Perception-based information is intrinsically imperfect and, as a result, is usually described in NL or in form of visual images. In this chapter we provide solutions of decision making problems with perception-based imperfect information by applying decision theories suggested in the previous chapters. The considered decision making problems include benchmark problems and real-world decision problems in the areas of business, economics, production and medicine. The NL-based and visual imperfect information in these problems makes the use of the existing decision theories inapplicable to solve them.

8.1 Benchmark Decision Making Problems

Zadeh's Two Boxes Problem

Prof. Lotfi Zadeh suggested a benchmark problem of decision making under imperfect information [5] which is described in Chapter 6. Let us solve this problem by applying the suggested decision theory with imperfect information.

At first let us consider solving Case 4 and Cases 2 and 3 as its special cases. These cases fall within the category of decision problems for which fuzzy logic-based decision theory suggested in Chapter 4 is developed. Let us denote boxes as A and B and colors of balls as w (white) and b (black).

The set of possible events will be represented as $\{Aw, Bw, Ab, Bb\}$, where Aw means "a white ball picked from box A", Bb means "a black ball picked from box B" etc. Then the set of the states of nature is: $S = \{s_1, s_2, s_3, s_4\}$, where $s_1 = (Aw, Bw)$, $s_2 = (Aw, Bb)$, $s_3 = (Ab, Bw)$, $s_4 = (Ab, Bb)$.

Denote probabilities of the events as $\tilde{P}(Aw), \tilde{P}(Bw), \tilde{P}(Ab), \tilde{P}(Bb)$. Then the probabilities of the states are defined as:

R.A. Aliev: *Fuzzy Logic-Based Generalized Theory of Decisions*, STUDFUZZ 293, pp. 265–319.
DOI: 10.1007/ 978-3-642-34895-2_8

$$\tilde{P}_1 = \tilde{P}(Aw, Bw) = \tilde{P}(Aw)\tilde{P}(Bw);$$
$$\tilde{P}_2 = \tilde{P}(Aw, Bb) = \tilde{P}(Aw)\tilde{P}(Bb);$$
$$\tilde{P}_3 = \tilde{P}(Ab, Bw) = \tilde{P}(Ab)\tilde{P}(Bw);$$
$$\tilde{P}_4 = \tilde{P}(Ab, Bb) = \tilde{P}(Ab)\tilde{P}(Bb).$$

Denote the outcomes of the events as $\tilde{X}(Aw), \tilde{X}(Bw), \tilde{X}(Ab), \tilde{X}(Bb)$, where $\tilde{X}(Aw)$ denotes the outcome faced when a white ball is picked from box A etc.

The alternatives are: $\tilde{f}_1$ (means choosing box A), $\tilde{f}_2$ (means choosing box B). As alternatives map states to outcomes, we write:

$$\tilde{f}_1(s_1) = \tilde{X}(Aw);$$
$$\tilde{f}_1(s_2) = \tilde{X}(Aw);$$
$$\tilde{f}_1(s_3) = \tilde{X}(Ab);$$
$$\tilde{f}_1(s_4) = \tilde{X}(Ab);$$

$$\tilde{f}_2(s_1) = \tilde{X}(Bw);$$
$$\tilde{f}_2(s_2) = \tilde{X}(Bb);$$
$$\tilde{f}_2(s_3) = \tilde{X}(Bw);$$
$$\tilde{f}_2(s_4) = \tilde{X}(Bb).$$

Assume that imperfect decision-relevant information is treated by using fuzzy outcomes and fuzzy probabilities given in form of the following triangular and trapezoidal fuzzy numbers, respectively:

$$\tilde{X}(Aw) = \$(15, 20, 25),$$
$$\tilde{X}(Ab) = \$(-10, -5, 0),$$
$$\tilde{X}(Bw) = \$(80, 100, 120),$$
$$\tilde{X}(Bb) = \$(-25, -20, -15).$$

$$\tilde{P}(Aw) = (0.25, 0.35, 0.5, 0.6),\ \tilde{P}(Bw) = (0.1, 0.2, 0.25, 0.35).$$

Then the unknown probabilities will be as follows:

$$\tilde{P}(Ab) = (0.4, 0.5, 0.65, 0.75),\ \tilde{P}(Bb) = (0.65, 0.75, 0.8, 0.9).$$

The probabilities of the states of nature are computed as follows:

$\tilde{P}(Aw,Bw)=(0.25,0.35,0.5,0.6)\cdot(0.1,0.2,0.25,0.35)=(0.025,0.07,0.125,0.21)$;

$\tilde{P}(Ab,Bw)=(0.4,0.5,0.65,0.75)\cdot(0.1,0.2,0.25,0.35)=(0.04,0.1,0.1625,0.2625)$;

$\tilde{P}(Ab,Bb)=(0.4,0.5,0.65,0.75)\cdot(0.65,0.75,0.8,0.9)=(0.26,0.375,0.52,0.675)$.

An overall utility for an alternative $\tilde{f}_j$, $j=1,2$ will be determined as a fuzzy-valued Choquet integral (4.1):

$$\tilde{U}(\tilde{f}_j)=(\tilde{u}(\tilde{f}_j(s_{(1)}))-_h\tilde{u}(\tilde{f}_j(s_{(2)})))\tilde{\eta}_{\tilde{P}^l}\left(\{s_{(1)}\}\right)+$$
$$+(\tilde{u}(\tilde{f}_j(s_{(2)}))-_h\tilde{u}(\tilde{f}_j(s_{(3)})))\tilde{\eta}_{\tilde{P}^l}\left(\{s_{(1)},s_{(2)}\}\right)+$$
$$+(\tilde{u}(\tilde{f}_j(s_{(3)}))-_h\tilde{u}(\tilde{f}_j(s_{(4)})))\tilde{\eta}_{\tilde{P}^l}\left(\{s_{(1)},s_{(2)},s_{(3)}\}\right)+$$
$$+(\tilde{u}(\tilde{f}_j(s_{(4)})))\tilde{\eta}_{\tilde{P}^l}\left(\{s_{(1)},s_{(2)},s_{(3)},s_{(4)}\}\right)$$

Here (i) means that the states are ordered such that $\tilde{u}(\tilde{f}_j(s_{(1)}))\geq \tilde{u}(\tilde{f}_j(s_{(2)}))\geq\tilde{u}(\tilde{f}_j(s_{(3)}))\geq\tilde{u}(\tilde{f}_j(s_{(4)}))$, $\tilde{u}(\tilde{f}_j(s_{(i)}))$ denotes utility of an outcome we face taking action $\tilde{f}_j$ at a state $s_{(i)}$, $\tilde{\eta}_{\tilde{P}^l}(\cdot)$ is a fuzzy number-valued fuzzy measure. For simplicity, we define $\tilde{u}(\tilde{f}_j(s_i))$ to be numerically equal to the corresponding outcomes $\tilde{f}_j(s_i)$. Then the overall utilities of the alternatives are determined as follows:

$$\tilde{U}(\tilde{f}_1)=(25,25,25)\tilde{\eta}_{\tilde{P}^l}\left(\{s_1,s_2\}\right)+(-10,-5,0);$$
$$\tilde{U}(\tilde{f}_2)=(105,120,135)\tilde{\eta}_{\tilde{P}^l}\left(\{s_1,s_3\}\right)+(-25,-20,-15).$$

The α-cuts of $\tilde{\eta}_{\tilde{P}^l}\left(\{s_1,s_2\}\right)$, $\tilde{\eta}_{\tilde{P}^l}\left(\{s_1,s_3\}\right)$ are found as numerical solutions to problem (4.13)-(4.14) (Chapter 4):

$$\eta^{\alpha}_{\tilde{P}^l}(\{s_1,s_2\})=$$
$$=\inf\left\{p(s_1)+p(s_2)\middle|(p(s_1),p(s_2),p(s_3))\in P_1^{\alpha}\times P_2^{\alpha}\times P_3^{\alpha},p(S_1)+p(S_2)+p(S_3)=1\right\}$$
$$\eta^{\alpha}_{\tilde{P}^l}(\{S_1,S_3\})=$$
$$=\inf\left\{p(s_1)+p(s_3)\middle|(p(s_1),p(s_2),p(s_3))\in P_1^{\alpha}\times P_2^{\alpha}\times P_3^{\alpha},p(S_1)+p(S_2)+p(S_3)=1\right\}.$$

The values of $\tilde{\eta}_{\tilde{P}^l}(\{s_1, s_2\}), \tilde{\eta}_{\tilde{P}^l}(\{s_1, s_3\})$ we found are triangular fuzzy numbers $\tilde{\eta}_{\tilde{P}^l}(\{s_1, s_2\}) = (0.25, 0.35, 0.35)$, $\tilde{\eta}_{\tilde{P}^l}(\{s_1, s_3\}) = (0.1, 0.2, 0.2)$.

Now we can compute fuzzy overall utilities $\tilde{U}(\tilde{f}_1), \tilde{U}(\tilde{f}_2)$. The computed $\tilde{U}(\tilde{f}_1), \tilde{U}(\tilde{f}_2)$ approximated by triangular fuzzy numbers are shown in Fig. 8.1.

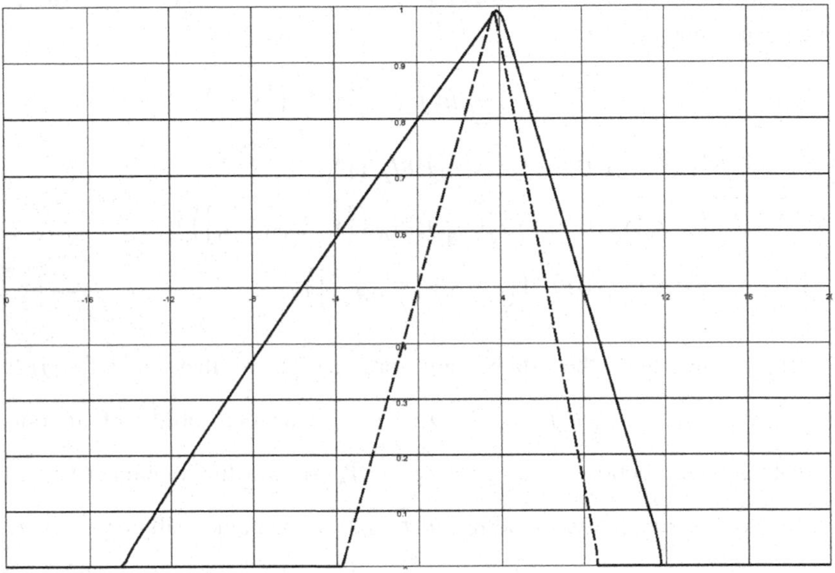

Fig. 8.1 Computed fuzzy overall utilities $\tilde{U}(\tilde{f}_1)$ (dashed), $\tilde{U}(\tilde{f}_2)$ (solid)

Applying Jaccard comparison method [13] we find that:

$$\tilde{U}(\tilde{f}_1) \geq \tilde{U}(\tilde{f}_2) \text{ with degree } 0.83$$

$$\tilde{U}(\tilde{f}_2) \geq \tilde{U}(\tilde{f}_1) \text{ with degree } 0.59$$

We determine the linguistic degree of the preference as it is shown in Fig. 8.2 below. According to Fig. 8.2, $\tilde{f}_1$ (choosing box A) has medium preference over $\tilde{f}_2$ (choosing box B).

The decision relevant information we considered in this problem is characterized by imperfect (fuzzy) probabilities, imperfect (fuzzy) outcomes and fuzzy utilities. Such type of decision-relevant information is presented in Cell 32 of Table 1 (see Chapter 3). Below we shortly present the results of solving the Zadeh's problem for its special cases – Cases 2 and 3 when only probabilities or outcomes are imprecise:

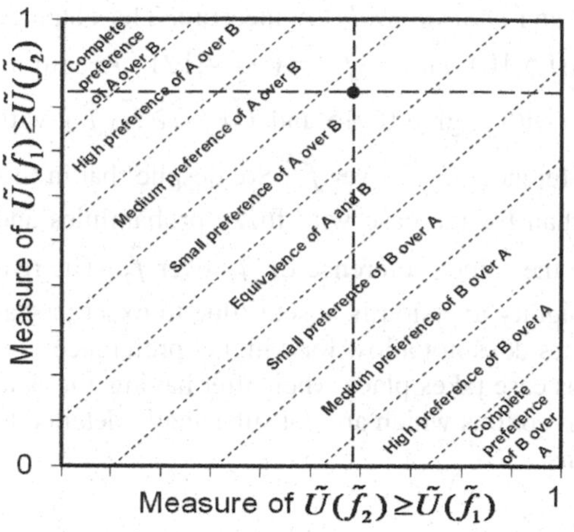

Fig. 8.2 Linguistic preferences

1. The case of imprecise probabilities described as fuzzy numbers $\tilde{P}(Aw) = T(0.25, 0.35, 0.5, 0.6)$, $\tilde{P}(Bw) = T(0.1, 0.2, 0.25, 0.35)$ and exactly known outcomes X(Aw)=\$20, X(Ab)=-\$5, X(Bw)=\$100, X(Bb)=-\$20. This is the situation presented in Cell 4 in Table 3.1 (see Chapter 3) – information on probabilities is imperfect and information on outcomes and utilities is precise (as we determine utilities to be numerically equal to outcomes). In this case the suggested fuzzy utility model degenerates to a fuzzy-valued utility described as Choquet integral with *numeric* integrand and *fuzzy number-valued* fuzzy measure. The obtained fuzzy overall utilities for this case are $\tilde{U}(f_1) = (1.25, 3.75, 3.75)$ and $\tilde{U}(f_2) = (-8, 4, 4)$. According to the Jaccard index [13], $\tilde{U}(f_1) \geq \tilde{U}(f_2)$ with degree 0.708 and $\tilde{U}(f_2) \geq \tilde{U}(f_1)$ with degree 0.266, which implies the medium preference of f_1 over f_2. In this case we have more strong preference of f_1 over f_2 than that obtained for the case of fuzzy probabilities and fuzzy outcomes (small preference of f_1 over f_2). The reason for this is that in the present case information is more precise (outcomes are exactly known) and also ambiguity aversion resulted from fuzzy probabilities is more clearly observed (whereas in the previous case it is weaken behind impreciseness of outcomes).

2. The case of imprecise outcomes described as fuzzy numbers $\tilde{X}(Aw) = \$(15, 20, 25)$, $\tilde{X}(Ab) = \$(-10, -5, 0)$, $\tilde{X}(Bw) = \$(80, 100, 120)$, $\tilde{X}(Bb) = \$(-25, -20, -15)$ and precise probabilities $P(Aw) = 0.4$, $P(Bw) = 0.2$. This is the situation presented in Cell 29 in Table 3.1 – information on probabilities is precise and information on outcomes and utilities is imperfect (as we determine utilities to be numerically equal to outcomes). In this case the suggested fuzzy utility model degenerates to a fuzzy-valued expected utility

with *fuzzy number-valued* integrand and *crisp probability* measure. The obtained fuzzy overall utilities are $\tilde{U}(\tilde{f}_1)=(0,5,10)$ and $\tilde{U}(\tilde{f}_2)=(-4,4,12)$. According to the Jaccard index, $\tilde{U}(\tilde{f}_1)\geq\tilde{U}(\tilde{f}_2)$ with degree 0.915 and $\tilde{U}(\tilde{f}_2)\geq\tilde{U}(\tilde{f}_1)$ with degree 0.651, which implies low preference of $\tilde{f}_1$ over $\tilde{f}_2$. So, despite that in this case the situation is less uncertain than for the case with fuzzy probabilities and fuzzy outcomes, we have in essence the same preference of $\tilde{f}_1$ over $\tilde{f}_2$. The reason for this is that for this case ambiguity aversion is absent (due to exact probabilities) and impreciseness of outcomes does not allow for a higher preference.

Consider now solving Case 5. This case takes place when after having a look at boxes (Fig. 8.3), a DM has visual perceptions which are not sufficiently detailed to be described by membership functions.

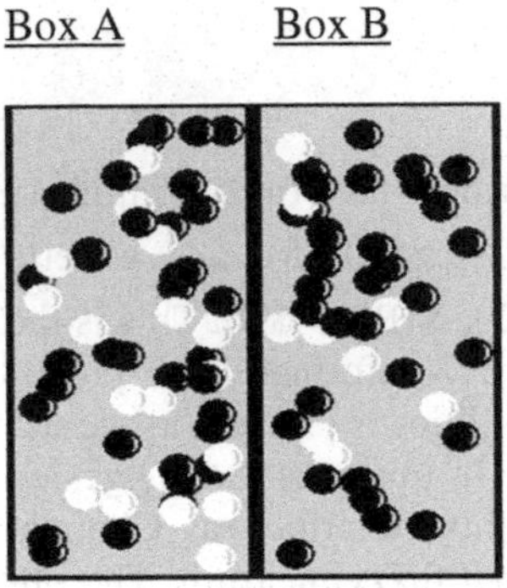

Fig. 8.3 Two boxes with large number of black and white balls

A DM can only graphically evaluate the probabilities as F-marks. Such essence of the relevant information requires the use of fuzzy geometry-based decision theory suggested in Chapter 6 for solution of Case 5.

Assuming that P_{i1}, P_{i2}, P_{i3} and P_{i4} are implementations of *P*(*Aw*), *P*(*Ab*), *P*(*Bw*), and *P*(*Bb*), respectively, and U_{i1} and U_{i2} are the corresponding assessments of the alternatives $\tilde{f}_1$ and $\tilde{f}_2$ respectively in an *i*-th rule, $i=1,2$, a rule base collected by a DM can be such as:

(If $(P_1 \text{ is } P_{11})$ and $(P_2 \text{ is } P_{12})$ and $(P_3 \text{ is } P_{13})$ and $(P_4 \text{ is } P_{14})$ Then $\mathrm{U}_1=U_{11}, \mathrm{U}_2=U_{12}$, 0.95)

(If $(P_1 \text{ is } P_{21})$ and $(P_2 \text{ is } P_{22})$ and $(P_3 \text{ is } P_{23})$ and $(P_4 \text{ is } P_{24})$ Then $\mathrm{U}_1=U_{21}, \mathrm{U}_2=U_{22}$, 0.95).

For simplicity, we used only 2 rules.

In our experiments, we tried to reproduce the conditions used in [5] in an attempt to obtain similar solution.

The values of generated P_{ij} and U_{ik} ($i=\overline{1,2},\ j=\overline{1,4},\ k=\overline{1,2}$) used in the rules are given in the Table 8.1 The user input is given in Table 8.2.

Table 8.1 The If-Then rules containing values of P_{ij} and U_{ik} described by f-marks

Rule #	P_{i1}	P_{i2}	P_{i3}	P_{i4}	U_{i1}	U_{i2}
1		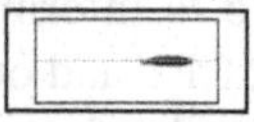		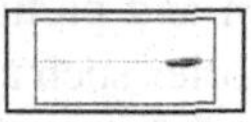		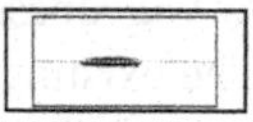
2			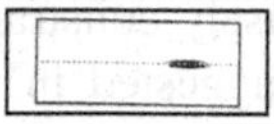	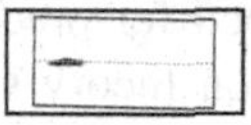		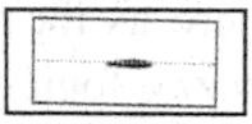

Table 8.2. User's graphically entered probabilities $P_i, i=\overline{1,4}$

P_1	P_2	P_3	P_4
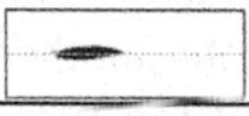			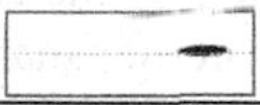

The values of computed integrated outcomes can be represented approximately as:

$$U_1 = v_1 \approx M\left((0.55,0),(0.85,0),0.05\right),$$

$$U_2 = v_2 \approx M\left((0.33,0),(0.61,0),0.05\right).$$

Their comparison gives:

$$g_{\geq}(v_1,v_2)=\frac{1}{2}\left(\frac{0.3}{0.3}+\frac{0.28}{0.28}\right)=1$$

$$g_{\geq}(v_2,v_1)=\frac{1}{2}\left(\frac{0.06}{0.52}+\frac{0.06}{0.52}\right)\approx 0.12.$$

This is quite in line with the result presented in [5]: the preference of one of the actions over the other and vice versa in classic model was 0.708/0.266 and in FG-based model 1/0.12.

8.2 Application in Medicine

In this example we will consider selection of an optimal treatment under imperfect information on stages of the disease and possible results of treatment [1]. In the considered problem, probabilities of the two most expressed stages of the disease are assigned linguistically (imprecisely) by the dentist but for the rest and the least expressed stage the dentist finds difficultly in evaluation of its occurrence probability.

Case history N583 – patient Vidadi Cabrayil Cabrayilov is entered with symptoms of the heavy stage of the periodontitis disease to the polyclinic of Medical University. He was examined and it was defined that intoxication indicators are approximately of 10%-20% level and exudative indicators are of 60%-80% level. The problem is to define an optimal treatment for the considered patient.

Such problem is essentially characterized by linguistic relevant information due to vagueness, imprecision and partial truth related to information on the disease. The existing decision theories such as SEU, CEU, CPT and others are inapplicable here as they are developed for precisely constrained information. We will apply fuzzy logic-based decision theory suggested in Chapter 4 which is able to deal with linguistic information of the considered problem.

Formal Description of the Problem [1]. 1) *States of nature.* States of nature are represented by the stages of the disease. During the patient examination it is very important to properly identify the actual stage of the disease. Without doubts, the "boundaries" of phases are not sharply defined, and one phase slips into another. Taking this into account, it is adequately to describe the phases by using fuzzy sets. The set of the fuzzy states of nature is

$$\mathcal{S}=\{\tilde{S}_1,\tilde{S}_2,\tilde{S}_3\},$$

where $\tilde{S}_1$-intoxication phase (1st stage of the disease), $\tilde{S}_2$-exudative stage (2nd stage of the disease), $\tilde{S}_3$-heavy phase (3rd stage of the disease). Membership functions for $\tilde{S}_1,\tilde{S}_2,\tilde{S}_3$ used to describe intoxication, exudative and heavy phases are shown in Fig.8.4.

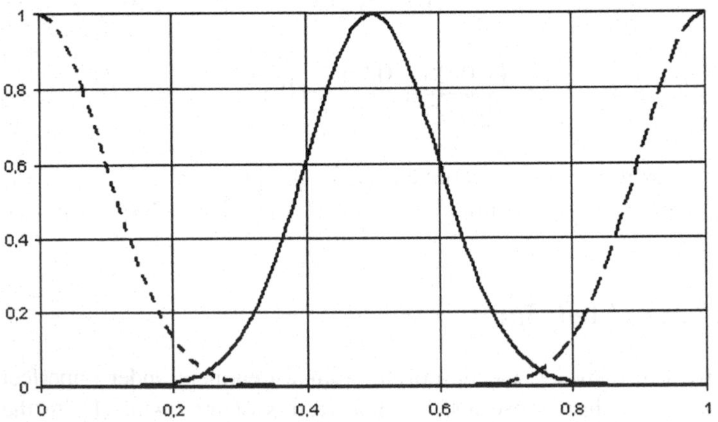

Fig. 8.4 Membership functions of $\tilde{S}_1$ (dotted), $\tilde{S}_2$ (solid) $\tilde{S}_3$ (dashed)

During a patient examination the development level of the disease is represented by a dentist's linguistic (imprecise) evaluations of likelihoods of various stages of the disease. We consider these approximate evaluations as linguistic probabilities. So, we have linguistic probability distribution $\tilde{P}^l$ over the fuzzy states of nature:

$$\tilde{P}^l = \tilde{P}_1/\tilde{S}_1 + \tilde{P}_2/\tilde{S}_2 + \tilde{P}_3/\tilde{S}_3,$$

where $\tilde{P}_1$ is the linguistic probability of the intoxication phase occurrence, $\tilde{P}_2$ is the linguistic probability of the exudative phase occurrence and $\tilde{P}_3$ is unknown imprecise probability of the heavy phase occurrence. The membership functions of the linguistic probabilities $\tilde{P}_1$, $\tilde{P}_2$ are shown in the Fig.8.5.

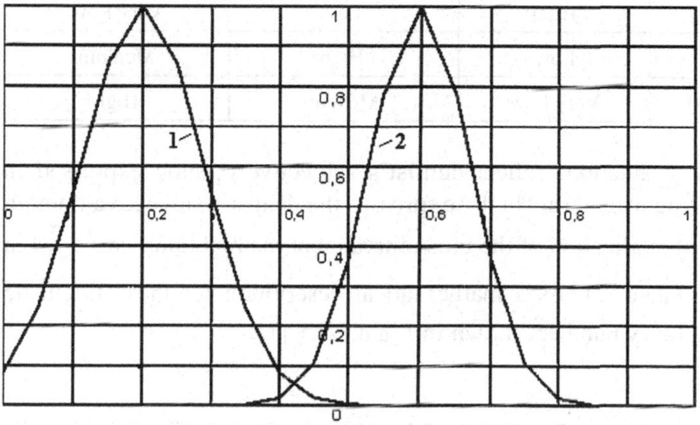

Fig. 8.5 Membership functions of linguistic probabilities $\tilde{P}_1$ (curve 1), $\tilde{P}_2$ (curve 2)

2) *Alternatives.* Alternatives are represented by the available treatment methods. The effectiveness of application of the available treatment methods at the various phases of the disease can be adequately determined in terms of dentist's linguistic (imprecise) evaluations. In view of this the alternatives should be considered as fuzzy functions [9]. The set of the fuzzy alternatives is as follows:

$$\mathcal{A} = \{\tilde{f}_1, \tilde{f}_2, \tilde{f}_3\},$$

where $\tilde{f}_1$ - closed treatment method, $\tilde{f}_2$ - open treatment method, $\tilde{f}_3$ - surgical entrance and tooth removal.

3) *Utilities.* Utility of an alternative $\tilde{f}_j$ taken at a state $\tilde{S}_i$ is considered as an effectiveness of a corresponding treatment method applied at a corresponding phase of the disease. Without doubts, due to uncertainty involved, effectiveness of

application of a considered treatment method at a considered phase of disease can be adequately described by a dentist only in terms of linguistic(imprecise) evaluations. So, utility of an alternative $\tilde{f}_j$ taken at a state $\tilde{S}_i$ will be considered as a fuzzy value $\tilde{u}\left(\tilde{f}_j\left(\tilde{S}_i\right)\right)$ of a fuzzy number-valued utility function $\tilde{u}$.

Let the dentist evaluate the effectiveness of applications of the treatment methods at various stages of the disease by using the following linguistic terms (Table 8.3):

Table 8.3 Linguistic evaluations of effectiveness of the treatment methods at the various stages of periodontitis

	Phase 1 ($\tilde{S}_1$)	Phase 2 ($\tilde{S}_2$)	Phase 3 ($\tilde{S}_3$)
Alternative 1 ($\tilde{f}_1$)	High	Low	Very Low
Alternative 2 ($\tilde{f}_2$)	Low	High	Medium
Alternative 3 ($\tilde{f}_3$)	Very Low	Medium	High

These linguistic evaluations reflect dentist's subjective opinion expressed in NL. For example, linguistic term "high" expresses the dentist's subjective opinion concerning utility of application of the closed treatment method (alternative $\tilde{f}_1$) in intoxication phase (state $\tilde{S}_1$). As a mathematical description for these linguistic evaluations we use fuzzy numbers shown in Fig.8.6-8.8 [1].

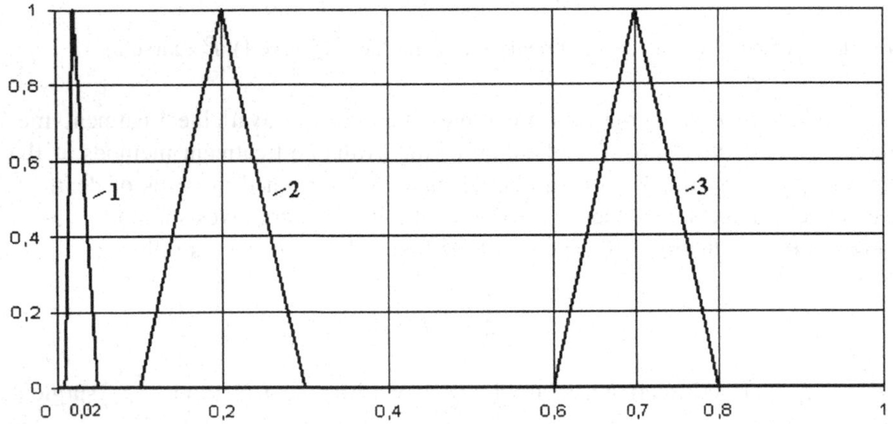

Fig. 8.6 Membership functions of linguistic utilities of the treatment methods application at the intoxication phase: Very Low (curve 1), Low (curve 2), Medium (curve 3)

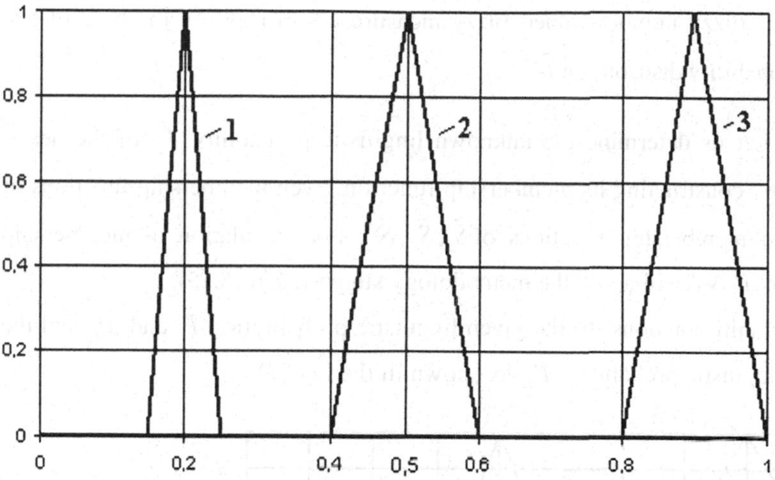

Fig. 8.7 Membership functions of linguistic utilities of the treatment methods application at the exudative phase: Low (curve 1), Medium (curve 2), High (curve 3)

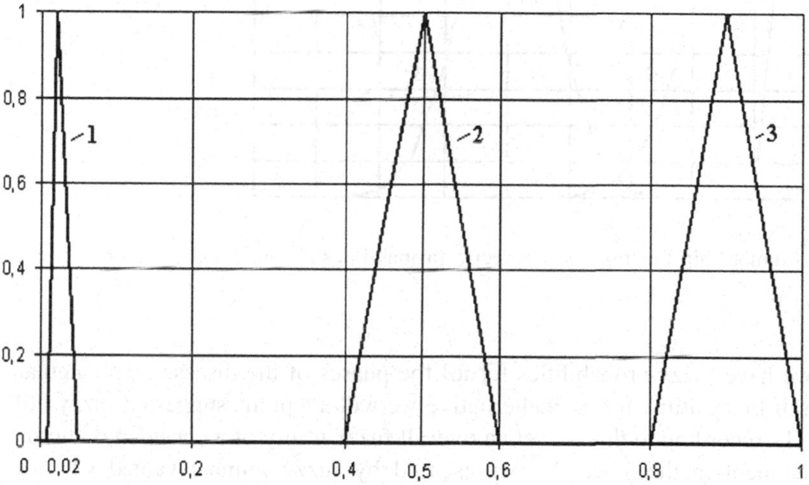

Fig. 8.8 Membership functions of linguistic utilities of the treatment methods application at the heavy phase: Very Low (curve 1), Medium (curve 2), High (curve 3)

So, on the base of the suggested theory, we will formulate the selection of the optimal treatment method of the periodontitis as a determination of a treatment method with the highest overall fuzzy utility represented by a fuzzy number-valued Choquet integral over $\mathcal{S}$:

$$\text{Find } \tilde{f}^* \in \mathcal{A} \text{ such that } \quad \tilde{U}(\tilde{f}^*) = \max_{\tilde{f}_j \in \mathcal{A}} \int_{\mathcal{S}} \tilde{u}(\tilde{f}_j(\tilde{S}_i))d\tilde{\eta}_{\tilde{P}^l},$$

where $\tilde{\eta}_{\tilde{P}^l}$ - fuzzy number-valued fuzzy measure constructed on the base of linguistic probability distribution $\tilde{P}^l$.

Solution. Let us determine the unknown linguistic probability $\tilde{P}_3$ of the heavy stage $\tilde{S}_3$ by constructing its membership function given membership functions of $\tilde{P}_1, \tilde{P}_2$ and membership functions of $\tilde{S}_1, \tilde{S}_2, \tilde{S}_3$. For calculation of membership function of $\tilde{P}_3$ we will apply the methodology suggested in [6,15].

Membership functions for the given linguistic probabilities $\tilde{P}_1$ and $\tilde{P}_2$ and the obtained linguistic probability $\tilde{P}_3$ are shown in the Fig.8.9.

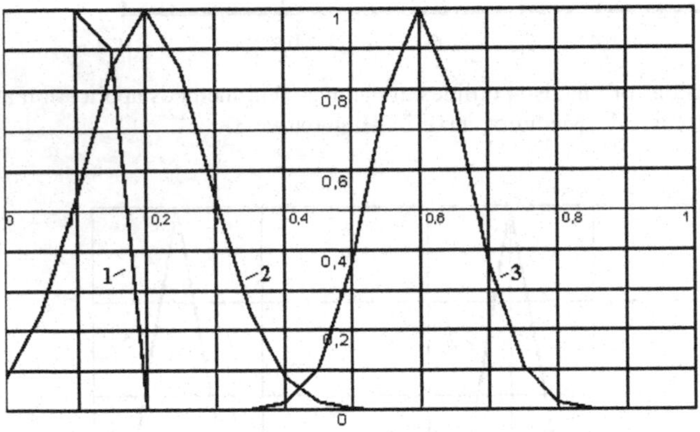

Fig. 8.9 Membership functions of imprecise probabilities $\tilde{P}_1$ (curve 1), $\tilde{P}_2$ (curve 2), and $\tilde{P}_3$ (curve 3)

So, we have fuzzy probabilities for all the phases of the disease. To calculate the overall fuzzy utility for each alternative we will adopt the suggested fuzzy utility model. According to this model an overall fuzzy utility of a considered alternative (treatment method) will be represented by fuzzy number-valued Choquet integral with the respect to fuzzy number-valued fuzzy measure as follows:

$$\tilde{U}(\tilde{f}_j) = \sum_{i=1}^{3} \left(\tilde{u}(\tilde{f}_j(\tilde{S}_{(i)})) -_h \tilde{u}(\tilde{f}_j(\tilde{S}_{(i+1)})) \right) \cdot \tilde{\eta}_{\tilde{P}^l}(\mathcal{H}_{(i)}),$$

where (i) means that utilities are ranked such that $\tilde{u}(\tilde{f}_j(\tilde{S}_{(1)})) \geq \ldots \geq \tilde{u}$, $(\tilde{f}_j(\tilde{S}_{(n)}))$ $\mathcal{H}_{(i)} = \left\{ \tilde{S}_{(1)}, \ldots, \tilde{S}_{(i)} \right\}, \tilde{u}(\tilde{f}_j(\tilde{S}_{(n+1)})) = 0$.

At first it is needed to rank $\tilde{u}(\tilde{f}_j(\tilde{S}_i)) = \tilde{u}_{ji}$ for each alternative $\tilde{f}_j$. For each alternative we have:

$$\text{Alternative } \tilde{f}_1:\ \tilde{u}_{11} = high > \tilde{u}_{12} = low > \tilde{u}_{13} = very\,low$$

$$\text{Alternative } \tilde{f}_2:\ \tilde{u}_{22} = high > \tilde{u}_{23} = medium > \tilde{u}_{21} = low$$

$$\text{Alternative } \tilde{f}_3:\ \tilde{u}_{33} = high > \tilde{u}_{32} = medium > \tilde{u}_{31} = very\,low$$

So, fuzzy utilities are ranked as it is shown below.

Alternative $\tilde{f}_1$ (closed treatment method):

$$\tilde{u}_{1(1)} = \tilde{u}_{11} = high\,; \tilde{u}_{1(2)} = \tilde{u}_{12} = low\,; \tilde{u}_{1(3)} = \tilde{u}_{13} = very\ low$$

Alternative $\tilde{f}_2$ (open treatment method):

$$\tilde{u}_{2(1)} = \tilde{u}_{22} = high\,; \tilde{u}_{2(2)} = \tilde{u}_{23} = medium\,; \tilde{u}_{2(3)} = \tilde{u}_{21} = low$$

Alternative $\tilde{f}_3$ (Surgical entrance or tooth removal method):

$$\tilde{u}_{3(1)} = \tilde{u}_{33} = high\,; \tilde{u}_{3(2)} = \tilde{u}_{32} = medium\,; \tilde{u}_{3(3)} = \tilde{u}_{31} = very\ low$$

So, overall fuzzy utilities for alternatives $\tilde{f}_1, \tilde{f}_2, \tilde{f}_3$ will be described as follows:

$$U(\tilde{f}_1) = (\tilde{u}_{11} -_h \tilde{u}_{12})\tilde{\eta}_{\tilde{P}^l}(\{\tilde{S}_1\}) + (\tilde{u}_{12} -_h \tilde{u}_{13})\tilde{\eta}_{\tilde{P}^l}(\{\tilde{S}_1, \tilde{S}_2\}) + \tilde{u}_{13}\tilde{\eta}_{\tilde{P}^l}(\{\tilde{S}_1, \tilde{S}_2, \tilde{S}_3\});$$

$$U(\tilde{f}_2) = (\tilde{u}_{22} -_h \tilde{u}_{23})\tilde{\eta}_{\tilde{P}^l}(\{\tilde{S}_2\}) + (\tilde{u}_{23} -_h \tilde{u}_{21})\tilde{\eta}_{\tilde{P}^l}(\{\tilde{S}_2, \tilde{S}_3\}) + \\ + \tilde{u}_{21}\tilde{\eta}_{\tilde{P}^l}(\{\tilde{S}_1, \tilde{S}_2, \tilde{S}_3\});$$

$$U(\tilde{f}_3) = (\tilde{u}_{33} -_h \tilde{u}_{32})\tilde{\eta}_{\tilde{P}^l}(\{\tilde{S}_3\}) + (\tilde{u}_{32} -_h \tilde{u}_{31})\tilde{\eta}_{\tilde{P}^l}(\{\tilde{S}_2, \tilde{S}_3\}) + \\ + \tilde{u}_{31}\tilde{\eta}_{\tilde{P}^l}(\{\tilde{S}_1, \tilde{S}_2, \tilde{S}_3\});$$

We have constructed fuzzy number-valued fuzzy measure from linguistic probability distribution P^l as its lower prevision on the base of the methodology suggested in [6,15]. Finally, we calculated overall fuzzy utility for each alternative and the obtained values approximated as triangular fuzzy numbers are the following:

$$\tilde{U}(\tilde{f}_1) = (0.16;\ 0.2;\ 0.24)\,;$$

$$\tilde{U}(\tilde{f}_2) = (0.37;\ 0.38;\ 0.39)\,; \tilde{U}(\tilde{f}_3) = (0.21;\ 0.27;\ 0.33)$$

Membership functions of these fuzzy utilities are shown in Fig. 8.10.

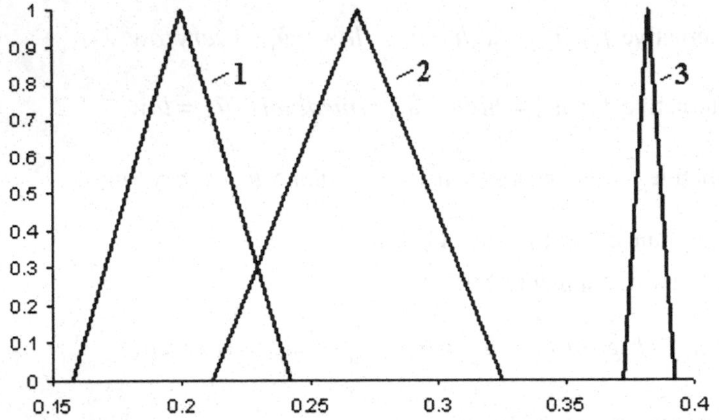

Fig. 8.10 Membership functions of $\tilde{U}(\tilde{f}_1)$ (curve 1), $\tilde{U}(\tilde{f}_2)$ (curve 3), $\tilde{U}(\tilde{f}_3)$ (curve 2)

Now it is necessary to compare $\tilde{U}(\tilde{f}_j),\ j=\overline{1,3}$ and determine the optimal treatment method as one with the highest overall fuzzy utility. Comparing fuzzy utilities $\tilde{U}(\tilde{f}_j),\ j=\overline{1,3}$ we found out that the best alternative is $\tilde{f}_2$ as one with the highest overall fuzzy utility. This means that the optimal treatment method for considered patient is an open treatment method. Below we present experimental results of applying the suggested approach to determine an optimal treatment method for patients.

Experimental Results. We examined 62 patients suffering from acute periodontitis at the age of 17-65. There were 38 men and 24 women from among them. In order to differentiate chronic periodontitis and to define diagnosis for the patients we held radiological examinations. To differentiate 3 phases of acute periodontitis one from another and to define the diagnosis of each phase together with radiological examinations we evaluated clinical situations on the base of acute periodontitis symptoms (such as pain, hiperemia, palpation, percussion and others) and defined clinical situations.

In oder to define the effectiveness of treatment during 3 different phases of acute periodontitis, together with conduction of the methods of radiological examination we observed the dynamics of changes of some clinical symptoms (pain, hiperemia, palpation, percussion and others) and evaluated the improvements of clinical evidence during the mentioned stages.

The results of our research on cure rates of the given treatment methods depending on 3 different phases of acute periodontitis are given in Table 8.4.

Table 8.4 Results of experimental investigation

Different phases of acute periodontitis development	The number of treated patients	Cure rate for the given methods (in number of patients)		
		"closed " treatment method	"open" treatment method	Surgical operation and the tooth removal
Intoxication phase	23	16	6	1
Exudative phase	23	-	21	2
Heavy phase	16	-	3	13

Using various methods of treatment during 3 phases of acute periodontitis, we have got different findings. The comparative results of our research show that in different stages of acute periodontitis "open" and "close" methods were the most effective. So, "open" teatment method proved to be more effective and superior for 21 patients. "Close" treatment method was effective for 16 patients, 6 patients used "open" treatment method as it positively effected on their treatment. The tooth of one patient was removed .

As to the treatment during heavy phase, it's necessary to note that at this phase periodontitis processes are irreversible as it reached the highest level and conservative treatment is less effective(3 patients); most teeth (13 patients), mainly multirooted teeth, were removed by surgical method. During the exudative phase the surgical method was unsatisfactory for 2 patients among the 23. During the intoxication phase for 21 of 23 and during the heavy phase 13 of 16 patients the offered method was satisfactory.

8.3 Applications in Production

Decision Making on the Base of a Two-Level Hierarchical Model

Let us consider decision making for an oil refinery plant which includes three units: preliminary distillation unit; cat cracker unit; cocker unit (see Fig.8.11).

Let us provide short description for the manufacturing that is schematically shown in Fig. 8.11. Preliminary distillation unit produces eight products: fraction OP-85 ($\tilde{f}_{11}$), fraction OP-85-180 ($\tilde{f}_{12}$), kerosene ($\tilde{f}_{13}$), diesel oil ($\tilde{f}_{14}$), tar ($\tilde{f}_{15}$), liquid petroleum gas ($\tilde{f}_{16}$), scrubber gas ($\tilde{f}_{17}$), vacuum gasoil ($\tilde{f}_{18}$). Tar enters cocker unit which produces gasoil ($\tilde{f}_{21}$), coke ($\tilde{f}_{22}$), heavy gasoil ($\tilde{f}_{23}$),

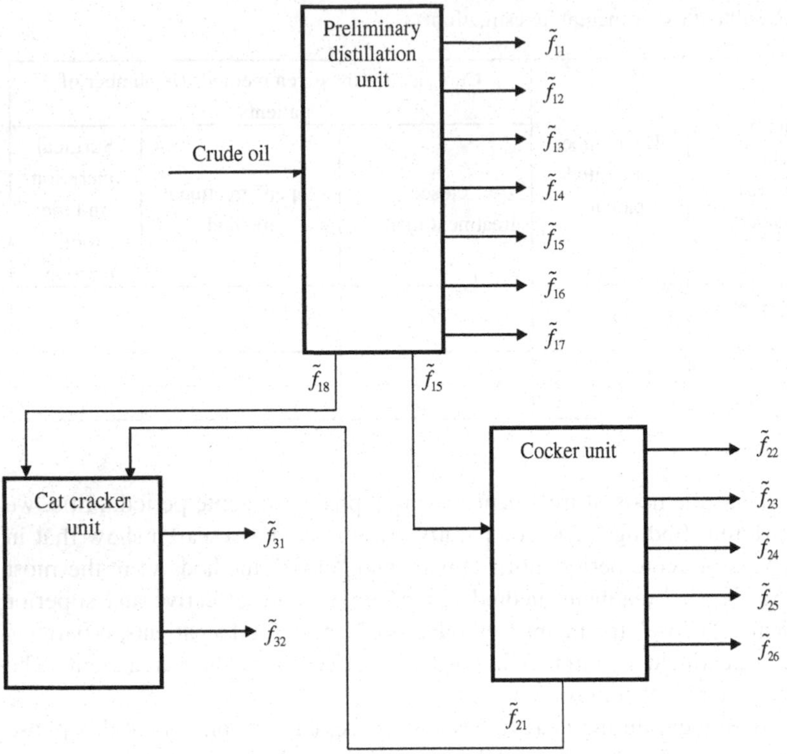

Fig. 8.11 Manufacturing scheme of oil refinery

coke light gasoil ($\tilde{f}_{24}$), waste ($\tilde{f}_{25}$), dry gas ($\tilde{f}_{26}$). Vacuum gasoil and gasoil enter cat cracker unit which produces petrol ($\tilde{f}_{31}$) and reflux ($\tilde{f}_{32}$).

The considered decision making problem is maximization of an overall amount of petroleum fractions produced at the plant. The following solutions to a Pareto-optimization problem were obtained for each unit at the micro-level, i.e. for each unit:

Table 8.5 Preliminary distillation unit

Preliminary distillation unit	1st variant			2nd variant			3rd variant		
	Center	Left	Right	Center	Left	Right	Center	Left	Right
fraction OP-85 ($\tilde{f}_{11}$)	0,0378	0,0367	0,0390	0,0224	0,0217	0,0231	0,0335	0,0325	0,0345
OP-85-180 $\tilde{f}_{12}$)	0,0750	0,0727	0,0772	0,1023	0,0992	0,1054	0,0916	0,0889	0,0944
Kerosene ($\tilde{f}_{13}$)	0,2852	0,2767	0,2938	0,3115	0,3022	0,3209	0,3385	0,3283	0,3486

Table 8.6 Cocker

Cocker unit	1st variant			2nd variant		
	Center	Left	Right	Center	Left	Right
Gas oil ($\tilde{f}_{21}$)	0,4164	0,4039	0,4228	0,4075	0,3953	0,4198
Coke ($\tilde{f}_{22}$)	0,2357	0,2287	0,2428	0,2450	0,2376	0,2523

Table 8.7 Cat cracker

Cat cracker unit	1st variant			2nd variant		
	Center	Left	Right	Center	Left	Right
Petrol ($\tilde{f}_{31}$)	0,1525	0,1479	0,1571	0,1438	0,1395	0,1481
Reflux ($\tilde{f}_{32}$)	0,3834	0,3719	0,3949	0,3917	0,3887	0,4117

For solving this decision making problem on coordination of the functioning of the three units we use the conventional approach to decision making in multi-agent systems [3,4] explained in Section 4.2. A DM expresses $\tilde{H}_0$ and $\tilde{H}_1$ according to (4.33)-(4.34):

$\tilde{H}_0$ is an overall amount of petroleum fractions that should be about $\tilde{b}_0 = (0.65, 0.58, 0.715)$,

$\tilde{H}_1$ is an overall amount of gas oil fractions that should be about $\tilde{b}_1 = (0.78, 0.702, 0.858)$.

At the next step $v_0(\bar{\lambda})$, $v_1(\bar{\lambda})$ and $v(\bar{\lambda}) = \pi_0 v_1 + \pi_1 v_2$ functions are constructed. The importance weights π_0 and π_1 are obtained according to the procedure of an involvement of a DM into a decision making process that was described above. In our case $\pi_0 = 0.6$ and $\pi_1 = 0.4$.

A vector λ is determined by solving the maximization problem: $\max_{\bar{\lambda}}(v(\bar{\lambda}))$. The results of solving this problem are given below:

$$\lambda_{11} = 0.09; \lambda_{12} = 0.45; \lambda_{13} = 0.46; \lambda_{21} = 0.81; \lambda_{22} = 0.19; \lambda_{31} = 0.75; \lambda_{32} = 0.25$$

On the base of calculated λ_{ij} the scheduled tasks desired in terms of an overall goal are calculated for each unit:

$$\tilde{f}_{11} = (0.0345, 0.0335, 0.0356); \tilde{f}_{12} = (0.0850, 0.0824, 0.0875);$$
$$\tilde{f}_{13} = (0.3119, 0.3026, 0.3213);$$
$$\tilde{f}_{21} = (0.1509, 0.1463, 0.1554); \tilde{f}_{22} = (0.23, 0.2070, 0.253);$$
$$\tilde{f}_{31} = (0.4142, 0.4017, 0.4220); \tilde{f}_{32} = (0.2380, 0.2309, 0.2452)$$

The obtained values of $\tilde{f}_{ij}$ are sent to the micro-level to be implemented by solving goal programming problems that completes solving the considered problem. The vector of technological parameters x_i determined by solving goal programming problems is sent further for realization.

Decision Making on Oil Extraction under Imperfect Information [6]

Assume that a manager of an oil-extracting company needs to make a decision on oil extraction at a potentially oil-bearing region. Knowledge about oil occurrence the manager has is described in NL and has the following form:

"probability of "occurrence of commercial oil deposits" is lower than medium"

The manager can make a decision based on this information, or at first having conducted seismic investigation of the region. Concerning the seismic investigation used, its accuracy is such that it with the probability "*very high*" confirms occurrence of commercial oil deposits and with the probability "*high*" confirms absence of commercial oil deposits. The manager has a set of alternative actions to choose from. The goal is to find the optimal action.

The considered problem comes within the same information framework as the problems considered in Sections 8.1 – this problem is characterized by imperfect information described in NL. We will use the theory suggested in Chapter 4 for solving of this problem.

Let us develop a general formal description of the problem. The set of the fuzzy states of nature is

$$\mathcal{S} = \{\tilde{S}_1, \tilde{S}_2\}$$

where $\tilde{S}_1$ denotes "occurrence of commercial oil deposits" and $\tilde{S}_2$ denotes "absence of commercial oil deposits". The states $\tilde{S}_1$ and $\tilde{S}_2$ are represented by triangular fuzzy numbers $\tilde{S}_1 = (1;1;0)$, $\tilde{S}_2 = (0;1;1)$.

The linguistic probability distribution $\tilde{P}^l$ over the states of nature that corresponds the knowledge of the manager is

$$\tilde{P}^l = \tilde{P}_1 / \tilde{S}_1 + \tilde{P}_2 / \tilde{S}_2,$$

where $\tilde{P}_1$ is a triangular fuzzy number $\tilde{P}_1 = (0.3;0.4;0.5)$ that represents linguistic term "lower than medium" and $\tilde{P}_2$ is unknown.

Taking into account the opportunities available to the manager, we consider the following set of the manager's possible actions: $\mathcal{A} = \{\tilde{f}_1, \tilde{f}_2, \tilde{f}_3, \tilde{f}_4, \tilde{f}_5, \tilde{f}_6\}$. The NL-based description of the manager actions $\tilde{f}_i$, $i = \overline{1,6}$ is given below in Table 8.8.

Table 8.8 Possible actions of the manager

Notation	NL-based description
$\tilde{f}_1$	Conduct seismic investigation and extract oil if seismic investigation shows occurrence of commercial oil deposits
$\tilde{f}_2$	Conduct seismic investigation and do not extract oil if seismic investigation shows occurrence of commercial oil deposits
$\tilde{f}_3$	Conduct seismic investigation and extract oil if seismic investigation shows absence of commercial oil deposits
$\tilde{f}_4$	Conduct seismic investigation and do not extract oil if seismic investigation shows absence of commercial oil deposits
$\tilde{f}_5$	Extract oil without seismic investigation
$\tilde{f}_6$	Abandon seismic investigation and oil extraction

In the problem, we have two types of events: geological events (states of the nature) - "occurrence of commercial oil deposits" ($\tilde{S}_1$) and "absence of commercial oil deposits" ($\tilde{S}_2$) and two seismic events (results of seismic investigation) - "seismic investigation shows occurrence of commercial oil deposits" (B_1) and "seismic investigation shows absence of commercial oil deposits" (B_2). Below we list possible combinations of geological and seismic events with fuzzy probabilities of their occurrence by taking into account NL-described information about accuracy of results of seismic investigation:

$B_1 / \tilde{S}_1$ - there are indeed commercial oil deposits and seismic investigation confirms their occurrence, $\tilde{P}(B_1 / \tilde{S}_1) = (0.7;0.8;0.9)$

$B_2 / \tilde{S}_1$ - there are indeed commercial oil deposits but seismic investigation shows their absence, $\tilde{P}(B_2 / \tilde{S}_1)$ is unknown;

$B_1/\tilde{S}_2$ - there are almost no commercial oil deposits but seismic investigation shows their occurrence, $\tilde{P}(B_1/\tilde{S}_2)$ is unknown;

$B_2/\tilde{S}_2$ - there are almost no commercial oil deposits and seismic investigation shows their absence, $\tilde{P}(B_2/\tilde{S}_2)=(0.6;0.7;0.8)$

According to (4.9)-(4.10) in Chapter 4 we have obtained unknown conditional probabilities $\tilde{P}(B_2/\tilde{S}_1)=(0.1;0.2;0.3)$ and $\tilde{P}(B_1/\tilde{S}_2)=(0.2;0.3;0.4)$.

Seismic investigation allows updating the prior knowledge about actual state of the nature with the purpose to obtain more credible information. Given a result of seismic investigation, the manager can revise prior probabilities of the states of the nature on the base of linguistic probabilities $\tilde{P}(B_j/\tilde{S}_k), k=\overline{1,2}, j=\overline{1,2}$ of possible combinations $\tilde{S}_k/B_j$ of geological and seismic events. These combinations are shown in Table 8.9.

Table 8.9 Possible combinations of seismic and geological events

Seismic events	Geological events	Notation
Seismic investigation shows occurrence of commercial oil deposits	occurrence of commercial oil deposits	$\tilde{S}_1/B_1$
Seismic investigation shows absence of commercial oil deposits	occurrence of commercial oil deposits	$\tilde{S}_1/B_2$
Seismic investigation shows occurrence of commercial oil deposits	absence of commercial oil deposits	$\tilde{S}_2/B_1$
Seismic investigation shows absence of commercial oil deposits	absence of commercial oil deposits	$\tilde{S}_2/B_2$

To revise probability of a state $\tilde{S}_k$ given seismic investigation result B_j we obtain a fuzzy posterior probability $\tilde{P}(\tilde{S}_k/B_j)$ of $\tilde{S}_k$ based on the fuzzy Bayes' formula (in α-cuts):

$$\tilde{P}^{\alpha}(\tilde{S}_k/B_j)=\left\{\frac{p_{jk}p_k}{\sum_{k=1}^{K}p_{jk}p_k}\,\middle|\,p_k\in P_k^{\alpha},p_{jk}\in P_{jk}^{\alpha},\sum_{k=1}^{K}p_k=1\right\}$$

The calculated $\tilde{P}(\tilde{S}_1/B_j)$, $j=\overline{1,2}$ are shown in Fig. 8.12 and 8.13.

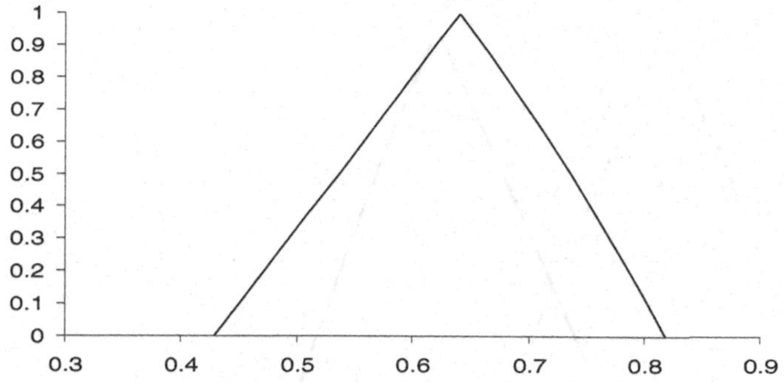

Fig. 8.12 Posterior probability $\tilde{P}(\tilde{S}_1 / B_1)$

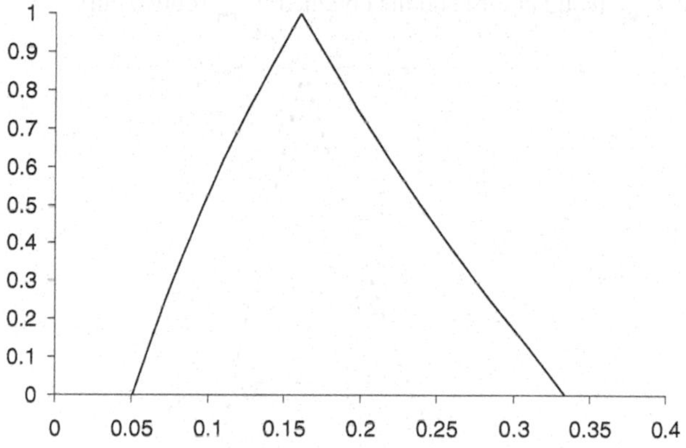

Fig. 8.13 Posterior probability $\tilde{P}(\tilde{S}_1 / B_2)$

Now we have new, revised (posterior) linguistic probabilities $\tilde{P}(\tilde{S}_1 / B_1)$ and $\tilde{P}(\tilde{S}_1 / B_2)$ for the state $\tilde{S}_1$ obtained on the basis of possible seismic investigation results B_1, B_2, respectively. We will denote them by $\tilde{P}(\tilde{S}_1 / B_1)$ and $\tilde{P}(S_1 / B_2)$ by $\tilde{P}_{1rev}^{B_1}$ and $\tilde{P}_{1rev}^{B_2}$, respectively. For these cases of unknown probabilities of absence of commercial oil deposits $\tilde{P}_{2rev}^{B_1}$ and $\tilde{P}_{2rev}^{B_2}$ we have obtained given seismic investigation results B_1, B_2. The membership functions for $\tilde{P}_{1rev}^{B_1}$ and $\tilde{P}_{2rev}^{B_1}$ are shown in the Fig. 8.14 and for $\tilde{P}_{1rev}^{B_2}$ and $\tilde{P}_{2rev}^{B_2}$ in Fig. 8.15. The membership functions for $\tilde{P}_{1rev}^{B_1}$ and $\tilde{P}_{2rev}^{B_1}$ are shown in the Fig. 8.14 and for $\tilde{P}_{1rev}^{B_2}$ and $\tilde{P}_{2rev}^{B_2}$ in Fig. 8.15.

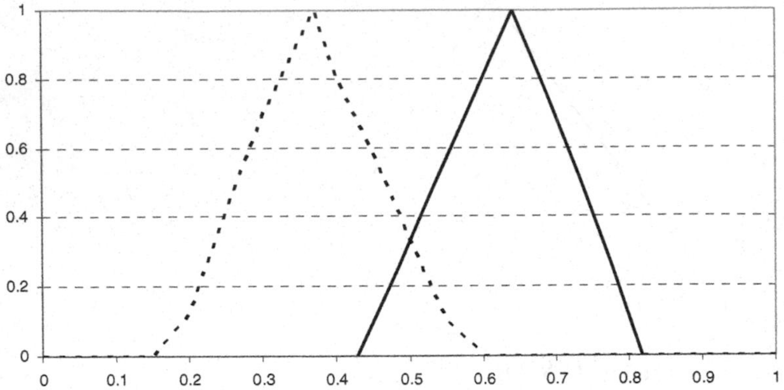

Fig. 8.14 Posterior probability $\tilde{P}_{1rev}^{B_1}$ (solid curve) and the obtained $\tilde{P}_{2rev}^{B_1}$ (dotted curve)

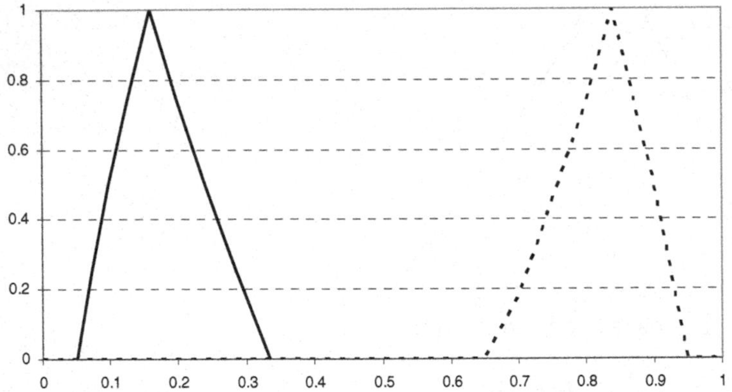

Fig. 8.15 Posterior probability $\tilde{P}_{1rev}^{B_2}$ (solid curve) and the obtained $\tilde{P}_{2rev}^{B_2}$ (dotted curve)

For actions $\tilde{f}_1, \tilde{f}_2, \tilde{f}_3, \tilde{f}_4$ depending on seismic investigations, the manager will use $\tilde{P}_{1rev}^{B_1}$ and $\tilde{P}_{2rev}^{B_1}$ or $\tilde{P}_{1rev}^{B_2}$ and $\tilde{P}_{2rev}^{B_2}$ instead of prior $\tilde{P}_1$.

For action $\tilde{f}_5$ the unknown $\tilde{P}_2$ is obtained from the $\tilde{P}_1$ as triangular fuzzy number (0.5;0.6;0.7) [5,15].

Now we have all required probabilities of the states $\tilde{S}_1$ and $\tilde{S}_2$. Assume that the manager evaluates utilities for various actions taken at various states of the nature from some scale. Because of incomplete and uncertain information about possible values of profit from oil sale and possible costs for seismic investigation and drilling of a well, the manager linguistically evaluates utilities for various actions taken at various states of the nature. Assume that the manager's linguistic utility evaluations for various actions taken at various states of the nature are as shown in the Table 8.10:

Table 8.10 Lingustic utility evaluations

	$\tilde{f}_1$	$\tilde{f}_2$	$\tilde{f}_3$	$\tilde{f}_4$	$\tilde{f}_5$	$\tilde{f}_6$
$\tilde{S}_1$	Positive significant	Negative very high	High	Negative very high	Positive high	0
$\tilde{S}_2$	Negative low	Negative very low	Negative low	Negative very low	Negative insignificant	

Below we give the representation of linguistic utilities for actions $\tilde{f}_i$ made at the states $\tilde{S}_k$ by triangular fuzzy numbers $\tilde{u}(\tilde{f}_i(\tilde{S}_k)) = \tilde{u}_{ik}$ defined on the scale $[-1,1]$:

$$\tilde{u}_{11} = (0.65;0.75;0.85)\,;\ \tilde{u}_{12} = (-0.11;-0.1;-0.09)\,;$$
$$\tilde{u}_{21} = (-0.88;-0.85;-0.82)\,;\ \tilde{u}_{22} = (-0.07;-0.04;-0.01)\,;$$
$$\tilde{u}_{31} = (0.65;0.75;0.85)\,;$$
$$\tilde{u}_{32} = (-0.11;-0.1;-0.09)\,;\ \tilde{u}_{41} = (-0.88;-0.85;-0.82)\,;$$
$$\tilde{u}_{42} = (-0.07;-0.04;-0.01)\,;$$
$$\tilde{u}_{51} = (0.7;0.8;0.9)\,;\ \tilde{u}_{52} = (-0.08;-0.07;-0.06)\,;\ \tilde{u}_6 = 0$$

To find the optimal action based on the methodology suggested in Section 4.1 we first calculate for each action $\tilde{f}_i$ its utility as a fuzzy-valued Choquet integral

$$\tilde{U}(\tilde{f}_i) = \int_S \tilde{u}(\tilde{f}_i(\tilde{S}))d\tilde{\eta}_{\tilde{P}^l}\,,$$

where $\tilde{\eta}_{\tilde{P}^l}$ is a fuzzy-valued fuzzy measure obtained from the linguistic probability distribution as a solution to the problem (4.13) – (4.14) (Chapter 4) based on the neuro-fuzzy-evolutionary technique covered in [5,15]. Let us note that depending upon actions, a fuzzy-valued measure will be constructed by considering either prior or posterior probability distributions. For actions $\tilde{f}_1, \tilde{f}_2$, fuzzy-valued measure will be constructed on the basis of $\tilde{P}^{B_1}_{rev}$ and for actions $\tilde{f}_3, \tilde{f}_4$ fuzzy-valued measure will be constructed based on $\tilde{P}^{B_2}_{rev}$ (as the seismic investigation has been involved). For action $\tilde{f}_5$ a fuzzy measure will be constructed on the basis of prior distribution. For action $\tilde{f}_6$ its utility, i.e. Choquet integral, is obviously equal to zero.

Fuzzy measures $\tilde{\eta}_1$ and $\tilde{\eta}_2$ defined on the base of $\tilde{P}^{B_1}_{rev}$ and $\tilde{P}^{B_2}_{rev}$ respectively are shown in Table 8.11:

Table 8.11 Fuzzy number-valued measures obtained from the posterior probabilities

$\mathcal{H} \subset \mathcal{S}$	$\{\tilde{S}_1\}$	$\{\tilde{S}_2\}$	$\{\tilde{S}_1, \tilde{S}_2\}$
$\tilde{\eta}_1(\mathcal{H})$	(0.43, 0.64, 0.64)	(0.18, 0.36, 0.36)	1
$\tilde{\eta}_2(\mathcal{H})$	(0.05, 0.16, 0.16)	(0.67, 0.84, 0.84)	1

The fuzzy-valued measure $\tilde{\eta}$ obtained on the base of prior probability is shown in Table 8.12:

Table 8.12 Fuzzy measure (approximated to triangular fuzzy numbers) obtained from the prior probabilities

$\mathcal{H} \subset \mathcal{S}$	$\{\tilde{S}_1\}$	$\{\tilde{S}_2\}$	$\{\tilde{S}_1, \tilde{S}_2\}$
$\tilde{\eta}(\mathcal{H})$	(0.3, 0.4,	(0.5, 0.6,	1

As utilities for $\tilde{u}_{ik}$ are fuzzy numbers, the corresponding values of Choquet integrals will also be fuzzy. We calculate a fuzzy utility for every action $\tilde{f}_i$ as a fuzzy value of a Choquet integral. Then using the fuzzy Jaccard compatibility-based ranking method, we find an action with the highest value of a fuzzy utility as an optimal one. The form of a Choquet integral for action $\tilde{f}_1$ reads as

$$\tilde{U}(\tilde{f}_1) = \sum_{i=1}^{2} \left(\tilde{u}(\tilde{f}_1(\tilde{S}_{(i)})) -_h \tilde{u}(\tilde{f}_1(\tilde{S}_{(i+1)}))\right)\tilde{\eta}_{\tilde{P}^l}\left(\mathcal{H}_{(i)}\right) =$$
$$= \left(\tilde{u}(\tilde{f}_1(\tilde{S}_{(1)})) -_h \tilde{u}(\tilde{f}_1(\tilde{S}_{(2)}))\right)\tilde{\eta}_1\left(\{\tilde{S}_{(1)}\}\right) + \tilde{u}(\tilde{f}_1(\tilde{S}_{(2)}))\tilde{\eta}_1\left(\{\tilde{S}_{(1)}, \tilde{S}_{(2)}\}\right)$$

As

$$\tilde{u}(\tilde{f}_1(\tilde{S}_{(1)})) = \tilde{u}_{11} = (0.65;0.75;0.85),\ \tilde{u}(\tilde{f}_1(\tilde{S}_2)) = \tilde{u}_{12} = (-0.11;-0.1;-0.09),$$

we find that $\tilde{u}_{11} \geq \tilde{u}_{12}$. Then $\tilde{u}(\tilde{f}_1(\tilde{S}_{(1)})) = \tilde{u}_{11}$, $\tilde{u}(\tilde{f}_1(\tilde{S}_{(2)})) = \tilde{u}_{12}$ and $\tilde{S}_{(1)} = \tilde{S}_1$, $\tilde{S}_{(2)} = \tilde{S}_2$. The Choquet integral for action f_1 is equal to

$$\tilde{U}(\tilde{f}_1) = \left(\tilde{u}(\tilde{f}_1(\tilde{S}_1)) -_h \tilde{u}(\tilde{f}_1(\tilde{S}_2))\right)\tilde{\eta}_1\left(\{\tilde{S}_1\}\right) + \tilde{u}(\tilde{f}_1(\tilde{S}_2))\tilde{\eta}_1\left(\{\tilde{S}_1, \tilde{S}_2\}\right) =$$
$$= (\tilde{u}_{11} -_h \tilde{u}_{12})\tilde{\eta}_1\left(\{\tilde{S}_1\}\right) + \tilde{u}_{12}\tilde{\eta}_1\left(\{\tilde{S}_1, \tilde{S}_2\}\right) = (\tilde{u}_{11} -_h \tilde{u}_{12})\tilde{\eta}_1\left(\{\tilde{S}_1\}\right) + \tilde{u}_{12} =$$
$$= \left((0.65;0.75;0.85) -_h (-0.11;-0.1;-0.09)\right)\cdot(0.43,\ 0.64,\ 0.64) + (-0.11;-0.1;-0.09)$$

The obtained result is approximated by a triangular fuzzy number comes as

$$\tilde{U}(\tilde{f}_1) = (0.2168, 0.444, 0.5116).$$

Based on this procedure, we have computed the fuzzy values of Choquet integrals also for the other actions $f_i, i = \overline{1,5}$ obtaining the following results:

$$\tilde{U}(\tilde{f}_2) = \sum_{i=1}^{2} \left(\tilde{u}(\tilde{f}_2(\tilde{S}_{(i)})) -_h \tilde{u}(\tilde{f}_2(\tilde{S}_{(i+1)})) \right) \tilde{\eta}_{\tilde{P}^l}\left(\mathcal{H}_{(i)}\right) =$$
$$= \left(\tilde{u}(\tilde{f}_2(\tilde{S}_{(1)})) -_h \tilde{u}(\tilde{f}_2(\tilde{S}_{(2)})) \right) \tilde{\eta}_1\left(\{\tilde{S}_{(1)}\}\right) + \tilde{u}(\tilde{f}_2(\tilde{S}_{(2)}))\tilde{\eta}_1\left(\{\tilde{S}_{(1)}, \tilde{S}_{(2)}\}\right) =$$
$$= \left(\tilde{u}(\tilde{f}_2(\tilde{S}_2)) -_h \tilde{u}(\tilde{f}_2(\tilde{S}_1)) \right) \tilde{\eta}_1\left(\{\tilde{S}_2\}\right) + \tilde{u}(\tilde{f}_2(\tilde{S}_1))\tilde{\eta}_1\left(\{\tilde{S}_1, \tilde{S}_2\}\right) =$$
$$= \left(\tilde{u}_{22} -_h \tilde{u}_{21} \right) \tilde{\eta}_1\left(\{\tilde{S}_2\}\right) + \tilde{u}_{21}\tilde{\eta}_1\left(\{\tilde{S}_1, \tilde{S}_2\}\right) =$$
$$= \left((-0.07;-0.04;-0.01) -_h (-0.88;-0.85;-0.82) \right) \cdot (0.67, 0.84, 0.84) +$$
$$+ (-0.88;-0.85;-0.82)$$

$$\tilde{U}(\tilde{f}_3) = \sum_{i=1}^{2} \left(\tilde{u}(\tilde{f}_3(\tilde{S}_{(i)})) -_h \tilde{u}(\tilde{f}_3(\tilde{S}_{(i+1)})) \right) \tilde{\eta}_{\tilde{P}^l}\left(\mathcal{H}_{(i)}\right) =$$
$$= \left(\tilde{u}(\tilde{f}_3(\tilde{S}_{(1)})) -_h \tilde{u}(\tilde{f}_3(\tilde{S}_{(2)})) \right) \tilde{\eta}_2\left(\{\tilde{S}_{(1)}\}\right) + \tilde{u}(\tilde{f}_3(\tilde{S}_{(2)}))\tilde{\eta}_2\left(\{\tilde{S}_{(1)}, \tilde{S}_{(2)}\}\right) =$$
$$= \left(\tilde{u}(\tilde{f}_3(\tilde{S}_1)) -_h \tilde{u}(\tilde{f}_3(\tilde{S}_2)) \right) \tilde{\eta}_2\left(\{\tilde{S}_1\}\right) + \tilde{u}(\tilde{f}_3(\tilde{S}_2))\tilde{\eta}_2\left(\{\tilde{S}_1, \tilde{S}_2\}\right) =$$
$$= \left(\tilde{u}_{31} -_h \tilde{u}_{32} \right) \tilde{\eta}_2\left(\{\tilde{S}_1\}\right) + \tilde{u}_{32}\tilde{\eta}_2\left(\{\tilde{S}_1, \tilde{S}_2\}\right) =$$
$$= \left((0.65;0.75;0.85) -_h (-0.11;-0.1;-0.09) \right) \cdot (0.43, 0.64, 0.64) +$$
$$+ (-0.11;-0.1;-0.09)$$

$$\tilde{U}(\tilde{f}_4) = \sum_{i=1}^{2} \left(\tilde{u}(\tilde{f}_4(\tilde{S}_{(i)})) -_h \tilde{u}(\tilde{f}_4(\tilde{S}_{(i+1)})) \right) \tilde{\eta}_{\tilde{P}^l}\left(\mathcal{H}_{(i)}\right) =$$
$$= \left(\tilde{u}(\tilde{f}_4(\tilde{S}_{(1)})) -_h \tilde{u}(\tilde{f}_4(\tilde{S}_{(2)})) \right) \tilde{\eta}_2\left(\{\tilde{S}_{(1)}\}\right) + \tilde{u}(\tilde{f}_4(\tilde{S}_{(2)}))\tilde{\eta}_2\left(\{\tilde{S}_{(1)}, \tilde{S}_{(2)}\}\right) =$$
$$= \left(\tilde{u}(\tilde{f}_4(\tilde{S}_2)) -_h \tilde{u}(\tilde{f}_4(\tilde{S}_1)) \right) \tilde{\eta}_2\left(\{\tilde{S}_2\}\right) + \tilde{u}(\tilde{f}_4(\tilde{S}_1))\tilde{\eta}_2\left(\{\tilde{S}_1, \tilde{S}_2\}\right) =$$
$$= \left(\tilde{u}_{42} -_h \tilde{u}_{41} \right) \tilde{\eta}_2\left(\{\tilde{S}_2\}\right) + \tilde{u}_{21}\tilde{\eta}_2\left(\{\tilde{S}_1, \tilde{S}_2\}\right) =$$
$$= \left((-0.07;-0.04;-0.01) -_h (-0.88;-0.85;-0.82) \right) \cdot (0.67, 0.84, 0.84) +$$
$$+ (-0.88;-0.85;-0.82)$$

$$\tilde{U}(\tilde{f}_5)=\sum_{i=1}^{2}\left(\tilde{u}(\tilde{f}_5(\tilde{S}_{(i)}))-_h\tilde{u}(\tilde{f}_5(\tilde{S}_{(i+1)}))\right)\tilde{\eta}_{\tilde{P}^l}\left(\mathcal{H}_{(i)}\right)=$$
$$=\left(\tilde{u}(\tilde{f}_5(\tilde{S}_{(1)}))-_h\tilde{u}(\tilde{f}_5(\tilde{S}_{(2)}))\right)\tilde{\eta}\left(\left\{\tilde{S}_{(1)}\right\}\right)+\tilde{u}(\tilde{f}_5(\tilde{S}_{(2)}))\tilde{\eta}\left(\left\{\tilde{S}_{(1)},\tilde{S}_{(2)}\right\}\right)=$$
$$=\left(\tilde{u}(\tilde{f}_5(\tilde{S}_1))-_h\tilde{u}(\tilde{f}_5(\tilde{S}_2))\right)\tilde{\eta}\left(\left\{\tilde{S}_1\right\}\right)+\tilde{u}(\tilde{f}_5(\tilde{S}_2))\tilde{\eta}\left(\left\{\tilde{S}_1,\tilde{S}_2\right\}\right)=$$
$$=\left(\tilde{u}_{51}-_h\tilde{u}_{52}\right)\tilde{\eta}\left(\left\{\tilde{S}_1\right\}\right)+\tilde{u}_{52}\tilde{\eta}\left(\left\{\tilde{S}_1,\tilde{S}_2\right\}\right)=$$
$$=\left((0.7;0.8;0.9)-_h(-0.08;-0.07;-0.06)\right)\cdot(0.3, 0.4, 0.4)+$$
$$+(-0.08;-0.07;-0.06)$$

The obtained results approximated by triangular fuzzy numbers are

$$\tilde{U}(\tilde{f}_2)=(\text{-}0.5317,\text{-}0.3316,\text{-}0.3016)\,;\tilde{U}(\tilde{f}_3)=(-0.072,0.036,0.0604)\,;$$
$$\tilde{U}(\tilde{f}_4)=(\text{-}0.8395,\text{-}0.7204,\text{-}0.6904)\,;\ \tilde{U}(\tilde{f}_5)=(0.154,0.278,0.324)\,.$$

These fuzzy numbers are shown in Fig.8.16

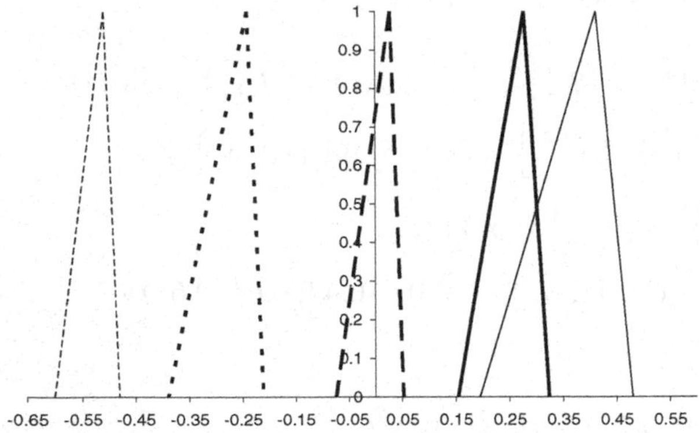

Fig. 8.16 Fuzzy values of Choquet integral for possible actions (for $\tilde{f}_1$ - thin solid line, for $\tilde{f}_2$ thick dotted line, for $\tilde{f}_3$ - thick dashed line, for $\tilde{f}_4$ - thin dashed line, for $\tilde{f}_5$ - thick solid line)

As it can be seen the highest fuzzy utilities are those of alternatives $\tilde{f}_1$ and $\tilde{f}_5$. The application of the fuzzy Jaccard compatibility-based ranking method [13] to compare the fuzzy utility values for $\tilde{f}_1$ and $\tilde{f}_5$ gave rise to the following results:

$\tilde{U}(\tilde{f}_1)\geq\tilde{U}(\tilde{f}_5)$ is satisfied with the degree 0.8748;

$\tilde{U}(\tilde{f}_5)\geq\tilde{U}(\tilde{f}_1)$ is satisfied with the degree 0.1635;

The best action is $\tilde{f}_1$ "Conduct seismic investigation and extract oil if seismic investigation shows occurrence of commercial oil deposits" as one with the highest fuzzy utility value being equal to $\tilde{U}(\tilde{f}_1) = (0.1953, 0.412, 0.4796)$.

We also considered the possibility of applying the classical (non-fuzzy) Choquet expected utility model to solve this problem. But as this model cannot take into account NL-described information about utilities, states, probabilities etc., we had to use only numerical information. Assume that the manager assigns numeric subjective probabilities and numerical utilities. Let us suppose that the manager assigned the following numerical values:

$$u(f_1(s_1)) = 0.6\,;\, u(f_1(s_2)) = -0.15\,;\, u(f_2(s_1)) = -0.86\,;\, u(f_2(s_2)) = -0.038$$
$$u(f_3(s_1)) = 0.6\,;\, u(f_3(s_2)) = -0.15\,;\, u(f_4(s_1)) = -0.86\,;\, u(f_4(s_2)) = -0.038\,;$$
$$u(f_5(s_1)) = 0.9\,;\, u(f_5(s_2)) = -0.08\,;\, P(s_1) = 0.35,\, P(b_1 / s_1) = 0.8\,;$$
$$P(b_2 / s_2) = 0.7\,;\, P(b_2 / s_1) = 0.2\,;\, P(b_1 / s_2) = 0.3\,.$$

To find the best alternative we used Choquet expected utility with possibility measure and obtained the following results for utilities of actions:

$$U(f_1) = 0.2775\,;\, U(f_2) = -0.51\,;\, U(f_3) = -0.05\,;\, U(f_4) = \text{-}0.05\,;$$
$$U(f_5) = 0.2825\,.$$

The best action is f_5 – "Extract oil without seismic investigation" as one with the highest utility value $U(f_5) = 0.2825$. This result differs from the above one we obtained when applying the suggested model given the NL-described information. "Extract oil without seismic investigation" appeared to be the best alternative despite of the fact that the probability of occurrence of commercial oil deposits is not high: $P(s_1) = 0.35$. The reason for this is that an assignment of subjective numeric values to probabilities and utilities leads to the loss of important partial information. In turn, this loss of information may result in choosing a decision that may not be suitable.

We also solved the considered problem with other initial information provided by the manager. When the information is *"probability of "occurrence of commercial oil deposits" is low"*, where *"low"* is described by $\tilde{P}_1 = (0.3;0.4;0.5)$ the obtained results being approximated by triangular fuzzy numbers are as follows:

$$\tilde{U}(\tilde{f}_1) = (-0.11, 0.082857, 0.1134)\,;\, \tilde{U}(\tilde{f}_2) = (\text{-}0.294, \text{-}0.168, \text{-}0.138)\,;$$
$$\tilde{U}(\tilde{f}_3) = (-0.11, \text{-}0.07538, \text{-}0.0626)\,;\, \tilde{U}(\tilde{f}_4) = (\text{-}0.1322, \text{-}0.0572, \text{-}0.0272)\,;$$
$$\tilde{U}(\tilde{f}_5) = (-0.08, 0.017, 0.036)\,.$$

These fuzzy numbers are shown in Fig. 8.17:

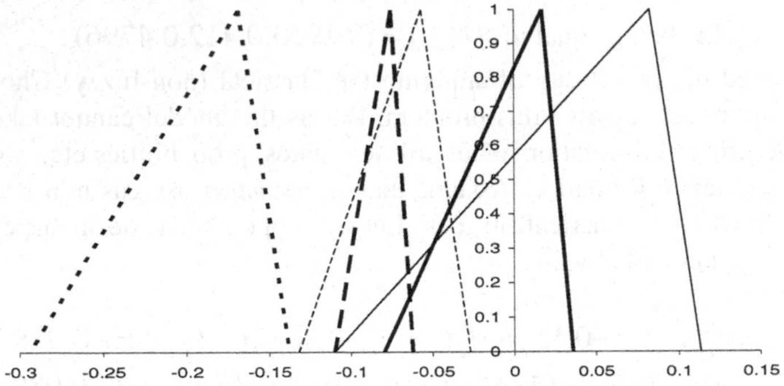

Fig. 8.17 Fuzzy values of Choquet integral for possible actions (for $\tilde{f}_1$ - thin solid line, for $\tilde{f}_2$ - thick dotted line, for $\tilde{f}_3$ - thick dashed line, for $\tilde{f}_4$ - thin dashed line, for $\tilde{f}_5$ - thicksolid line)

As can be seen, the best actions are $\tilde{f}_1$ and $\tilde{f}_5$. The application of the fuzzy Jaccard compatibility-based ranking method to compare the fuzzy utility values for $\tilde{f}_1$ and $\tilde{f}_5$ resulted in the following:

$\tilde{U}(\tilde{f}_1) \geq \tilde{U}(\tilde{f}_5)$ is satisfied with the degree 0.9902;
$\tilde{U}(\tilde{f}_5) \geq \tilde{U}(\tilde{f}_1)$ is satisfied with the degree 0.4193;

Here the best action is $\tilde{f}_1$ "Conduct seismic investigation and extract oil if seismic investigation shows occurrence of commercial oil deposits" as the one with the highest fuzzy utility value $\tilde{U}(\tilde{f}_1) = (0.1953, 0.412, 0.4796)$. This is due to the fact that the probability of occurrence of commercial oil deposits is so low that it is better to begin to extract oil only if seismic investigation shows their occurrence.

When the *"probability of "oil's occurrence" is high"* with *"high"* described as $\tilde{P}_1 = (0.8; 0.9; 1)$ the obtained results approximated by triangular fuzzy numbers are the following:

$$\tilde{U}(\tilde{f}_1) = (0.539, 0.668, 0.764);$$
$$\tilde{U}(\tilde{f}_2) = (-0.63, -0.5776, -0.5476);$$
$$\tilde{U}(\tilde{f}_3) = (0.127, 0.476, 0.5508);$$
$$\tilde{U}(\tilde{f}_4) = (-0.63, -0.4432, -0.4132);$$
$$\tilde{U}(\tilde{f}_5) = (0.544, 0.713, 0.804).$$

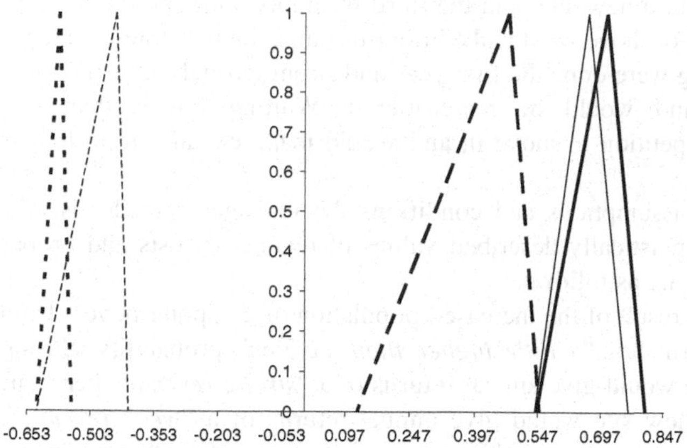

Fig. 8.18 Fuzzy values of Choquet integral for possible actions (for $\tilde{f}_1$ - thin solid line, for $\tilde{f}_2$ - thick dotted line, for $\tilde{f}_3$ - thick dashed line, for $\tilde{f}_4$ - thin dashed line, for $\tilde{f}_5$ - thick solid line)

Here again the highest fuzzy utilities are those of $\tilde{f}_1$ and $\tilde{f}_5$ alternatives. The use of the fuzzy Jaccard compatibility-based ranking method [13] to compare the fuzzy utility values for $\tilde{f}_1$ and $\tilde{f}_5$ produced the following results:

$\tilde{U}(\tilde{f}_1) \geq \tilde{U}(\tilde{f}_5)$ is satisfied with the degree 0.6164;

$\tilde{U}(\tilde{f}_5) \geq \tilde{U}(\tilde{f}_1)$ is satisfied with the degree 0.9912;

The best action is $\tilde{f}_5$ "Extract oil without seismic investigation" as one with the highest fuzzy utility value $\tilde{U}(\tilde{f}_5) = (0.544, 0.713, 0.804)$. This is due to the fact that the probability of occurrence of commercial oil deposits is so high that it is more reasonable not to spend money on seismic investigation and to begin to extract oil.

8.4 Applications in Business and Economics

Business Development For a Computer Firm [2,15]

A manager of a computer firm needs to make a decision concerning his business over the next five years. There has been good sales growth over the past couple of years. The owner sees three options. The first is to enlarge the current store, the second is to move to a new site, and the third is simply wait and do nothing. The decision to expand or move would take little time, and, so, the store would not lose revenue. If nothing were done the first year and strong growth occurred, then the decision to expand would be reconsidered. Waiting longer than one year would allow competition to move in and would make expansion no longer feasible.

The description of assumptions and conditions the manager consider is NL-based and includes linguistically described values of revenues, costs and probabilities for the problem are as follows:

"Strong growth as a result of the increased population of computer buyers from the electronics new firm has "*a little higher than medium*" probability. Strong growth with a new site would give annual returns of a "*strong revenue*" per year. Weak growth with a new site would give annual returns of a "*weak revenue*". Strong growth with an expansion would give annual returns of an "*about strong revenue*" per year. Weak growth with an expansion would mean annual returns of a "*lower than weak*". At the existing store with no changes, there would be returns of a "*medium revenue*" per year if there is strong growth and of a "*higher than weak revenue*" per year if growth is weak. Expansion at the current site would cost "*low*". The move to the new site would cost "*high*". If growth is strong and the existing site is enlarged during the second year, the cost would still be "*low*". Operating costs for all options are equal."

The considered decision problem is characterized by linguistically described imperfect information on all its elements: alternatives, outcomes, states of nature and probabilities. Really, in a broad variety of economic problems decision relevant information can not be represented by precise evaluations. In contrast, we deal with evaluations with blurred, not sharply defined boundaries. Moreover, the considered problem is characterized by a second-order uncertainty represented as a mix of probabilistic and fuzzy uncertainties in form of fuzzy probabilities. Such essence of the considered problem makes it impossible to apply the existing decision theories for solving it. In order to apply any existing theory, one needs to precisiate the linguistic relevant information to precise numbers that leads to distortion and loss of information. As a result, one will arrive at a decision problem which is not equivalent to the original one and the resulted solution will not be trustful. For solving this problem it is more adequate to apply decision theory suggested in Chapter 4 which is able to deal with linguistic information on all elements of decision problems. Below we consider application of this theory to solving the considered problem.

Below it is given representation of linguistic evaluations of revenues and costs in form of triangular fuzzy numbers.

Revenues: "strong revenue" = (175.5;195;214.5); "weak revenue" = (103.5; 115; 126.5); "about strong revenue" = (171; 190; 209); lower than weak" = =(90;100;110); "medium revenue" = (153;170;187); "higher than weak revenue"= (94.5;105;115.5).

Costs: "high"=(189;210;231); "low"=(78.3; 87; 95.7); "zero"=(0;0;5)

In Table 8.13. it is given linguistic description of revenues, costs and final values (outcomes) for each alternative decision.

Fuzzy values of the outcomes, calculated using the formula **"Value of an outcome = revenue – cost"** on the base of the fuzzy values of revenues and costs are given below.

Values of outcomes: "about large" = (646.5;765;841.5); "about medium" = (286.5;365;401.5); "a little large"= (759.3;863;949.3); "medium" = (354.3; 413;454.3); "large2"= (741.3;843;927;3); "large3"= (760;850;930); "higher than medium" = (472.5;525;577.5)

The set of the fuzzy states of the nature is $\mathcal{S}=\{\tilde{S}_1,\tilde{S}_2\}$, where $\tilde{S}_1$ - "strong growth", $\tilde{S}_2$ - "weak growth". The membership functions of $\tilde{S}_1$ and $\tilde{S}_2$ are shown in the Fig. 8.19:

The linguistic probability distribution $\tilde{P}^l$ over the states of the nature that corresponds the knowledge the manager has is:

$$\tilde{P}^l = \tilde{P}_1/\tilde{S}_1 + \tilde{P}_2/\tilde{S}_2$$

$\tilde{P}_1 =$ "a little higher than medium", described by triangular fuzzy number (0.45;0.55;0.65) and $\tilde{P}_2$ is unknown.

Table 8.13 Linguistic description of revenues, costs and final values

Alternative	Revenue	Cost	Value of an outcome = revenue – cost
Move to new location, strong growth	"strong revenue" x 5 yrs	"high"	"about large"
Move to new locations, weak growth	"weak revenue" x 5 yrs	"high"	"about medium"
Expand store, strong growth	"about strong revenue" x 5 yrs	"low"	"large1"
Expand store, weak growth	"lower than weak" x 5 yrs	"low"	"medium"
Do nothing now, strong growth, expand next year	"medium revenue" x 1 yr + + "about strong revenue" x 4 yrs	"low"	"large3"
Do nothing now, strong growth, do not expand next year	"medium revenue" x 5 yr	"zero"	"large2"
Do nothing now, wegrowth	"higher than weak revenue"	"zero"	"higher than medium"

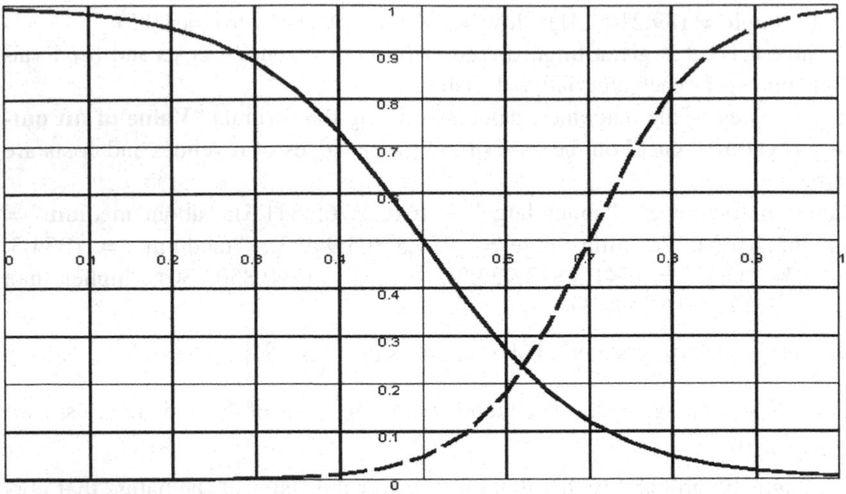

Fig. 8.19 Fuzzy states of the nature: $\tilde{S}_1$ (dashed), $\tilde{S}_2$ (solid)

The set of the manager's possible actions is $\mathcal{A}=\{\tilde{f}_1,\tilde{f}_2,\tilde{f}_3\}$, where $\tilde{f}_1$ denotes "move" decision, $\tilde{f}_2$ denotes "expand" decision, $\tilde{f}_3$ denotes "do nothing" decision.

Fuzzy probabilities $\tilde{P}_1$(given) and $\tilde{P}_2$ (obtained on the base of solving the problem (4.9)-(4.10) (Chapter 4) are given in the Fig.8.20

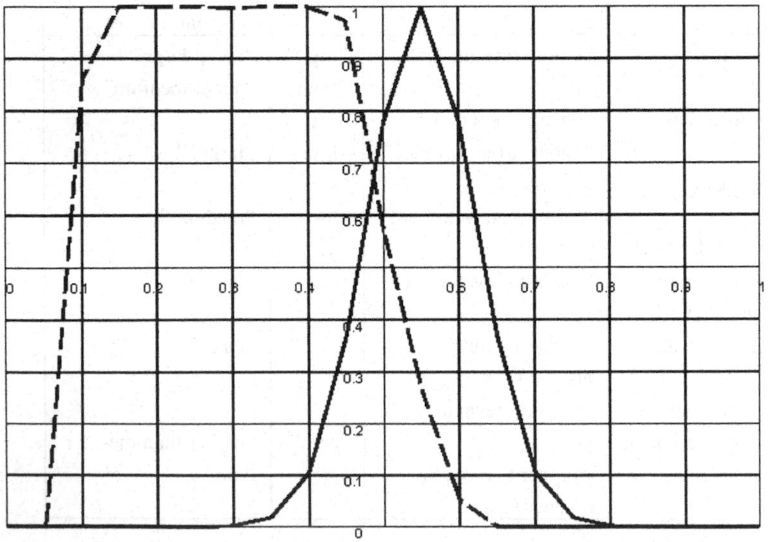

Fig. 8.20 The given (solid curve) and the obtained (dashed curve) fuzzy probabilities

The defuzzified values of the fuzzy-valued measure $\tilde{\eta}_{\tilde{P}^l}$ obtained from $\tilde{P}^l$ on the base of formulas (4.20)-(4.21) (Chapter 4) by using the approach suggested in [5,12] are given in Table 8.14. The defuzzified values are used for simplicity of further calculations.

Utility values are calculated as follows:

$$\tilde{U}(\tilde{f})=\tilde{\eta}_{\tilde{P}^l}(\{\tilde{S}_{(1)},\tilde{S}_{(2)}\})\tilde{u}(\tilde{f}(\tilde{S}_{(2)}))+\tilde{\eta}_{\tilde{P}^l}(\{\tilde{S}_{(1)}\})(\tilde{u}(\tilde{f}(\tilde{S}_{(1)}))-\tilde{u}(\tilde{f}(\tilde{S}_{(2)})))$$

Table 8.14 Fuzzy measure $\tilde{\eta}_{\tilde{P}^l}$

$\mathcal{H}$	$\{\tilde{S}_1\}$	$\{\tilde{S}_2\}$	$\{\tilde{S}_1,\tilde{S}_2\}$
$\tilde{\eta}_{\tilde{P}^l}(\mathcal{H})$	0.47	0.38	1

The calculated fuzzy utility values of $\tilde{f}_i, i=\overline{1,3}$ are

$$\tilde{U}(\tilde{f}_1)=(419.52;\ 516.2;\ 612.8)\,;\tilde{U}(\tilde{f}_2)=(480.5;\ 583.1;\ 684,7)\,;$$
$$\tilde{U}(\tilde{f}_3)=(559.4;\ 647.85;\ 750.65)\,.$$

These fuzzy utility values are shown in Fig. 8.21 below:

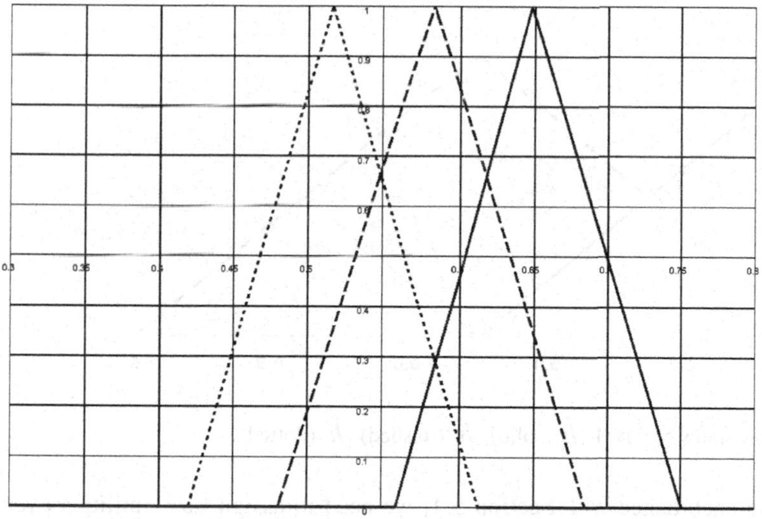

Fig. 8.21 Fuzzy utility values of the manager's possible actions: $\tilde{U}(\tilde{f}_1)$ (dotted), $\tilde{U}(\tilde{f}_2)$ (dashed), $\tilde{U}(\tilde{f}_3)$ (solid)

Using the Fuzzy Jaccard compatibility-based ranking method we got the following results on ranking the alternatives:

$$\tilde{U}(\tilde{f}_1) \geq \tilde{U}(\tilde{f}_2) \text{ is satisfied with the degree 0.377;}$$
$$\tilde{U}(\tilde{f}_2) \geq \tilde{U}(\tilde{f}_1) \text{ is satisfied with the degree 0.651;}$$
$$\tilde{U}(\tilde{f}_2) \geq \tilde{U}(\tilde{f}_3) \text{ is satisfied with the degree 0.3032;}$$
$$\tilde{U}(\tilde{f}_3) \geq \tilde{U}(\tilde{f}_2) \text{ is satisfied with the degree 0.6965.}$$

As can be seen, the best alternative is $\tilde{f}_3$ - “Do nothing”.

Behavioral Decision Making in Investment Problem. In this section we will consider behavioral decision making conditioned by risk attitudes in an investment problem. We will consider a DM’s states as fuzzy sets to describe impreciseness of concepts of “risk seeking”, “risk averse” and “risk neutrality” as these concepts are rather qualitative and are a matter of a degree.

Let us denote the considered three fuzzy states of a DM as follows: $\tilde{h}_1$ – risk aversion, $\tilde{h}_2$ – risk neutrality, and $\tilde{h}_3$ – risk seeking. Fuzziness of these states reflects impreciseness of the information on the each of the risk attitudes levels. The fuzzy states $\tilde{h}_1$, $\tilde{h}_2$, $\tilde{h}_3$ are given in the Fig.8.22:

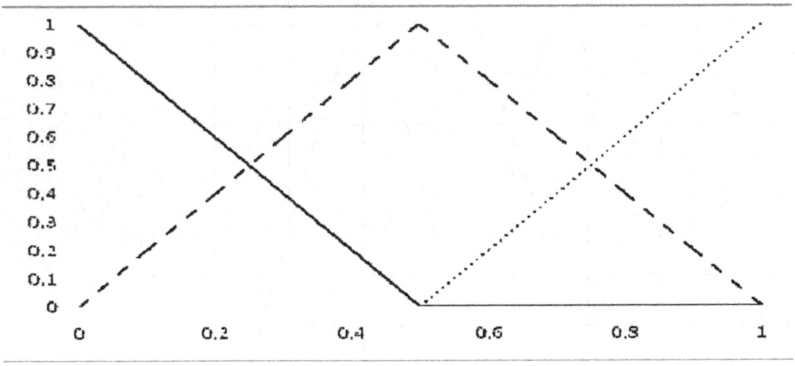

Fig. 8.22 Fuzzy states of a DM: $\tilde{h}_1$ (solid), $\tilde{h}_2$ (dashed), $\tilde{h}_3$ (dotted)

Further, in accordance with Section 5.1, we need to assign fuzzy utilities over combined states space Ω. We obtain fuzzy utilities $\tilde{u}(\tilde{f}(\tilde{S}_i, \tilde{h}_j))$ on the base of the Zadeh’s extension principle and the ideas used in Prospect theory.

The next step in solving the considered problem is to determine FJPs for the combined states from fuzzy marginal probabilities over the states of nature and the DM's states. As the majority of people are risk averse, we consider a DM with $\tilde{h}_1$ (risk aversion) as the most probable state. Let the fuzzy probabilities of fuzzy states $\tilde{h}_1$, $\tilde{h}_2$, $\tilde{h}_3$ be defined as follows: $\tilde{P}(\tilde{h}_1)$ ("high" – given), $\tilde{P}(\tilde{h}_2)$ ("small" – given), $\tilde{P}(\tilde{h}_3)$ ("very small" –computed) – see Fig.8.23:

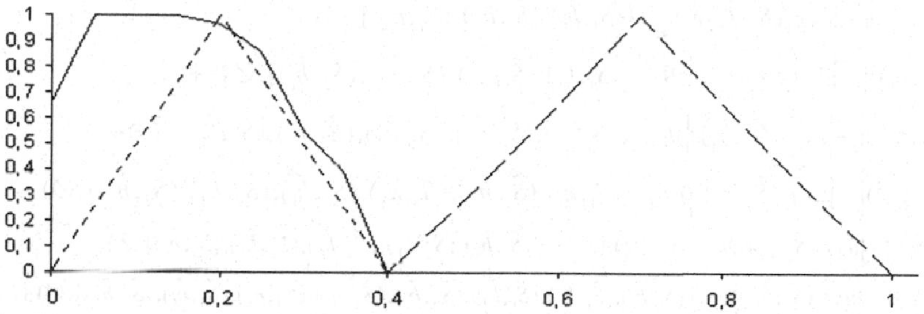

Fig. 8.23 Fuzzy probabilities: $\tilde{P}(\tilde{h}_1)$ (dashed), $\tilde{P}(\tilde{h}_2)$ (dotted), $\tilde{P}(\tilde{h}_3)$ (solid)

We determine FJPs of the combined states $(\tilde{S}_i, \tilde{h}_j)$ from the fuzzy marginal probabilities on the base of the notions of dependence between events by assuming positive dependence between "risk aversion" state and gains and also between "risk seeking" state and losses, negative dependence – between "risk aversion" state and losses and also between "risk seeking" state and gains, and independence between "risk neutrality" and both gains and losses. For example, the FJPs of combined states for the case of $\tilde{f}_1$ action (common bonds) are described by the following trapezoidal fuzzy numbers:

$$\tilde{P}(\tilde{S}_1,\tilde{h}_1)=(0.12,0.35,0.5,0.7),\tilde{P}(\tilde{S}_2,\tilde{h}_1)=(0.04,0.21,0.3,0.5),$$
$$\tilde{P}(\tilde{S}_3,\tilde{h}_1)=(0,0.105,0.15,0.3),\tilde{P}(\tilde{S}_4,\tilde{h}_1)=(0,0.035,0.15,0.3);$$

$$\tilde{P}(\tilde{S}_1,\tilde{h}_2)=(0,0.1,0.1,0.28),\tilde{P}(\tilde{S}_2,\tilde{h}_2)=(0,0.06,0.06,0.2),$$
$$\tilde{P}(\tilde{S}_3,\tilde{h}_2)=(0,0.03,0.03,0.12),\tilde{P}(\tilde{S}_4,\tilde{h}_2)=(0,0.01,0.03,0.12);$$

$$\tilde{P}(\tilde{S}_1,\tilde{h}_3)=(0,0,0.05,0.28),\tilde{P}(\tilde{S}_2,\tilde{h}_3)=(0,0,0.03,0.2),$$
$$\tilde{P}(\tilde{S}_3,\tilde{h}_3)=(0,0,0.015,0.12),\tilde{P}(\tilde{S}_4,\tilde{h}_3)=(0,0,0.015,0.12).$$

Given these FJPs, in accordance with Chapter 5, we need to determine fuzzy-valued bi-capacity $\tilde{\eta}_{\tilde{P}^l}(\cdot,\cdot)$ over subsets of the combined states space to be used in determination of the fuzzy overall utility for each alternative. The fuzzy overall utility $\tilde{U}(\tilde{f}_1)$ is expressed as follows:

$$\tilde{U}(\tilde{f}_1)=(\left|\tilde{u}(\tilde{f}_1(\tilde{S}_1,\tilde{h}_1))\right|-\left|\tilde{u}(\tilde{f}_1(\tilde{S}_1,\tilde{h}_2))\right|)(\tilde{\eta}_{\tilde{P}^l}(\{(\tilde{S}_1,\tilde{h}_1)\},\varnothing))+$$

$$(\left|\tilde{u}(\tilde{f}_1(\tilde{S}_1,\tilde{h}_2))\right|-\left|\tilde{u}(\tilde{f}_1(\tilde{S}_1,\tilde{h}_3))\right|)(\tilde{\eta}_{\tilde{P}^l}(\{(\tilde{S}_1,\tilde{h}_1),(\tilde{S}_1,\tilde{h}_2)\},\varnothing))$$

$$(\left|\tilde{u}(\tilde{f}_1(\tilde{S}_1,\tilde{h}_3))\right|-\left|\tilde{u}(\tilde{f}_1(\tilde{S}_2,\tilde{h}_1))\right|)\tilde{\eta}_{\tilde{P}^l}(\{(\tilde{S}_1,\tilde{h}_1),(\tilde{S}_1,\tilde{h}_2),(\tilde{S}_1,\tilde{h}_3)\},\varnothing)+$$

$$(\left|\tilde{u}(\tilde{f}_1(\tilde{S}_2,\tilde{h}_1))\right|-\left|\tilde{u}(\tilde{f}_1(\tilde{S}_2,\tilde{h}_2))\right|)\tilde{\eta}_{\tilde{P}^l}(\{(\tilde{S}_1,\tilde{h}_1),(\tilde{S}_1,\tilde{h}_2),(\tilde{S}_1,\tilde{h}_3),(\tilde{S}_2,\tilde{h}_1)\},\varnothing\})+$$

$$(\left|\tilde{u}(\tilde{f}_1(\tilde{S}_2,\tilde{h}_2))\right|-\left|\tilde{u}(\tilde{f}_1(\tilde{S}_2,\tilde{h}_3))\right|)\tilde{\eta}_{\tilde{P}^l}(\{(\tilde{S}_1,\tilde{h}_1),(\tilde{S}_1,\tilde{h}_2),(\tilde{S}_1,\tilde{h}_3),(\tilde{S}_2,\tilde{h}_1),(\tilde{S}_2,\tilde{h}_2)\},\varnothing)+$$

$$(\left|\tilde{u}(\tilde{f}_1(\tilde{S}_2,\tilde{h}_3))\right|-\left|\tilde{u}(\tilde{f}_1(\tilde{S}_3,\tilde{h}_1))\right|)\tilde{\eta}_{\tilde{P}^l}(\{(\tilde{S}_1,\tilde{h}_1),(\tilde{S}_1,\tilde{h}_2),(\tilde{S}_1,\tilde{h}_3),(\tilde{S}_2,\tilde{h}_1),(\tilde{S}_2,\tilde{h}_2),(\tilde{S}_2,\tilde{h}_3)\},\varnothing)+$$

$$(\left|\tilde{u}(\tilde{f}_1(\tilde{S}_3,\tilde{h}_1))\right|-\left|\tilde{u}(\tilde{f}_1(\tilde{S}_3,\tilde{h}_2))\right|)\tilde{\eta}_{\tilde{P}^l}(\{(\tilde{S}_1,\tilde{h}_1),(\tilde{S}_1,\tilde{h}_2),(\tilde{S}_1,\tilde{h}_3),(\tilde{S}_2,\tilde{h}_1),(\tilde{S}_2,\tilde{h}_2),(\tilde{S}_2,\tilde{h}_3),(\tilde{S}_3,\tilde{h}_1)\},\varnothing)+$$

$$(\left|\tilde{u}(\tilde{f}_1(\tilde{S}_3,\tilde{h}_2))\right|-\left|\tilde{u}(\tilde{f}_1(\tilde{S}_3,\tilde{h}_3))\right|)\tilde{\eta}_{\tilde{P}^l}(\{\tilde{S}_1,\tilde{h}_1),(\tilde{S}_1,\tilde{h}_2),(\tilde{S}_1,\tilde{h}_3),(\tilde{S}_2,\tilde{h}_1),(\tilde{S}_2,\tilde{h}_2),(\tilde{S}_2,\tilde{h}_3),(\tilde{S}_3,\tilde{h}_1),(\tilde{S}_3,\tilde{h}_2)\},\varnothing)+$$

$$(\left|\tilde{u}(\tilde{f}_1(\tilde{S}_3,\tilde{h}_1))\right|-\left|\tilde{u}(\tilde{f}_1(\tilde{S}_3,\tilde{h}_2))\right|)\tilde{\eta}_{\tilde{P}^l}(\{(\tilde{S}_1,\tilde{h}_1),(\tilde{S}_1,\tilde{h}_2),(\tilde{S}_1,\tilde{h}_3),(\tilde{S}_2,\tilde{h}_1),(\tilde{S}_2,\tilde{h}_2),(\tilde{S}_2,\tilde{h}_3),(\tilde{S}_3,\tilde{h}_1)\},\varnothing+$$

$$(\left|\tilde{u}(\tilde{f}_1(\tilde{S}_3,\tilde{h}_2))\right|-\left|\tilde{u}(\tilde{f}_1(\tilde{S}_3,\tilde{h}_3))\right|)\tilde{\eta}_{\tilde{P}^l}(\{\tilde{S}_1,\tilde{h}_1),(\tilde{S}_1,\tilde{h}_2),(\tilde{S}_1,\tilde{h}_3),(\tilde{S}_2,\tilde{h}_1),(\tilde{S}_2,\tilde{h}_2),(\tilde{S}_2,\tilde{h}_3),(\tilde{S}_3,\tilde{h}_1),(\tilde{S}_3,\tilde{h}_2)\},\varnothing+$$

$$(\left|\tilde{u}(\tilde{f}_1(\tilde{S}_3,\tilde{h}_3))\right|-\left|\tilde{u}(\tilde{f}_1(\tilde{S}_4,\tilde{h}_1))\right|)\tilde{\eta}_{\tilde{P}^l}(\{\tilde{S}_1,\tilde{h}_1),(\tilde{S}_1,\tilde{h}_2),(\tilde{S}_1,\tilde{h}_3),(\tilde{S}_2,\tilde{h}_1),(\tilde{S}_2,\tilde{h}_2),(\tilde{S}_2,\tilde{h}_3),(\tilde{S}_3,\tilde{h}_1),(\tilde{S}_3,\tilde{h}_2),(\tilde{S}_3,\tilde{h}_3)\},\varnothing+$$

$$(\left|\tilde{u}(\tilde{f}_1(\tilde{S}_4,\tilde{h}_1))\right|-\left|\tilde{u}(\tilde{f}_1(\tilde{S}_4,\tilde{h}_2))\right|\tilde{\eta}_{\tilde{P}^l}(\{(\tilde{S}_1,\tilde{h}_1),(\tilde{S}_1,\tilde{h}_2),(\tilde{S}_1,\tilde{h}_3),(\tilde{S}_2,\tilde{h}_1),(\tilde{S}_2,\tilde{h}_2),(\tilde{S}_2,\tilde{h}_3),(\tilde{S}_3,\tilde{h}_1),(\tilde{S}_3,\tilde{h}_2),(\tilde{S}_3,\tilde{h}_3),(\tilde{S}_4,\tilde{h}_1)\},\varnothing=$$

$$(\left|\tilde{u}(\tilde{f}_1(\tilde{S}_4,\tilde{h}_2))\right|-\left|\tilde{u}(\tilde{f}_1(\tilde{S}_4,\tilde{h}_3))\right|\tilde{\eta}_{\tilde{P}^l}(\{(\tilde{S}_1,\tilde{h}_1),(\tilde{S}_1,\tilde{h}_2),(\tilde{S}_1,\tilde{h}_3),(\tilde{S}_2,\tilde{h}_1),(\tilde{S}_2,\tilde{h}_2),(\tilde{S}_2,\tilde{h}_3),(\tilde{S}_3,\tilde{h}_1),(\tilde{S}_3,\tilde{h}_2),(\tilde{S}_3,\tilde{h}_3),(\tilde{S}_4,\tilde{h}_1),(\tilde{S}_4,\tilde{h}_2)\},\varnothing=$$

$$\left|\tilde{u}(\tilde{f}_1(\tilde{S}_4,\tilde{h}_3))\right|\tilde{\eta}_{\tilde{P}^l}(\{(\tilde{S}_1,\tilde{h}_1),(\tilde{S}_1,\tilde{h}_2),(\tilde{S}_1,\tilde{h}_3),(\tilde{S}_2,\tilde{h}_1),(\tilde{S}_2,\tilde{h}_2),(\tilde{S}_2,\tilde{h}_3),(\tilde{S}_3,\tilde{h}_1),(\tilde{S}_3,\tilde{h}_2),(\tilde{S}_3,\tilde{h}_3),(\tilde{S}_4,\tilde{h}_1),(\tilde{S}_4,\tilde{h}_2),(\tilde{S}_4,\tilde{h}_3)\},\varnothing=$$

$$=(\left|\tilde{u}(\tilde{f}_1(\tilde{S}_1,\tilde{h}_3))\right|-\left|\tilde{u}(\tilde{f}_1(\tilde{S}_2,\tilde{h}_1))\right|)\tilde{\eta}_{\tilde{P}^l}(\{(\tilde{S}_1,\tilde{h}_1),(\tilde{S}_1,\tilde{h}_2),(\tilde{S}_1,\tilde{h}_3)\},\varnothing+$$

$$+(\left|\tilde{u}(\tilde{f}_1(\tilde{S}_2,\tilde{h}_3))\right|-\left|\tilde{u}(\tilde{f}_1(\tilde{S}_3,\tilde{h}_1))\right|)\tilde{\eta}_{\tilde{P}^l}(\{(\tilde{S}_1,\tilde{h}_1),(\tilde{S}_1,\tilde{h}_2),(\tilde{S}_1,\tilde{h}_3),(\tilde{S}_2,\tilde{h}_1),(\tilde{S}_2,\tilde{h}_2),(\tilde{S}_2,\tilde{h}_3)\},\varnothing+$$

$$+(\left|\tilde{u}(\tilde{f}_1(\tilde{S}_3,\tilde{h}_3))\right|-\left|\tilde{u}(\tilde{f}_1(\tilde{S}_4,\tilde{h}_1))\right|)\tilde{\eta}_{\tilde{P}^l}(\{\tilde{S}_1,\tilde{h}_1),(\tilde{S}_1,\tilde{h}_2),(\tilde{S}_1,\tilde{h}_3),(\tilde{S}_2,\tilde{h}_1),(\tilde{S}_2,\tilde{h}_2),(\tilde{S}_2,\tilde{h}_3),(\tilde{S}_3,\tilde{h}_1),(\tilde{S}_3,\tilde{h}_2),(\tilde{S}_3,\tilde{h}_3)\},\varnothing+$$

$$\left|\tilde{u}(\tilde{f}_1(\tilde{S}_4,\tilde{h}_3))\right|\tilde{\eta}_{\tilde{P}^l}(\{(\tilde{S}_1,\tilde{h}_1),(\tilde{S}_1,\tilde{h}_2),(\tilde{S}_1,\tilde{h}_3),(\tilde{S}_2,\tilde{h}_1),(\tilde{S}_2,\tilde{h}_2),(\tilde{S}_2,\tilde{h}_3),(\tilde{S}_3,\tilde{h}_1),(\tilde{S}_3,\tilde{h}_2),(\tilde{S}_3,\tilde{h}_3),(\tilde{S}_4,\tilde{h}_1),(\tilde{S}_4,\tilde{h}_2),(\tilde{S}_4,\tilde{h}_3)\},\varnothing)$$

We will determine the fuzzy-valued bi-capacity $\tilde{\eta}_{\tilde{P}^l}(\cdot,\cdot)$ as $\tilde{\eta}_{\tilde{P}^l}(\mathcal{V},\mathcal{W})=\tilde{\eta}_{\tilde{P}^l}(\mathcal{V})-\tilde{\eta}_{\tilde{P}^l}(\mathcal{W})$, $\mathcal{V},\mathcal{W}\in\Omega$, where $\tilde{\eta}_{\tilde{P}^l}$ is the fuzzy-valued lower probability constructed from the FJPs (as it was done in the previous example). The calculated fuzzy overall utilities $\tilde{U}(\tilde{f}_1),\tilde{U}(\tilde{f}_2),\tilde{U}(\tilde{f}_3)$ approximated by TFNs are given in the Fig.8.24:

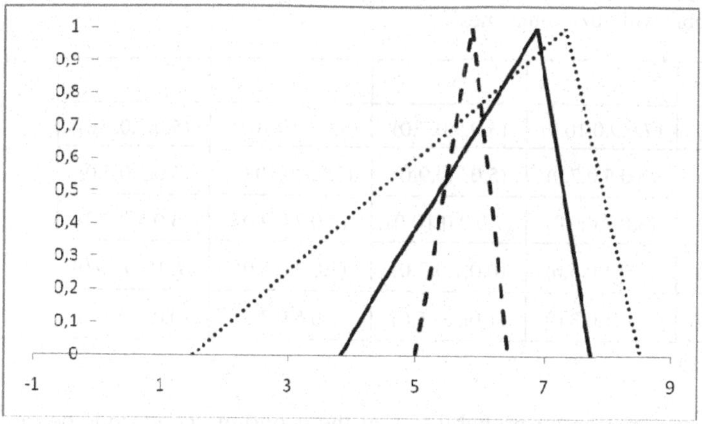

Fig. 8.24 Fuzzy overall utilities $\tilde{U}(\tilde{f}_1)$ (solid) $\tilde{U}(\tilde{f}_2)$ (dashed), $\tilde{U}(\tilde{f}_3)$ (dotted)

Conducting pairwise comparison by using formula (5.2) (see Chapter 5), we got the following results:

$Deg(\tilde{f}_1 \tilde{\succsim}_l \tilde{f}_2) = 0.038$;
$Deg(\tilde{f}_2 \tilde{\succsim}_l \tilde{f}_1) = 0.0247$;
$Deg(\tilde{f}_1 \tilde{\succsim}_l \tilde{f}_3) = 0.026$;
$Deg(\tilde{f}_3 \succ_l \tilde{f}_1) = 0.01$;
$Deg(\tilde{f}_2 \succ_l \tilde{f}_3) = 0.035$;
$Deg(\tilde{f}_3 \succ_l \tilde{f}_2) = 0.027$.

So, as $Deg(\tilde{f}_1 \succ_l \tilde{f}_i) > Deg(\tilde{f}_i \succ_l \tilde{f}_1),\ i = 2,3$, then the best solution is $\tilde{f}_1$.

The Supplier Selection Problem. Consider a problem of decision making with imperfect information as a problem of a supplier selection [14] by taking into account possible economic conditions. The set of alternatives is represented by a set of five suppliers: $\mathcal{A} = \{\tilde{f}_1, \tilde{f}_2, \tilde{f}_3, \tilde{f}_4, \tilde{f}_5\}$. The set of states of nature is represented by five possible economic conditions: $S = \{s_1, s_2, s_3, s_4, s_5\}$. Each economic condition s_j is characterized by requirements to profitability, relationship closeness, technological capability, conformance quality and conflict resolution aspects of a supplier. For simplicity, states of nature are considered in classical sense. Payoff table containing fuzzy evaluations of outcomes of the alternatives under the economic conditions is given below (Table 8.15):

Table 8.15 Payoff table with fuzzy outcomes

	s_1	s_2	s_3	s_4	s_5
$\tilde{f}_1$	(5.0,7.0,9.0)	(7.0,9.0,10.0)	(3.0,5.0,7.0)	(9.0,10.0,10.0)	(5.0,7.0,9.0)
$\tilde{f}_2$	(1.0,3.0,5.0)	(3.0,5.0,7.0)	(5.0,7.0,9.0)	(7.0,9.0,10.0)	(1.0,3.0,5.0)
$\tilde{f}_3$	(3.0,5.0,7.0)	(5.0,7.0,9.0)	(7.0,9.0,10.0)	(5.0,7.0,9.0)	(3.0,5.0,7.0)
$\tilde{f}_4$	(0.0,1.0,3.0)	(1.0,3.0,5.0)	(0.0,1.0,3.0)	(1.0,3.0,5.0)	(7.0,9.0,10.0)
$\tilde{f}_5$	(7.0,9.0,10.0)	(0.0,1.0,3.0)	(1.0,3.0,5.0)	(3.0,5.0,7.0)	(0.0,1.0,3.0)

Let linguistic information on probabilities of the economic conditions be described as follows:

$$\tilde{P}^l = (0.2,\ 0.3,\ 0.4)/s_1 + (0.1,\ 0.2,\ 0.3)/s_2 + (0.0,\ 0.1,\ 0.2)/s_3 + \\ +(0.3,\ 0.3,\ 0.5)/s_4 + (0.0,\ 0.1,\ 0.4)/s_5$$

It is more convenient to apply fuzzy optimality-based approach to decision making under imperfect information without a utility function which is presented in Chapter 4. Application of this approach to the considered problem will provide valid results at a notably less complex and time consuming computations.

By applying fuzzy optimality concept-based approach, the considered problem can be solved as follows. At first, according to (4.50)-(4.52), we calculated nbF, neF, nwF:

$$nbF = \begin{bmatrix} 0 & 0.28667 & 0.21 & 0.55667 & 0.42667 \\ 0.02 & 0 & 0.056667 & 0.35667 & 0.28 \\ 0.033333 & 0.14667 & 0 & 0.41333 & 0.31 \\ 0.02 & 0.086667 & 0.053333 & 0 & 0.15667 \\ 0.046667 & 0.16667 & 0.10667 & 0.31333 & 0 \end{bmatrix},$$

$$neF = \begin{bmatrix} 5 & 4.6933 & 4.7567 & 4.4233 & 4.5267 \\ 4.6933 & 5 & 4.7967 & 4.5567 & 4.5533 \\ 4.7567 & 4.7967 & 5 & 4.5333 & 4.5833 \\ 4.4233 & 4.5567 & 4.5333 & 5 & 4.53 \\ 4.5267 & 4.5533 & 4.5833 & 4.53 & 5 \end{bmatrix},$$

$$nwF = \begin{bmatrix} 0 & 0.02 & 0.033333 & 0.02 & 0.046667 \\ 0.28667 & 0 & 0.14667 & 0.086667 & 0.16667 \\ 0.21 & 0.056667 & 0 & 0.053333 & 0.10667 \\ 0.55667 & 0.35667 & 0.41333 & 0 & 0.31333 \\ 0.42667 & 0.28 & 0.31 & 0.15667 & 0 \end{bmatrix}.$$

Next we calculated $\mu_{\tilde{D}}$ and $d(\tilde{f}_i, \tilde{f}_k)$ (see formulas (4.8),(4.6)):

$$\mu_{\tilde{D}} = \begin{bmatrix} 0.5 & 0.47333 & 0.48233 & 0.44633 & 0.462 \\ 0.52667 & 0.5 & 0.509 & 0.473 & 0.48867 \\ 0.51767 & 0.491 & 0.5 & 0.464 & 0.47967 \\ 0.55367 & 0.527 & 0.536 & 0.5 & 0.51567 \\ 0.538 & 0.51133 & 0.52033 & 0.48433 & 0.5 \end{bmatrix},$$

$$d(\tilde{f}_i, \tilde{f}_k) = \begin{bmatrix} 0 & 0.93023 & 0.84127 & 0.96407 & 0.89062 \\ 0 & 0 & 0 & 0.75701 & 0.40476 \\ 0 & 0.61364 & 0 & 0.87097 & 0.65591 \\ 0 & 0 & 0 & 0 & 0 \\ 0 & 0 & 0 & 0.5 & 0 \end{bmatrix}.$$

Finally we calculated degree of optimality for each of the considered alternatives (see formula (4.56)):

$$do = \begin{bmatrix} 1 \\ 0.069767 \\ 0.15873 \\ 0.035928 \\ 0.10938 \end{bmatrix}.$$

So, the preferences obtained are: $\tilde{f}_1 \succsim \tilde{f}_3 \succsim \tilde{f}_5 \succsim \tilde{f}_2 \succsim \tilde{f}_4$. The degrees of preferences are the following:

$$Deg(\tilde{f}_1 \succsim_l \tilde{f}_3) = 0.84,$$
$$Deg(\tilde{f}_3 \succsim_l \tilde{f}_5) = 0.05,$$
$$Deg(\tilde{f}_5 \succsim_l \tilde{f}_2) = 0.04,$$
$$Deg(\tilde{f}_2 \succsim_l \tilde{f}_4) = 0.034.$$

We have also solved this problem by applying the method suggested in [14].

The preference obtained by this method is: $\tilde{f}_1 \succsim \tilde{f}_3 \succsim \tilde{f}_2 \succsim \tilde{f}_5 \succsim \tilde{f}_4$. This is almost the same ordering as that obtained by the method suggested in this paper. However, the suggested method, as compared to the method in [14] has several advantages. The first is that our method not only determines ordering among alternatives, but also determines to what degree the considered alternative is optimal. This degree is an overall extent to which the considered alternative is better than all the other. The second when one has alternatives which are equivalent according to the method in [14] , by using our approach it is possible to differentiate them into more and less optimal ones by determining the corresponding value of k_F . Given these two advantages, the method suggested in Section 4.3 is of almost the same computational complexity as the method in [14].

Decision Making in a Hotel Management [7]

A management of a hotel should make a decision concerning a construction of an additional wing. The alternatives are buildings with 30 (f_1), 40 (f_2) and 50 (f_3) rooms. The results of each decision depend on a combination of local government legislation and competition in the field. With respect to this, three states of nature are considered: positive legislation and low competition (s_1), positive legislation and strong competition (s_2), no legislation and low competition (s_3). The outcomes (results) of each decision are values of anticipated payoffs (in percentage) described by Z-numbers. The problem is to find how many rooms to build in order to maximize the return on investment.

Z -information for the utilities of the each act taken at various states of nature and probabilities on states of nature are provided in Table 8.16, Table 8.17, respectively.

Table 8.16 The utility values of actions under various states

	s_1	s_2	s_3
f_1	(high; likely)	(below than high; likely)	(medium; likely)
f_2	(below than high; likely)	(low; likely)	(below than high; likely)
f_3	(below than high; likely)	(high; likely)	(medium; likely)

Table 8.17 The values of probabilities of states of nature

$P(s_1)$ =(medium; quite sure)	$P(s_2)$ =(more than medium; quite sure)	$P(s_2)$ =(low; quite sure)

Here $\tilde{Z}_{v_{sj}(f_i(s_j))} = (\tilde{v}_{s_j}(f_i(s_j)), \tilde{R}_1)$, where the outcomes and corresponding reliability are the trapezoidal fuzzy numbers:

$Z_{11} = (A_{11}, B_{11}) = \tilde{Z}_{v_{s1}(f_1(s_1))} = (\tilde{v}_{s_1}(f_1(s_1)), \tilde{R}_1) =$ {high; likely} =[(0.0, 0.8, 0.9, 1.0), (0.0, 0.7, 0.7, 0.8)],

$Z_{12} = (A_{12}, B_{12}) = \tilde{Z}_{v_{s2}(f_1(s_2))} = (\tilde{v}_{s_2}(f_1(s_2)), \tilde{R}_1) =$ {below than high; likely} =[(0.0, 0.7, 0.8, 1.0), (0.0, 0.7, 0.7, 1.0)],

$Z_{13} = (A_{13}, B_{13}) = \tilde{Z}_{v_{s3}(f_1(s_3))} = (\tilde{v}_{s_3}(f_1(s_3)), \tilde{R}_1) =$ {medium; likely} = =[(0.0, 0.5, 0.6, 1.0), (0.0, 0.7, 0.7, 1.0)],

$Z_{21} = (A_{21}, B_{21}) = \tilde{Z}_{v_{s1}(f_2(s_1))} = (\tilde{v}_{s_1}(f_2(s_1)), \tilde{R}_1) =$ {below than high; likely} =[(0.6, 0.7, 0.8, 1.0), (0.0, 0.7, 0.7, 1.0)],

$Z_{22} = (A_{22}, B_{22}) = \tilde{Z}_{v_{s2}(f_2(s_2))} = (\tilde{v}_{s_2}(f_2(s_2)), \tilde{R}_1) =$ {low; likely} = =[(0.0, 0.4, 0.5, 1.0), (0.0, 0.7, 0.7, 1.0)],

$Z_{23} = (A_{23}, B_{23}) = \tilde{Z}_{v_{s3}(f_2(s_3))} = (\tilde{v}_{s_3}(f_2(s_3)), \tilde{R}_1) =$ {below than high; likely} =[(0.0, 0.7, 0.8, 1.0), (0.0, 0.7, 0.7, 1.0)],

$Z_{31} = (A_{31}, B_{31}) = \tilde{Z}_{v_{s1}(f_3(s_1))} = (\tilde{v}_{s_1}(f_3(s_1)), \tilde{R}_1) =$ {below than high; likely} =[(0., 0.7, 0.8, 1.0), (0.0, 0.7, 0.7, 1.0)],

$Z_{32} = (A_{32}, B_{32}) = \tilde{Z}_{v_{s2}(f_3(s_2))} = (\tilde{v}_{s_2}(f_3(s_2)), \tilde{R}_1) =$ {high; likely} = =[(0.0, 0.8, 0.9, 1.0), (0.0, 0.7, 0.7, 1.0)],

$Z_{33} = (A_{33}, B_{33}) = \tilde{Z}_{v_{s3}(f_3(s_3))} = (\tilde{v}_{s_3}(f_3(s_3)), \tilde{R}_1) =$ {medium; likely} = =[(0.0, 0.5, 0.6, 1.0), (0.0, 0.7, 0.7, 1.0)].

Let the probabilities for s_1 and s_2 be Z-numbers $\tilde{Z}_{P(s_j)} = (\tilde{P}(s_j)), \tilde{R}_2)$, where the probabilities and the corresponding reliability are the triangular fuzzy numbers:

$Z_{41} = (A_{41}, B_{41}) = \tilde{Z}_{P(s_1)} = (\tilde{P}(s_1)), \tilde{R}_2)$ =(medium; quite sure) = =[(0.0, 0.3, 0.3, 1.0), (0.0, 0.9, 0.9, 1.0)].

$Z_{42} = (A_{42}, B_{42}) = \tilde{Z}_{P(s_2)} = (\tilde{P}(s_2)), \tilde{R}_2)$ =(more than medium quite sure) = =[(0.0, 0.4, 0.4, 1.0), (0.0, 0.9, 0.9, 1.0)].

In accordance with [3] we have calculated probability for s_3:

$Z_{43} = (A_{43}, B_{43}) = \tilde{Z}_{P(s_3)} = (\tilde{P}(s_3)), \tilde{R}_2)$ =(low; quite sure) = =[(0.0, 0.3, 0.3, 1.0), (0.0, 0.9, 0.9, 1)].

As it is shown in section 1.1. a Z-number (A,B) can be interpreted as $\text{Prob}(U\ is\ A)\ is\ B$.

This expresses that we do not know the true probability density over U, but have a constraint in form of a fuzzy subset P of the space $\mathbf{P}$ of all probability densities over U. This restriction induces a fuzzy probability B. Let p *be* density function over U. The probability $\text{Prob}_p(U\ is\ A)$ (probability that $U\ is\ A$) is determined on the base of the definition of the probability of a fuzzy subset as

$$\text{Prob}_\text{p}(U\ is\ A) = \int_{-\infty}^{+\infty} \mu_A(u)\, p_U(u)du\,.$$

Then the degree to which p satisfies the Z-valuation $\text{Prob}_p(U\ is\ A)\ is\ B$ is

$$\mu_P(p) = \mu_B(\text{Prob}_p(U\ is\ A)) = \mu_B\Big(\int_{-\infty}^{+\infty} \mu_A(u)\, p_U(u)du\Big)\,.$$

Here p is taken as some a parametric distribution. The density function of a normal distribution is

$$p_U(u) = normpdf(u,m,\sigma)) = \frac{1}{\sigma\sqrt{2\pi}}\exp\left(-\frac{(u-m)^2}{2\sigma^2}\right).$$

In this situation, for any m, σ we have

$$\text{Prob}_{m,\sigma}(U\ is\ A) = \int_{-\infty}^{+\infty} \mu_A(u)\, p_{m,\sigma}(u)du = \int_{-\infty}^{+\infty} \mu_A(u)\frac{1}{\sigma\sqrt{2\pi}}\exp\left(\frac{(u-m)^2}{2\sigma^2}\right)du =$$
$$= quad(trapmf(u,[a_1,a_{2,},a_3,a_4])*normpdf(u,m,\sigma),-\text{inf},+\text{inf})$$

Then the space $\mathbf{P}$ of probability distributions will be the class of all normal distributions each uniquely defined by its parameters m,σ.

Let $U = (A_U, B_U)$ and $V = (A_V, B_V)$ be two independent Z-numbers. Consider determination of $W = U + V$. First, we need compute $A_U + A_V$ using Zadeh's extension principle:

$$\mu_{(A_U+A_V)}(w) = \sup_u(\mu_{A_U}(u) \wedge \mu_{A_V}(w-u)),\ \ \wedge = \min\,.$$

As the sum of random variables involves the convolution of the respective density functions we can construct $\tilde{P}_W$, the fuzzy subset of $\mathbf{P}$, associated with the random variable W. Recall that the convolution of density functions p_1 and p_2 is defined as the density function

$$p = p_1 \oplus p_2$$

such that

$$p(w) = \int_{-\infty}^{+\infty} p_1(u)p_2(w-u)du = \int_{-\infty}^{+\infty} p_1(w-u)p_2(u)du$$

One can then find the fuzzy subset $\tilde{P}_W$. For any $p_W \in \mathbf{P}$, one obtains

$$\mu_{P_W}(p_W) = \max_{p_U, p_V}[\mu_{P_U}(p_U) \wedge \mu_{P_V}(p_V)],$$

subject to

$$p_W = p_U \oplus p_V,$$

that is,

$$p_W(w) = \int_{-\infty}^{+\infty} p_U(u)p_V(w-u)du = \int_{-\infty}^{+\infty} p_U(w-u)p_V(u)du.$$

Given $\mu_{P_U}(p_U) = \mu_{P_U}(m_U, \sigma_U)$ and $\mu_{P_V}(p_V) = \mu_{P_V}(m_V, \sigma_V)$ as

$$\mu_{P_U}(m_U, \sigma_U) = \mu_{B_U}\left(\int_{-\infty}^{+\infty} \mu_{A_U}(u)\frac{1}{\sigma_U\sqrt{2\pi}}\exp\left(\frac{(u-m_U)^2}{2\sigma_U{}^2}\right)du,\right.$$

$$\mu_{P_V}(m_V, \sigma_V) = \mu_{B_V}\left(\int_{-\infty}^{+\infty} \mu_{A_V}(u)\frac{1}{\sigma_V\sqrt{2\pi}}\exp\left(\frac{(u-m_V)^2}{2\sigma_V{}^2}\right)du\right.$$

one can define $\tilde{P}_W$ as follows

$$p_W = p_{m_U,\sigma_U} \oplus p_{m_V,\sigma_V},$$

$$p_W(w) = p_{m_W,\sigma_W} = normpdf[w, m_W, \sigma_W] =$$

$$= quad(normpdf(u, m_U, \sigma_U) * normpdf(w-u, m_V, \sigma_V), -\text{inf}, +\text{inf}) =$$

$$= \int_{-\infty}^{+\infty} \frac{1}{\sigma_U\sqrt{2\pi}}\exp\left(\frac{(u-m_U)^2}{2\sigma_U{}^2}\right)\frac{1}{\sigma_V\sqrt{2\pi}}\exp\left(\frac{(w-u-m_V)^2}{2\sigma_V{}^2}\right)du$$

where

$$m_W = m_U + m_V \text{ and } \sigma_W = \sqrt{\sigma_U^2 + \sigma_V^2},$$
$$\mu_{\tilde{P}_W}(p_W) = \sup(\mu_{\tilde{P}_U}(p_U) \wedge \mu_{\tilde{P}_V}(p_V))$$

subject to

$$p_W = p_{m_U,\sigma_U} \oplus p_{m_V,\sigma_V}$$

B_W is found as follows.

$$\mu_{B_W}(b_W) = \sup(\mu_{\tilde{P}_W}(p_W))$$

subject to

$$b_W = \int_{-\infty}^{+\infty} p_W(w)\mu_{A_W}(w)dw$$

Let us now consider determination of $W = U \cdot V$.

$A_U \cdot A_V$ is defined by:

$$\mu_{(A_U \cdot A_V)}(w) = \sup_u(\mu_{A_U}(u) \wedge \mu_{A_V}(\frac{w}{u})), \quad \wedge = \min .$$

the probability density p_W associated with W is obtained as

$$p_W = p_{m_U,\sigma_U} \otimes p_{m_V,\sigma_V},$$

$$p_W(w) = p_{m_W,\sigma_W} = \int_{-\infty}^{+\infty} \frac{1}{\sigma_U\sqrt{2\pi}} \exp\left(\frac{(u-m_U)^2}{2\sigma_U{}^2}\right) \frac{1}{\sigma_V\sqrt{2\pi}} \exp\left(\frac{(\frac{w}{u}-m_V)^2}{2\sigma_V{}^2}\right) du$$

where

$$m_W = \frac{m_U m_V}{\sigma_U \sigma_V} + r,$$

and

$$\sigma_W = \frac{\sqrt{m_U^2\sigma_V^2 + m_V^2\sigma_U^2 + 2m_U m_V \sigma_U \sigma_V r + \sigma_U^2\sigma_V^2 + \sigma_U^2\sigma_V^2 r^2}}{\sigma_U \sigma_V}.$$

Where r is correlation coefficient.

If U and V are two independent random variables, then

$$m_W = \frac{m_U m_V}{\sigma_U \sigma_V}, \text{ and } \sigma_W = \frac{\sqrt{m_U^2\sigma_V^2 + m_V^2\sigma_U^2 + \sigma_U^2\sigma_V^2}}{\sigma_U \sigma_V}.$$

If take into account compatibility conditions $\sigma_U \sigma_V = 1$.

The other steps are analogous to those of determination of $W = U + V$.

Following the proposed decision making method, and using given Z - information we get the values of utility and their reliabilites for acts f_1, f_2, f_3

$$f_1 \succ f_3 \succ f_2$$

Decision Making Problem of One-Product Dynamic Economic Model [11]

In the present study a problem of optimal control for a single-product dynamic macroeconomic model is considered. In this model gross domestic product is divided into productive consumption, gross investment and nonproductive consumption. The model is described by a fuzzy differential equation (FDE) to take into account imprecision inherent in the dynamics that may be naturally conditioned by influence of various external factors, unforeseen contingencies of future etc [8,9,11]. The considered problem is characterized by four criteria. Application of the classical Pareto optimality principle for solving such problems leads to a large Pareto optimal set that complicates determination of a solution. We applied fuzzy Pareto optimality (FPO) formalism to solve the considered problem that allows to softly narrow a Pareto optimal set by determining degrees of optimality for considered alternatives. Five optimal alternatives are obtained with their degrees of optimality and the alternative with the degree of optimality equal to one is taken as the solution of the considered problem stability. The applied approach is characterized by a low computational complexity as compared with the existing decision making methods for solving multiobjective optimal control problems.

Statement of the Problem. Let us consider a single-product dynamical macroeconomic model which reflects interaction between factors of production when a gross domestic product (GDP) is divided into productive consumption, gross investment and non-productive consumption as the performance of production activity. In its turn productive consumption is assumed to be completely consumed on capital formation and depreciation. These processes are complicated by a presence of possibilistic, that is, fuzzy uncertainty which is conditioned by imprecise evaluation of future trends, unforeseen contingencies and other vagueness and impreciseness inherent in economical processes. Under the above mentioned assumptions the considered dynamical economical model can be described by the following fuzzy differential equation:

$$\frac{d\tilde{K}}{dt}=\frac{1}{q}\left((1-a)\tilde{u}_1-\mu\tilde{K}-\tilde{u}_2\right) \tag{8.1}$$

Here $\tilde{K}$ is a fuzzy variable describing imprecise information on capital, i.e. fuzzy value of capital, $\tilde{u}_1$ is a fuzzy value of GDP (the first control variable), $\tilde{u}_2$ (the second control variable) – is a fuzzy value of a non-productive consumption, $a,\mu,q>0$ are coefficients related to the productive consumption, net capital formation and depreciation respectively.

Let us consider a multiobjective optimal control problem of (8.1) within the period of planning $[t_0,T]$ with four objective functions (criteria): profit ($\tilde{J}_1$), reduction of production expenditures of GDP($\tilde{J}_2$), a value of capital at the end of period $[t_0,T]$($\tilde{J}_3$), a discount sum of a direct consumption over $[t_0,T]$($\tilde{J}_4$). The considered fuzzy multiobjective optimal control problem is formulated below:

$$\begin{aligned}&\sup_{\tilde{\mathbf{u}}\in\mathcal{U}}(\tilde{J}_1(\tilde{\mathbf{u}})=\int_{t_0}^{T}p(t)\tilde{u}_2(t)dt,\tilde{J}_2(\tilde{\mathbf{u}})=-c\int_{t_0}^{T}|\tilde{u}_1(t)|dt,\tilde{J}_3(\tilde{\mathbf{u}})=\\&\tilde{K}(T),\tilde{J}_4(\tilde{\mathbf{u}})=\int_{t_0}^{T}\theta(t)\tilde{u}_2(t)dt))^T\end{aligned} \tag{8.2}$$

subject to

$$\begin{aligned}&\frac{d\tilde{K}}{dt}=\frac{1}{q}\left((1-a)\tilde{u}_1-\mu\tilde{K}-\tilde{u}_2\right),\tilde{K}(t)\in\mathcal{E}^1,t\in[t_0,T],\ \tilde{K}(t_0)=\tilde{K}_0,\\&\tilde{K}(T)\in\mathcal{K}(T),\\&\mathcal{K}(T)=\{\tilde{K}\in\mathcal{E}^1:\tilde{K}_*\le\tilde{K}(T)\le\tilde{K}^*\},\\&\tilde{\mathbf{u}}=(\tilde{u}_1,\tilde{u}_2)^T\in\mathcal{U}\subset\mathcal{E}^2,\\&\mathcal{U}=\mathcal{U}_1\times\mathcal{U}_2,\\&\mathcal{U}_1=\{\tilde{u}_1\in\mathcal{E}^1:\tilde{u}_{1*}\le\tilde{u}_1(t)\le\tilde{u}_1^*\},\\&\mathcal{U}_2=\{\tilde{u}_2\in\mathcal{E}^1:\tilde{u}_{2*}\le\tilde{u}_2(t)\le\tilde{u}_2^*\}\end{aligned} \tag{8.3}$$

Here $p(t)$ is the price of production unit, produced at the time t, $\theta(t)$ - discount function, $c = const > 0$, $[t_0, T]$ is a term of forecasting (or planning).

Method of Solution. The considered problem is solved as follows. At the first stage it is necessary to determine feasible area defined by (8.3). Each feasible solution is represented by two control actions $\tilde{u}_1(t,\tilde{K})$ and $\tilde{u}_2(t,\tilde{K})$ for which (8.3) are satisfied. We choose $\tilde{u}_1(t,\tilde{K})$ and $\tilde{u}_2(t,\tilde{K})$ as

$$\tilde{u}_1(t,\tilde{K}) = a_1\tilde{K} + b_1, \tag{8.4}$$

$$\tilde{u}_2(t,\tilde{K}) = a_2\tilde{K} + b_2, \tag{8.5}$$

and determine such a_1, b_1, a_2, b_2 for which (8.3) are satisfied. Taking into account (8.1) and (8.4)-(8.5), the solution for FDE (8.1) under the second case of the strongly generalized differentiability [8] is represented as follows:

$$\tilde{K}(t) = \bigcup_{\alpha \in (0,1]} \alpha[K_1^\alpha(t), K_2^\alpha(t)] \tag{8.6}$$

$$K_1^\alpha(t) = -\frac{\beta}{\gamma} + \left(K_1^\alpha(t_0) + \frac{\beta}{\gamma}\right)e^{\gamma t}$$

$$K_2^\alpha(t) = -\frac{\beta}{\gamma} + \left(K_2^\alpha(t_0) + \frac{\beta}{\gamma}\right)e^{\gamma t}.$$

where $\beta = b_2 - b_1(1-a)$, $\gamma = (1\text{-}a)a_1 - (\mu + a_2)$.

At the second stage, we need to calculate values of the criteria for the feasible solutions determined at the previous stage. We will consider discrete form of the criteria $\tilde{J}_i$, $i = 1,\ldots,4$:

$$\begin{aligned} J_{11}^\alpha &= \sum_{n=0}^{N} p\left(a_2\left(-\frac{\beta}{\gamma} + \left(K_{01}^\alpha + \frac{\beta}{\gamma}\right)e^{n\Delta}\right) + b_2\right)\Delta, \\ J_{12}^\alpha &= \sum_{n=0}^{N} p\left(a_2\left(-\frac{\beta}{\gamma} + \left(K_{02}^\alpha + \frac{\beta}{\gamma}\right)e^{n\Delta}\right) + b_2\right)\Delta \end{aligned} \tag{8.7}$$

$$J_{21}^{\alpha} = -c\sum_{n=0}^{N}\left|a_1(-\frac{\beta}{\gamma}+(\tilde{K}_0+\frac{\beta}{\gamma})e^{\gamma n\Delta}+b_1)\right|_2^{\alpha}\Delta,$$
$$J_{22}^{\alpha} = -c\sum_{n=0}^{N}\left|a_1(-\frac{\beta}{\gamma}+(\tilde{K}_0+\frac{\beta}{\gamma})e^{\gamma n\Delta}+b_1)\right|_1^{\alpha}\Delta \tag{8.8}$$

$$J_{31}^{\alpha} = \sum_{n=0}^{N}(-\frac{\beta}{\gamma}+(K_{01}^{\alpha}+\frac{\beta}{\gamma}))e^{\gamma n\Delta},\ J_{32}^{\alpha} = \sum_{n=0}^{N}(-\frac{\beta}{\gamma}+(K_{02}^{\alpha}+\frac{\beta}{\gamma}))e^{\gamma n\Delta} \tag{8.9}$$

$$J_{41} = \Delta\sum_{n=0}^{N-1}e^{rn\Delta}((-\frac{\beta}{\gamma}+(K_{01}^{\alpha}+\frac{\beta}{\gamma})e^{\gamma n\Delta})a_2+b_2),$$
$$J_{42} = \Delta\sum_{n=0}^{N-1}e^{rn\Delta}((-\frac{\beta}{\gamma}+(K_{02}^{\alpha}+\frac{\beta}{\gamma})e^{\gamma n\Delta})a_2+b_2) \tag{8.10}$$

At the third stage, given the values of the criteria calculated for feasible solutions at the previous stage we need to extract from Pareto optimal solutions them in terms of maximization of the criteria (8.7)-(8.10).

At the fourth stage, it is needed to calculate *nbF*, *neF*, *nwF* for each pair of alternatives $\tilde{\mathbf{u}}^{\mathbf{i}}$, $\tilde{\mathbf{u}}^{\mathbf{k}}$. Let us mention that FO-based formalism is suggested in [10] for multicriteria problems with non-fuzzy criteria. In contrast, in our problem we consider fuzzy-valued criteria. As a result, instead of calculation of μ_b, μ_e, μ_w suggested in [10] (see Section 4.3), we will calculate possibility measures of similarity of differences $\tilde{J}_j(\tilde{\mathbf{u}}^{\mathbf{i}}) - \tilde{J}_j(\tilde{\mathbf{u}}^{\mathbf{k}})$ to fuzzy sets $\tilde{A}_b, \tilde{A}_e, \tilde{A}_w$ describing linguistic evaluations "*better*", "*equivalent*" and "*worse*" respectively. Then *nbF, neF, nwF* will be calculated as follows:

$$nbF(\tilde{\mathbf{u}}^{\mathbf{i}},\tilde{\mathbf{u}}^{\mathbf{k}}) = \sum_{j=1}^{M}P_{\tilde{A}_b}((\tilde{J}_j(\tilde{\mathbf{u}}^{\mathbf{i}}) - \tilde{J}_j(\tilde{\mathbf{u}}^{\mathbf{k}})), \tag{8.11}$$

$$neF(\tilde{\mathbf{u}}^{\mathbf{i}},\tilde{\mathbf{u}}^{\mathbf{k}}) = \sum_{j=1}^{M}P_{\tilde{A}_e}((\tilde{J}_j(\tilde{\mathbf{u}}^{\mathbf{i}}) - \tilde{J}_j(\tilde{\mathbf{u}}^{\mathbf{k}})), \tag{8.12}$$

$$nwF(\tilde{\mathbf{u}}^{\mathbf{i}},\tilde{\mathbf{u}}^{\mathbf{k}}) = \sum_{j=1}^{M}P_{\tilde{A}w}((\tilde{J}_j(\tilde{\mathbf{u}}^{\mathbf{i}}) - \tilde{J}_j(\tilde{\mathbf{u}}^{\mathbf{k}})), \tag{8.13}$$

where

$$P_{\tilde{A}_b}((\tilde{J}_j(\tilde{\mathbf{u}}^{\mathbf{i}}) - \tilde{J}_j(\tilde{\mathbf{u}}^{\mathbf{k}})) = \frac{Poss\left((\tilde{J}_j(\tilde{\mathbf{u}}^{\mathbf{i}}) - \tilde{J}_j(\tilde{\mathbf{u}}^{\mathbf{k}}))\middle|\tilde{A}_b\right)}{Poss\left((\tilde{J}_j(\tilde{\mathbf{u}}^{\mathbf{i}}) - \tilde{J}_j(\tilde{\mathbf{u}}^{\mathbf{k}}))\middle|\tilde{A}_b\right) + Poss\left((\tilde{J}_j(\tilde{\mathbf{u}}^{\mathbf{i}}) - \tilde{J}_j(\tilde{\mathbf{u}}^{\mathbf{k}}))\middle|\tilde{A}_e\right) + Poss\left((\tilde{J}_j(\tilde{\mathbf{u}}^{\mathbf{i}}) - \tilde{J}_j(\tilde{\mathbf{u}}^{\mathbf{k}}))\middle|\tilde{A}_w\right)}$$

$$P_{\tilde{A}_e}((\tilde{J}_j(\tilde{\mathbf{u}}^{\mathbf{i}})-\tilde{J}_j(\tilde{\mathbf{u}}^{\mathbf{k}}))=\frac{Poss\left((\tilde{J}_j(\tilde{\mathbf{u}}^{\mathbf{i}})-\tilde{J}_j(\tilde{\mathbf{u}}^{\mathbf{k}}))\middle|\tilde{A}_e\right)}{Poss\left((\tilde{J}_j(\tilde{\mathbf{u}}^{\mathbf{i}})-\tilde{J}_j(\tilde{\mathbf{u}}^{\mathbf{k}}))\middle|\tilde{A}_b\right)+Poss\left((\tilde{J}_j(\tilde{\mathbf{u}}^{\mathbf{i}})-\tilde{J}_j(\tilde{\mathbf{u}}^{\mathbf{k}}))\middle|\tilde{A}_e\right)+Poss\left((\tilde{J}_j(\tilde{\mathbf{u}}^{\mathbf{i}})-\tilde{J}_j(\tilde{\mathbf{u}}^{\mathbf{k}}))\middle|\tilde{A}_w\right)}$$

$$P_{\tilde{A}_w}((\tilde{J}_j(\tilde{\mathbf{u}}^{\mathbf{i}})-\tilde{J}_j(\tilde{\mathbf{u}}^{\mathbf{k}}))=\frac{Poss\left((\tilde{J}_j(\tilde{\mathbf{u}}^{\mathbf{i}})-\tilde{J}_j(\tilde{\mathbf{u}}^{\mathbf{k}}))\middle|\tilde{A}_w\right)}{Poss\left((\tilde{J}_j(\tilde{\mathbf{u}}^{\mathbf{i}})-\tilde{J}_j(\tilde{\mathbf{u}}^{\mathbf{k}}))\middle|\tilde{A}_b\right)+Poss\left((\tilde{J}_j(\tilde{\mathbf{u}}^{\mathbf{i}})-\tilde{J}_j(\tilde{\mathbf{u}}^{\mathbf{k}}))\middle|\tilde{A}_e\right)+Poss\left((\tilde{J}_j(\tilde{\mathbf{u}}^{\mathbf{i}})-\tilde{J}_j(\tilde{\mathbf{u}}^{\mathbf{k}}))\middle|\tilde{A}_w\right)}$$

As

$$P_{\tilde{A}_b}((\tilde{J}_j(\tilde{\mathbf{u}}^{\mathbf{i}})-\tilde{J}_j(\tilde{\mathbf{u}}^{\mathbf{k}}))+P_{\tilde{A}_e}((\tilde{J}_j(\tilde{\mathbf{u}}^{\mathbf{i}})-\tilde{J}_j(\tilde{\mathbf{u}}^{\mathbf{k}}))+P_{\tilde{A}_w}((\tilde{J}_j(\tilde{\mathbf{u}}^{\mathbf{i}})-\tilde{J}_j(\tilde{\mathbf{u}}^{\mathbf{k}}))=1$$

always holds then

$$\begin{aligned}&nbF(\tilde{\mathbf{u}}^{\mathbf{i}},\tilde{\mathbf{u}}^{\mathbf{k}})+neF(\tilde{\mathbf{u}}^{\mathbf{i}},\tilde{\mathbf{u}}^{\mathbf{k}})+nwF(\tilde{\mathbf{u}}^{\mathbf{i}},\tilde{\mathbf{u}}^{\mathbf{k}})=\\&=\sum_{j=1}^{M}(P_{\tilde{A}_b}((\tilde{J}_j(\tilde{\mathbf{u}}^{\mathbf{i}})-\tilde{J}_j(\tilde{\mathbf{u}}^{\mathbf{k}}))+P_{\tilde{A}_e}((\tilde{J}_j(\tilde{\mathbf{u}}^{\mathbf{i}})-\tilde{J}_j(\tilde{\mathbf{u}}^{\mathbf{k}}))+P_{\tilde{A}_w}((\tilde{J}_j(\tilde{\mathbf{u}}^{\mathbf{i}})-\tilde{J}_j(\tilde{\mathbf{u}}^{\mathbf{k}})))=M\end{aligned} \quad (14)$$

The membership functions of $\tilde{A}_b,\tilde{A}_e,\tilde{A}_w$ are shown below (Fig.8.25.):

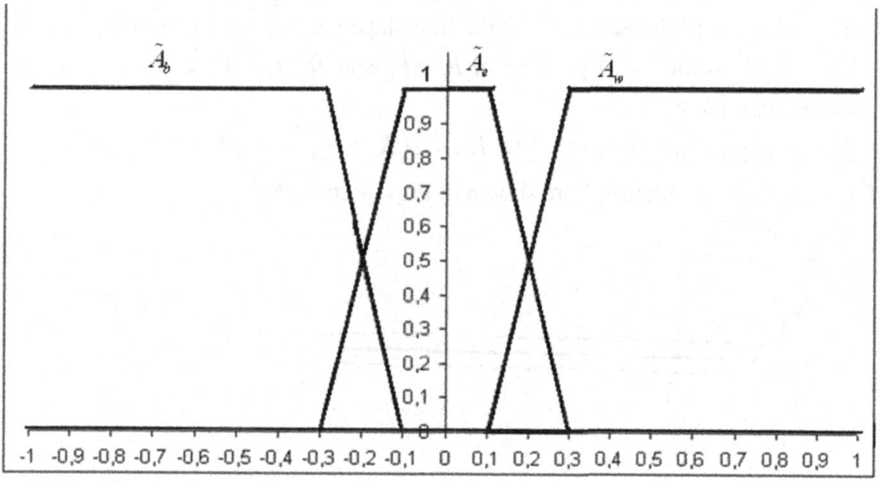

Fig. 8.25 The membership functions of $\tilde{A}_b,\tilde{A}_e,\tilde{A}_w$

At the fifth stage, in accordance with FO-based approach (see Section 4.3), on the base of $nbF(\tilde{\mathbf{u}}^{\mathbf{i}},\tilde{\mathbf{u}}^{\mathbf{k}})$, $neF(\tilde{\mathbf{u}}^{\mathbf{i}},\tilde{\mathbf{u}}^{\mathbf{k}})$, and $nwF(\tilde{\mathbf{u}}^{\mathbf{i}},\tilde{\mathbf{u}}^{\mathbf{k}})$ we need to calculate the greatest $kF=1-d(\tilde{\mathbf{u}}^{\mathbf{i}},\tilde{\mathbf{u}}^{\mathbf{k}})$ among all $\tilde{\mathbf{u}}^{\mathbf{k}}\in\mathcal{U}$, such that $\tilde{\mathbf{u}}^{\mathbf{i}}$ $(1-kF)$-dominates $\tilde{\mathbf{u}}^{\mathbf{k}}$, where $d(\tilde{\mathbf{u}}^{\mathbf{i}},\tilde{\mathbf{u}}^{\mathbf{k}})$ is defined as follows:

$$d(\tilde{\mathbf{u}}^{\mathrm{i}},\tilde{\mathbf{u}}^{\mathrm{k}}) = \begin{cases} 0, & \text{if } nbF(\tilde{\mathbf{u}}^{\mathrm{i}},\tilde{\mathbf{u}}^{\mathrm{k}}) \le \dfrac{M - neF(\tilde{\mathbf{u}}^{\mathrm{i}},\tilde{\mathbf{u}}^{\mathrm{k}})}{2} \\ \dfrac{2 \cdot nbF(\tilde{\mathbf{u}}^{\mathrm{i}},\tilde{\mathbf{u}}^{\mathrm{k}}) + neF(\tilde{\mathbf{u}}^{\mathrm{i}},\tilde{\mathbf{u}}^{\mathrm{k}}) - M}{nbF(\tilde{\mathbf{u}}^{\mathrm{i}},\tilde{\mathbf{u}}^{\mathrm{k}})}, & \text{otherwise} \end{cases}$$

At the final stage it is needed to determine for each $\tilde{\mathbf{u}}^{\mathrm{i}}$ its degree of optimality $do(\tilde{\mathbf{u}}^{\mathrm{i}}) = 1 - \max_{\tilde{\mathbf{u}} \in \mathcal{U}} d(\tilde{\mathbf{u}}^{\mathrm{i}},\tilde{\mathbf{u}})$ and choose an optimal alternative (solution) as a solution with the highest *do*.

Simulation Results

Let us consider solving of the problem (8.2)-(8.3) under the following data:

$$\tilde{K}_0 = (2000, 2020, 2040)$$

$$\tilde{K}_* = 2000, \tilde{K}^* = 3000\,; \tilde{u}_{1*} = 600, \tilde{u}_1^* = 800\,; \tilde{u}_{2*} = 450, \tilde{u}_2^* = 550\,;$$

$$p = 1500;\ c = 0.5;\ a = 0.05;\ q = 0.95;\ \mu = 0.08;$$

In accordance with the Section 4, at the first stage we obtained an approximate set of the feasible solutions $\tilde{u}_1(t,\tilde{K}) = a_1\tilde{K} + b_1$ and $\tilde{u}_2(t,\tilde{K}) = a_2\tilde{K} + b_2$ for the considered problem.

For example, the plots of $\tilde{u}_1(t,\tilde{K}) = a_1\tilde{K} + b_1$, $\tilde{u}_2(t,\tilde{K}) = a_2\tilde{K} + b_2$ and $\tilde{K}(t)$ for feasible solution 1 are shown in Figs. 8.26-8.28:

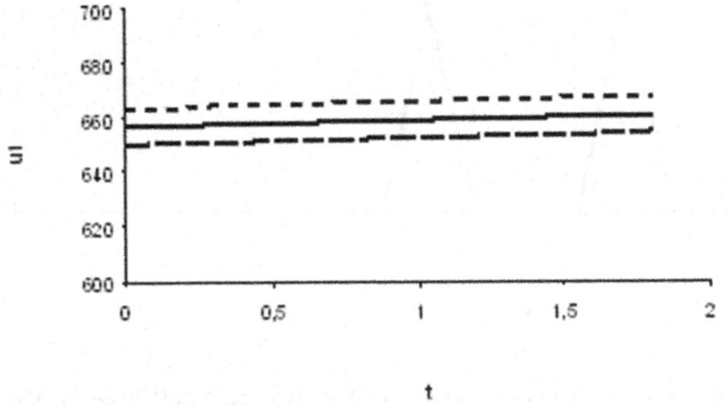

Fig. 8.26 Graphical representation of $\tilde{u}_1(t,\tilde{K})$ (the core and the support bounds)

Then we calculate values of criteria (8.7)-(8.10) for the obtained feasible solutions. The calculated fuzzy values as triangular fuzzy numbers are shown in Table 8.18.

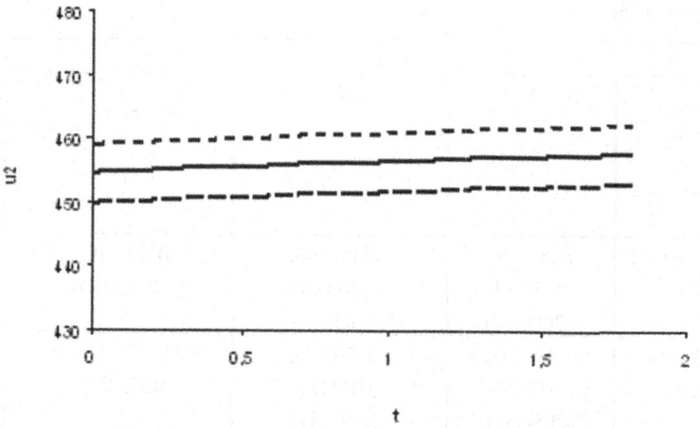

Fig. 8.27 Graphical representation of $\tilde{u}_2(t, \tilde{K})$ (the core and the support bounds)

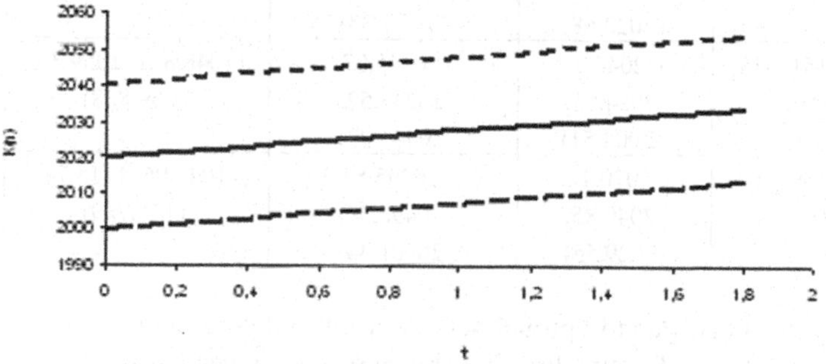

Fig. 8.28 Fuzzy value of capital (the core and the support bounds)

Table 8.18 Feasible solutions(in the space of the criteria)

Feasible solution	Criteria values			
	$\tilde{J}_1$	$\tilde{J}_2$	$\tilde{J}_3$	$\tilde{J}_4$
1	(1354567, 1368113, 1381658)	(-2046.9, -2026.83, -2006.77)	(20067.66, 20268.34, 20469.01)	(945.0581, 954.5087, 963.9593)
2	(1351675, 1365220, 1378766)	(-2047.62, -2027.55, -2007.48)	(20024.81, 20225.49, 20426.16)	(943.0032, 952.4537, 961.9043)
3	(1384002, 1397842, 1411682)	(-2091.38, -2070.88, -2050.37)	(20503.73, 20708.77, 20913.81)	(965.9727, 975.6324, 985.2922)

Now given the feasible solutions described in the space of criteria (8.7)-(8.10) we need to determine the corresponding Pareto optimal set. The Pareto optimal set is given in Table 8.19.

Table 8.19 Pareto optimal set

Feasible solution	Criteria values			
	$\tilde{J}_1$	$\tilde{J}_2$	$\tilde{J}_3$	$\tilde{J}_4$
1	(1354567, 1368113, 1381658)	(-2046.9, -2026.83, -2006.77)	(20067.66, 20268.34, 20469.01)	(945.0581, 954.5087, 963.9593)
3	(1384002, 1397842, 1411682)	(-2091.38, -2070.88, -2050.37)	(20503.73, 20708.77, 20913.81)	(965.9727, 975.6324, 985.2922)
6	(1456329, 1470892 1485456)	(-2063.13, -2042.91, -2022.68)	(20226.79, 20429.06, 20631.33)	(1016.202, 1026.364, 1036.526)
9	(1503380, 1518414, 1533448)	(-2044.6, -2024.55, -2004.51)	(20045.07, 20245.52, 20445.97)	(1048.861, 1059.35, 1069.838)
11	(1583052, 1598882, 1614713)	(-2070.14, -2049.85, -2029.55)	(20295.53, 20498.49, 20701.44)	(1104.696, 1115.743, 1126.789)

Now, given the obtained Pareto optimal set, we need to determine the alternative with the highest degree of optimality. For this purpose, we need at first to calculate values of nbF, neF, nwF [10]. The values of nbF, neF, nwF are given in Tables 8.20-8.22:

Table 8.20 nbF

$nbF(\tilde{\mathbf{u}}^i, \tilde{\mathbf{u}}^k)$	$\tilde{\mathbf{u}}^1$	$\tilde{\mathbf{u}}^2$	$\tilde{\mathbf{u}}^3$	$\tilde{\mathbf{u}}^4$	$\tilde{\mathbf{u}}^5$
$\tilde{\mathbf{u}}^1$	0	0.719504417	0.420087801	0.242182752	0.512271202
$\tilde{\mathbf{u}}^2$	1.587055972	0	0.783480605	1	0.726849113
$\tilde{\mathbf{u}}^3$	2.678826696	1.957762366	0	0.848204309	0.33129111
$\tilde{\mathbf{u}}^4$	2.280101146	2.740620486	1.365479496	0	0.545431948
$\tilde{\mathbf{u}}^5$	2.914733443	2.48284307	2.556180883	2.513137814	0

Table 8.21 neF

$neF(\tilde{\mathbf{u}}^i,\tilde{\mathbf{u}}^k)$	$\tilde{\mathbf{u}}^1$	$\tilde{\mathbf{u}}^2$	$\tilde{\mathbf{u}}^3$	$\tilde{\mathbf{u}}^4$	$\tilde{\mathbf{u}}^5$
$\tilde{\mathbf{u}}^1$	0	1.69343961	0.901085504	1.477716102	0.57299535
$\tilde{\mathbf{u}}^2$	1.69343961	0	1.258757029	0.259379514	0.790307817
$\tilde{\mathbf{u}}^3$	0.901085504	1.258757029	0	1.786316195	1.112528002
$\tilde{\mathbf{u}}^4$	1.477716102	0.259379514	1.786316195	0	0.941430238
$\tilde{\mathbf{u}}^5$	0.572995355	0.790307817	1.112528002	0.941430238	0

Table 8.22 nwF

$nwF(\tilde{\mathbf{u}}^i,\tilde{\mathbf{u}}^k)$	$\tilde{\mathbf{u}}^1$	$\tilde{\mathbf{u}}^2$	$\tilde{\mathbf{u}}^3$	$\tilde{\mathbf{u}}^4$	$\tilde{\mathbf{u}}^5$
$\tilde{\mathbf{u}}^1$	0	1.587055972	2.678826696	2.280101146	2.914733443
$\tilde{\mathbf{u}}^2$	0.719504417	0	1.957762366	2.740620486	2.48284307
$\tilde{\mathbf{u}}^3$	0.420087801	0.783480605	0	1.365479496	2.556180883
$\tilde{\mathbf{u}}^4$	0.242182752	1	0.848204309	0	2.513137814
$\tilde{\mathbf{u}}^5$	0.512271202	0.726849113	0.331291114	0.545431948	0

Given the calculated values of nbF, neF, nwF we need to calculate for each pair of alternatives $\tilde{\mathbf{u}}_i, \tilde{\mathbf{u}}_k$ the greatest kF such that $\tilde{\mathbf{u}}_i$ *(1-kF)* -dominates $\tilde{\mathbf{u}}_k$ as $1-d(\tilde{\mathbf{u}}_i,\tilde{\mathbf{u}}_k)$. The calculated $d(\tilde{\mathbf{u}}_i,\tilde{\mathbf{u}}_k)$ are given below:

$$d(u_i,\ u_j)=\begin{bmatrix} 0 & 0.546642066 & 0.843182166 & 0.893784207 & 0.824247667 \\ 0 & 0 & 0.599808118 & 0.635119125 & 0.707251287 \\ 0 & 0 & 0 & 0.378823109 & 0.87039606 \\ 0 & 0 & 0 & 0 & 0.782967752 \\ 0 & 0 & 0 & 0 & 0 \end{bmatrix}$$

Finally, for each $\tilde{u}$ we calculated its degree of optimality *do*:

$$do(\tilde{\mathbf{u}}^1) = 0.106215793, \quad do(\tilde{\mathbf{u}}^2) = 0.292748713, \quad do(\tilde{\mathbf{u}}^3) = 0.12960394,$$
$$do(\tilde{\mathbf{u}}^4) = 0.217032248, \quad do(\tilde{\mathbf{u}}^5) = 1$$

As one can see, the optimal alternative $\tilde{\mathbf{u}}^*$ is $\tilde{\mathbf{u}}^{\mathrm{i}}$ as it has the highest degree of optimality: $do(\tilde{\mathbf{u}}^5) = 1$.

Application of FPO formalism allowed obtaining intuitively meaningful solution for the considered problem complicated by four conflicting criteria and non-stochastic uncertainty intrinsic to real-world economic problems. This is due to the fact that FPO formalism develops Pareto optimality principle by differentiate between "less" and "more" Pareto optimal solutions. Fuzzy Pareto optimal solutions with different degrees of optimality provide an additional information about alternatives and a freedom for choosing an appropriate alternative (control actions) when the alternative with the highest degree of optimality may not be implementable in real-world situation. The obtained results show validity of the applied approach.

References

1. Aliev, R.A., Aliyev, B.F., Gardashova, L.A., Huseynov, O.H.: Selection of an Optimal Treatment Method for Acute Periodontitis Disease. Journal of Medical Systems 36(2), 639–646 (2012)
2. Aliev, R.A.: Decision Making Theory with Imprecise Probabilities. In: Proceedings of the Fifth International Conference on Soft Computing and Computing with Words in System Analysis, Decision and Control (ICSCCW 2009), p. 1 (2009)
3. Aliev, R.A., Krivosheev, V.P., Liberzon, M.I.: Optimal decision coordination in hierarchical systems. News of Academy of Sciences of USSR, Tech. Cybernetics 2, 72–79 (1982) (in English and Russian)
4. Aliev, R.A., Liberzon, M.I.: Coordination methods and algorithms for integrated manufacturing systems, p. 208. Radio I svyaz, Moscow (1987) (in Russian)
5. Aliev, R.A., Pedrycz, W., Fazlollahi, B., Huseynov, O.H., Alizadeh, A.V., Guirimov, B.G.: Fuzzy logic-based generalized decision theory with imperfect information. Information Sciences 189, 18–42 (2012)
6. Aliev, R.A., Pedrycz, W., Huseynov, O.H.: Decision theory with imprecise probabilities. International Journal of Information Technology & Decision Making 11(2), 271–306 (2012), doi:10.1142/S0219622012400032
7. Alizadeh, A.V., Aliev, R.R., Aliyev, R.A.: Operational Approach to Z-information-based decision making. In: Proceedings of the Tenth International Conference on Application of Fuzzy Systems and Soft Computing (ICAFS 2012), pp. 269–277 (2012)
8. Bede, B., Gal, S.G.: Generalizations of the differentiability of fuzzy-number-valued functions with applications to fuzzy differential equations. Fuzzy Sets and Systems 151, 581–599 (2005)

9. Diamond, P., Kloeden, P.: Metric spaces of fuzzy sets. Theory and applications. World Scientific, Singapoure (1994)
10. Farina, M., Amato, P.: A fuzzy definition of "optimality" for many-criteria decision-making and optimization problems. Submitted to IEEE Trans. on Sys. Man and Cybern. (2002)
11. Gurbanov, R.S., Gardashova, L.A., Huseynov, O.H., Aliyeva, K.R.: Operational Approach to Z-information-based decision making. In: Proceedings of the Tenth International Conference on Application of Fuzzy Systems and Soft Computing (ICAFS 2012), pp. 161–174 (2012)
12. Musayev, A.F., Alizadeh, A.V., Guirimov, B.G., Huseynov, O.H.: Computational framework for the method of decision making with imprecise probabilities. In: Proceedings of the Fifth International Conference on Soft Computing and Computing with Words in System Analysis, Decision and Control, ICSCCW, Famagusta, North Cyprus, pp. 287–290 (2009)
13. Setnes, M.: Compatibility-Based Ranking of Fuzzy numbers. In: Annual Meeting of the North American Fuzzy Information. Processing Society (NAFIPS 1997), pp. 305–310 (1997)
14. Vahdani, B., Zandieh, M.: Selecting suppliers using a new fuzzy multiple criteria decision model: the fuzzy balancing and ranking method. International Journal of Production Research 48(18), 5307–5326 (2010)
15. Zadeh, L.A., Aliev, R.A., Fazlollahi, B., Alizadeh, A.V., Guirimov, B.G., Huseynov, O.H.: Decision Theory with Imprecise Probabilities. In: Contract on Application of Fuzzy Logic and Soft Computing to Communications, Planning and Management of Uncertainty, Berkeley, Baku, p. 95 (2009)

Index